Understanding the Universe: An Introduction to Astronomy, 2nd Edition

Alex Filippenko, Ph.D.

PUBLISHED BY:

THE GREAT COURSES
Corporate Headquarters
4840 Westfields Boulevard, Suite 500
Chantilly, Virginia 20151-2299
Phone: 1-800-832-2412
Fax: 703-378-3819
www.thegreatcourses.com

Printed in the United States of America

Alex Filippenko, Ph.D.

Professor of Astronomy
University of California, Berkeley

Professor Alex Filippenko received his bachelor's degree in physics (1979) from the University of California, Santa Barbara, and his doctorate in astronomy (1984) from the California Institute of Technology. He subsequently became a Miller Postdoctoral Fellow for Basic Research in Science at the University of California, Berkeley. In 1986, he joined the faculty at UC Berkeley, where he has remained through the present time. A member of the International Astronomical Union, Dr. Filippenko has served as president of the Astronomical Society of the Pacific and as councilor of the American Astronomical Society.

An observational astronomer who makes frequent use of the Hubble Space Telescope and the Keck 10-meter telescopes, Dr. Filippenko's primary areas of research are exploding stars (supernovae), active galaxies, black holes, gamma-ray bursts, and cosmology. He and his collaborators recognized a new class of exploding star, obtained some of the best evidence for the existence of small black holes in our Milky Way Galaxy, and found that other galaxies commonly show vigorous activity in their centers that suggests the presence of supermassive black holes. His robotic telescope at Lick Observatory in California is the world's most successful search engine for relatively nearby supernovae, having discovered more than 800 of them since 1998. Dr. Filippenko also made major contributions to the discovery that the expansion rate of the universe is speeding up with time (the *accelerating universe*), driven by a mysterious form of dark energy—the top "Science Breakthrough of 1998," according to the editors of *Science* magazine.

Dr. Filippenko's research findings are documented in more than 600 published papers, and he is one of the world's most highly cited astronomers. He has been recognized with several major awards, including the Newton Lacy Pierce Prize of the American Astronomical Society (1992), the Robert M. Petrie Prize of the Canadian Astronomical Society (1997), and the Richtmyer

Memorial Award of the American Association of Physics Teachers (2007). A Fellow of the California Academy of Sciences, and an elected member of the National Academy of Sciences. Dr. Filippenko has also been a Guggenheim Foundation Fellow (2001) and a Phi Beta Kappa Visiting Scholar (2002). He has held distinguished visiting positions at numerous colleges and universities, including the Marlar Lecturer at Rice University and both the Spitzer Lecturer and Farnum Lecturer at Princeton University.

At the UC Berkeley campus, Dr. Filippenko has won the coveted Distinguished Teaching Award (1991) and the Donald S. Noyce Prize for Excellence in Undergraduate Teaching in the Physical Sciences (1991), each of which is generally given at most once per career. He was voted the "Best Professor" on campus seven times in student polls. Also, in 2002, he received the Distinguished Research Mentoring of Undergraduates Award, given by UC Berkeley.

Dr. Filippenko has delivered hundreds of public lectures on astronomy and has played a prominent role in science newscasts and television documentaries, such as "Mysteries of Deep Space," "Stephen Hawking's Universe," "Runaway Universe," and more than 20 episodes of "The Universe" on The History Channel. With Jay M. Pasachoff, Dr. Filippenko coauthored an introductory astronomy textbook, *The Cosmos: Astronomy in the New Millennium*, now in its 3rd edition, which won the 2001 Texty Excellence Award of the Text and Academic Authors Association for the best new textbook in the physical sciences. This 2006 edition of *Understanding the Universe* combines and updates Dr. Filippenko's previous introduction to astronomy courses, recorded for The Teaching Company in 1998 and 2003. He also recorded *Black Holes Explained* in 2009.

Dr. Filippenko was the recipient of the 2004 Carl Sagan Prize for Science Popularization from the Trustees of Wonderfest, the San Francisco Bay Area Festival of Science. The Carnegie Foundation for the Advancement of Teaching and the Council for Advancement and Support of Education honored him as the "Outstanding Doctoral and Research Universities Professor of the Year" in 2006. In 2010, he won the Astronomical Society of the Pacific's *Richard H. Emmons* award for undergraduate teaching.

Table of Contents

INTRODUCTION

Professor Biography ... i
Course Scope ... 1

LECTURE GUIDES

LECTURE 49
Solar Neutrinos—Probes of the Sun's Core ... 5

LECTURE 50
Brown Dwarfs and Free-Floating Planets ... 21

LECTURE 51
Our Sun's Brilliant Future ... 37

LECTURE 52
White Dwarfs and Nova Eruptions ... 51

LECTURE 53
Exploding Stars—Celestial Fireworks! ... 67

LECTURE 54
White Dwarf Supernovae—Stealing to Explode ... 82

LECTURE 55
Core-Collapse Supernovae—Gravity Wins ... 98

LECTURE 56
The Brightest Supernova in Nearly 400 Years ... 113

LECTURE 57
The Corpses of Massive Stars ... 129

LECTURE 58
Einstein's General Theory of Relativity ... 145

Table of Contents

LECTURE 59
Warping of Space and Time 160

LECTURE 60
Black Holes—Abandon Hope, Ye Who Enter 176

LECTURE 61
The Quest for Black Holes..193

LECTURE 62
Imagining the Journey to a Black Hole 208

LECTURE 63
Wormholes—Gateways to Other Universes? 225

LECTURE 64
Quantum Physics and Black-Hole Evaporation 241

LECTURE 65
Enigmatic Gamma-Ray Bursts 257

LECTURE 66
Birth Cries of Black Holes..272

LECTURE 67
Our Home—The Milky Way Galaxy...288

LECTURE 68
Structure of the Milky Way Galaxy 303

LECTURE 69
Other Galaxies—"Island Universes" 320

LECTURE 70
The Dark Side of Matter 336

LECTURE 71
Cosmology—The Really Big Picture 352

Table of Contents

LECTURE 72

Expansion of the Universe and the Big Bang 368

SUPPLEMENTAL MATERIAL

Useful Symbols...385
Universe Timeline..386
Solar System Timeline..388
Glossary 389
Biographical Notes 411
Bibliography...416

Understanding the Universe: An Introduction to Astronomy, 2nd Edition

Scope:

This visually rich course is designed to provide a nontechnical description of modern astronomy, including the structure and evolution of planets, stars, galaxies, and the Universe as a whole. It includes almost all of the material in my first two astronomy courses for The Teaching Company, produced in 1998 and 2003, but with a large number of new images, diagrams, and animations. The discoveries reported in the 2003 course are integrated throughout these new lectures, and more recent findings (through mid-2006) are included, as well. Much has happened in astronomy during the past few years; we will discuss the most exciting and important advances.

Astronomical objects have been explored with breathtaking data obtained by the Hubble Space Telescope, the Chandra X-Ray Observatory, the Keck 10-meter telescopes, planetary probes, and other modern instruments. We will explore amazing phenomena, such as quasars, exploding stars, neutron stars, and black holes, and we will see how they increase our understanding of the physical principles of nature. We will also investigate recent newsworthy topics, such as the *Cassini* mission to Saturn, evidence for liquid water on ancient Mars, the discovery of many small bodies beyond Neptune in our Solar System, the detection of numerous planets around other stars, the nonzero mass of ghostly neutrinos, enormously powerful gamma-ray bursts, the conclusive evidence for a supermassive black hole in the center of our Milky Way Galaxy, the determination of the age of the Universe, the discovery of a long-range repulsive effect accelerating the expansion of the Universe, and progress in the unification of nature's fundamental forces. Scientifically reasonable speculations regarding the birth of the Universe, the possibility of multiple universes, and the probability of extraterrestrial life are also included.

This course concentrates on the most exciting aspects of our fantastic Universe and on the methods astronomers have used to develop an understanding of it. The lectures present, in clear and simple terms, explanations of how the Universe "works," as well as the interrelationships among its different components. Reliance on basic mathematics and physics is minimal but appropriate in some sections to deepen the interested viewer's quantitative understanding of the material.

The course is divided into three major sections, each of which consists of several units. (These major sections are called "parts" during the lectures, but they are not to be confused with the eight 12-lecture "parts" used in packaging the lectures.) There are 24 lectures in section 1, entitled "Observing the Heavens." The first unit, "Celestial Sights for Everyone," describes simple daytime and nighttime observations that you can make to better appreciate the sky and what it contains. Various commonly observed phenomena, such as seasons, lunar phases, and eclipses, are also discussed. The second unit, "The Early History of Astronomy," covers the study of astronomy from the ancient Greeks through Newton, including the transition from geocentric (Earth-centered) to heliocentric (Sun-centered) models of the Universe. In the third unit, "Basic Concepts and Tools," we provide an overview of distance and time scales in the Universe to put our discussions in perspective. Because the study of light is of central importance to astronomy, we spend several lectures explaining its physical nature and utility. Modern telescopes, the main instruments used by astronomers, are also described.

Section 2, "The Contents of the Universe," consists of 46 lectures in 5 units. In the first unit, "Our Solar System," we discuss the major constituents of our own planetary system, including the Sun, planets and their moons, comets, asteroids, and Kuiper-belt objects. The discovery of a distant body larger than Pluto and the subsequent, highly controversial demotion of Pluto from planetary status have recently made worldwide headlines. The formation of other stars and planetary systems, as well as the discovery of such extrasolar planets, is the subject of the second unit, "Other Planetary Systems." During the past decade, about 200 planets have been found orbiting other stars, making this one of the most exciting areas of modern astronomy. The search for extraterrestrial life is also described.

In the third unit of section 2, "Stars and Their Lives," we learn about the properties of other stars and the various observations needed to deduce them. Nuclear reactions, the source of energy from the stars, are described, as well. We examine how stars eventually become red giants, subsequently shedding their outer layers to end up as dense white dwarfs, retired stars. The explosive fates of some rare types of stars are the subject of the fourth unit, "Stellar Explosions and Black Holes," and we explain how the heavy elements necessary for life are created. Bizarre stellar remnants include neutron stars and black holes, the realm of Einstein's general theory of relativity. We continue our exploration of black holes with such phenomena as black-hole evaporation and powerful gamma-ray bursts, as well as speculations that black holes are gateways to other universes. In the fifth unit, "The Milky Way and Other Galaxies," we extend our explorations to the giant collections of stars called galaxies, along the way examining evidence for mysterious dark matter.

Section 3, "Cosmology: The Universe as a Whole," comprises the final 26 lectures of the course in 3 units. The first unit, "Cosmic Expansion and Distant Galaxies," introduces the expansion of the Universe and shows how it is used to study the evolution of galaxies. We discuss active galaxies and quasars, in which matter is inferred to be falling into a central, supermassive black hole. In the second unit, "The Structure and Evolution of the Universe," aspects of the Universe, such as its age, geometry, and possible fate, are considered. We examine evidence for the stunning conclusion that the expansion of the Universe is currently accelerating. We also describe the cosmic microwave background radiation—the generally uniform afterglow of the Big Bang—as well as the tiny irregularities that reveal the presence of early density variations from which all of the large-scale structure of the Universe subsequently formed. The nature of dark energy accelerating the Universe is explored in terms of modern attempts to unify forces, such as string theory.

In the third and final unit, "The Birth of the Cosmos, and Other Frontiers," we examine the very early history of the Universe, showing how the lightest elements formed during a phase of primordial nucleosynthesis. The recognition of several troubling problems with the standard Big Bang theory led to a magnificent refinement—an inflationary epoch of expansion that lasted only a tiny fraction of a second. The possible connection between inflation and the currently accelerating expansion of space is also discussed. We then consider very speculative ideas regarding the birth of the Universe and the hypothesis of multiple universes. We end, in the last lecture, on a philosophical note, with some reflections on intelligent life in the cosmos and of our place in the grand scheme of things. ■

Overview of Course Organization

Major Section	Lectures	Units
Observing the Heavens	2–24	• Celestial Sights for Everyone (2–11) • The Early History of Astronomy (12–16) • Basic Concepts and Tools (17–24)
The Contents of the Universe	25–70	• Our Solar System (25–36) • Other Planetary Systems (37–42) • Stars and Their Lives (43–52) • Stellar Explosions and Black Holes (53–66) • The Milky Way and Other Galaxies (67–70)
Cosmology: The Universe as a Whole	71–96	• Cosmic Expansion and Distant Galaxies (71–78) • The Structure and Evolution of the Universe (79–90) • The Birth of the Cosmos, and Other Frontiers (91–96)

Solar Neutrinos—Probes of the Sun's Core
Lecture 49

"The detection of neutrinos and neutrino oscillations is one of the greatest achievements of the past few decades. It affects not just astrophysics, it also affects fundamental particle physics, throwing a giant wrench into the theory, and that's really exciting."

In the previous lecture, we learned that the Sun produces its energy through nuclear fusion. But we can't actually conduct physical experiments on the core of the Sun, so how do we know nuclear fusion is occurring now? The Sun could not burn from chemical reactions, such as flames burning paper or wood, because such a reaction couldn't produce enough energy to power the Sun. Also, we know that temperatures in the Sun are too high for chemical burning in the conventional sense. The Sun cannot burn through gravitational contraction because this process would allow the Sun to live only about 50 million years; the Sun has likely been shining at its present rate for at least 3 billion years. In addition, other studies prove that the Sun is not gravitationally contracting, at least not much. Physicists have concluded that nuclear fusion must be occurring now because there is no other source of energy for the Sun to use. Though earlier concepts about physics couldn't account for nuclear fusion, the advent of **quantum mechanics** changed our perception of how particles behave and answered some questions about nuclear fusion in the Sun.

Still, we cannot physically test the Sun for nuclear fusion reactions. We do know that photons emerge from the Sun, though this fact still doesn't prove that nuclear fusion occurs. Why not? When a photon is produced in the middle of the Sun, it encounters an opaque gas on its way to the Sun's surface (photosphere). This causes photons to move randomly on their way out, which on average, takes about 100,000 years. If nuclear fusion reactions in the Sun stopped right now, we wouldn't know it for 100,000 years because photons are already on their way to the surface. Thus, we can't rely on photons to tell us that nuclear fusion is occurring right now. We must rely on ghostly particles called *neutrinos*, which are produced during fusion reactions. Neutrinos have only a slight mass and hardly interact with other matter. It takes about 2 seconds for a neutrino to go from the core of the Sun to its surface and another 8.3 minutes to reach Earth. If we could detect neutrinos coming in great numbers from the Sun, we could prove that nuclear fusion is occurring right now.

Where do these neutrinos come from and how do they interact? Recall the first step of the proton-proton chain, in which two protons combine to form a deuteron. In the process, one of the protons turns into a neutron, a positron—an antielectron—and a neutrino. The positron annihilates an electron and produces photons. In the proton-proton chain, the neutrino—called an *electron neutrino* because it is associated with processes that form electrons and antielectrons—simply exits the Sun at a high speed. Every square centimeter of the Earth, every second, is bathed by about 60 billion such neutrinos from the Sun. With such high concentrations of neutrinos all around us, surely some would react with earthly elements and we could detect them.

In fact, neutrinos can combine with a neutron in an atomic nucleus, turning it into a proton and an electron. Specifically, a nucleus of chlorine can occasionally absorb an electron neutrino, turning it into a radioactive form of argon, which can be detected with a Geiger counter or some other device that detects radioactivity. One such experiment was conducted by Ray Davis, whose amazingly precise methodology allowed for the detection of a single radioactive argon nucleus in a 100,000-gallon tank of dry-cleaning solution. Another experiment by Masatoshi Koshiba made a similar discovery. Koshiba could actually confirm that the neutrinos were coming from the Sun rather than some other source.

The detection of neutrinos was a great breakthrough, but there was another puzzle. Davis's experiment detected only about one-third of the expected number of neutrinos, suggesting that our theory about the Sun was incorrect or that something else was at work. For example, perhaps the temperature in the Sun's core wasn't quite as high as we thought. Or maybe the Sun simply wasn't fusing much at all at the present time. Alternatively, perhaps the electron neutrinos turned into some other kind—officially known as "flavor"—of neutrinos, such as *muon neutrinos* or *tau neutrinos*.

A 1998 experiment measured muon neutrinos produced by the interaction of **cosmic rays** from space with molecules in the Earth's atmosphere. The experiment detected muons coming from the direction just above the detector (close to the Earth's surface), muons coming from the back side of Earth, and muons coming from other random directions. The number of neutrinos coming from the back side of Earth was smaller than that coming from close to the detector, suggesting that muon neutrinos passing through Earth turn into another flavor of neutrino on the way. This was the first real evidence that at least muon neutrinos can change their flavor; therefore, perhaps electron neutrinos from the Sun could change, as well.

Recent experiments in Canada showed that two types of reactions can occur when electron neutrinos travel through a certain form of heavy water (that is, water containing atoms of **deuterium** instead of normal hydrogen). In one reaction, sensitive only to electron neutrinos, a very energetic electron is formed. It travels through the water with a speed higher than the local speed of light (which is slower than that in a vacuum, due to an interaction of light with water), producing electromagnetic radiation called **Cerenkov radiation**, or Cerenkov light. This accounts for one-third of the expected number of neutrinos, as in previous experiments. A second type of reaction, however, was able to detect all three known flavors of neutrinos: electron, muon, and tau. Remarkably, the total rate at which neutrinos were detected matched theoretical expectations! This essentially proves the hypothesis that the Sun's source of energy is nuclear fusion. Electron neutrinos are produced in the Sun, but about two-thirds of them turn into other neutrino flavors during their journey to the Earth.

"In fundamental particle physics, neutrinos are supposed to be massless, and yet it turns out they cannot change flavors if they're massless."

The three observed neutrino flavors turn out to be different combinations of more fundamental neutrinos, called *type 1*, *type 2*, and *type 3*. For example (simplified to illustrate the physical principles), an electron neutrino might be a specific combination of type 1 and type 2 neutrinos. The quantum mechanical waves of type 1 and 2 neutrinos can sometimes be in phase and sometimes out of phase, thus creating the different flavors of observed neutrinos—electron, muon, and tau. Such *neutrino oscillations* (from one observed flavor to another) imply that neutrinos have nonzero (but very small) mass. Previously, scientists thought that neutrinos had no mass at all. In order to change their flavor, neutrinos must move slower than the speed of light—but to move slower than the speed of light, they must have mass. Particles with no mass have to travel at the speed of light; otherwise, they wouldn't exist.

This great discovery challenges previous theories about **particle physics**, which had asserted that neutrinos are massless. It affects our understanding of the ways in which particles fundamentally behave at the microscopic and submicroscopic scales. ■

Important Terms

Cerenkov radiation: Electromagnetic radiation emitted by a charged particle traveling at greater than the speed of light in a transparent medium. The blue light emitted is the electromagnetic equivalent of a sonic boom heard when an aircraft exceeds the speed of sound.

cosmic rays: High-energy protons and other charged particles, probably formed by supernovae and other violent processes.

deuterium: An isotope of hydrogen that contains one proton and one neutron.

particle physics: The study of the elementary constituents of nature.

quantum mechanics: A 20th-century theory that successfully describes the behavior of matter on very small scales (such as atoms) and radiation.

Suggested Reading

Golub and Pasachoff, *Nearest Star: The Surprising Science of Our Sun.*

Institute for Advanced Study, School of Natural Sciences, *John Bahcall*, www.sns.ias.edu/~jnb/ ("popular articles" link).

Pasachoff, *Astronomy: From the Earth to the Universe*, 6th ed.

Pasachoff and Filippenko, *The Cosmos: Astronomy in the New Millennium*, 3rd ed.

Sudbury Neutrino Observatory (SNO), www.sno.phy.queensu.ca/.

Sutton, *Spaceship Neutrino*.

Questions to Consider

1. Why do neutrinos give us different information about the Sun than light does?

2. Do you think astronomers were overly bold in predicting that the solar neutrino problem would be resolved by changes in our theories of fundamental particles, rather than by abandoning the standard model of solar energy production?

3. The solar neutrino experiments that preceded SNO were able to detect only electron neutrinos. Why was the ability to detect *all* flavors of neutrinos so important in helping to resolve the solar neutrino problem?

Solar Neutrinos—Probes of the Sun's Core

Lecture 49—Transcript

In the previous lecture, I discussed the process by which we think our Sun is producing energy. Basically, four protons—hydrogen nuclei—combine or fuse together, bound by the strong nuclear force, to form a helium nucleus. That helium nucleus has less mass than the original four protons did. Through $E = mc^2$, Einstein's famous equation, that mass difference between the four original protons and the final helium nucleus gets turned into energy. This is nuclear fusion. That's what we think is going on in the middle of the Sun. But you might ask, "How do we really know that's happening?" We can't stare into the middle of the Sun very easily. In fact, with optical light, or even with radio waves, the Sun is completely opaque. You can't get anywhere close to the middle of the Sun with electromagnetic radiation, as viewed from outside. Maybe the Sun actually shines through some other process.

Let's consider the possibilities. Could it be chemical burning? No, it can't. Chemical burning, like fire, flames of paper or wood burning, doesn't produce enough energy, by a gargantuan factor, to explain the observed luminosity or power of the Sun. It doesn't even come close. If you were to pack a lot of wood or paper in the Sun and just make it all go off very quickly to produce the observed luminosity of the Sun, then you'd use it all up really quickly, and you wouldn't have a very long-lasting Sun. Therefore, it just can't be happening. In fact, we know that temperatures in the Sun are too high in general for any chemical burning in the conventional sense—so basically there's no chemical burning.

Could it be gravitational contraction, like in a pre-main-sequence star? After all, a pre-main-sequence star shines quite brightly. But it does so not because of fusion, but rather because it's slowly gravitationally contracting, releasing its gravitational energy and turning it into light. A physicist named Lord Kelvin—after whom the absolute, or Kelvin, temperature scale is named—calculated that, through this process of gradual gravitational contraction, the Sun could have lived only about 50 million years. In other words, at the current rate at which it's generating energy, you couldn't have had it generating energy for more than about 50 million years; otherwise, the star would have contracted down to zero volume right now. Clearly, the Sun is pretty large—so there's only a certain amount of contraction that could have happened.

This so-called Kelvin contraction time, 50 million years, though much longer than the time the Sun could have shone if it were only chemically burning, is still nowhere near enough time. The Earth has been around for about 4.5 billion years, and we have evidence through studies of fossils that life forms existed easily up to 3 billion years ago, and they needed surface temperatures comparable to what we have right now. So, we have pretty good evidence that the Sun has been shining at about its present rate for at least 3 billion years, a far cry from just 50 million. We also know now through other studies that the Sun is simply not gravitationally contracting, at least not much. In fact, we think it's not contracting at all—so that hypothesis pretty much flew out the door.

Well, what could it be? Physicists, essentially by the process of elimination, decided that nuclear fusion must be what's going on. There's really no other source of energy that the Sun could be using. All the other potential ideas had been shot full of holes through various observations and physical reasoning. So, they said, "Okay, nuclear fusion is going on." In fact, they didn't even really know how that could be happening because, though the temperatures in a star are pretty high, and the protons get pretty close to one another, they still don't hit each other. You might wonder, "Well, how can the strong nuclear force, which has a very short range, even get these protons to bind together to form a deuteron if, in fact, the protons never get close enough to each other to feel each other, to hit each other?" Physicists didn't know how, at the relatively low temperatures of the Sun, you could do that.

If the temperature were a billion degrees, sure, the protons could slam into each other, and then the strong nuclear force could take over. But the temperatures are only 10 million degrees, and at such temperatures, the protons don't even get close enough together. Some physicists said, "No, it can't be nuclear fusion." Then quantum mechanics was developed, and our perception of particles changed. There are waves of some probability distribution of where they might be. It turns out these probability functions overlap ever so slightly, even when protons are pretty far apart from each other, and the strong nuclear force can take over. I don't want to get into the details of that, but basically, the nuclear fusion problem was solved with the development of quantum physics.

Physically, people said, "Okay, it can happen, but we still don't really know that it's happening. You don't see the middle of the Sun." You might say, "Well, yeah, but we know that photons are leaking out." The fact that photons are leaking out right now, you might think, means that fusion must be going on right now: or does it? Well, it doesn't really; it doesn't directly. That's because when a photon is produced in the middle of the Sun, it encounters a very opaque gas on its way out. The Sun is so opaque that the photon bounces around and takes, on average, 100,000 years to escape. It doesn't travel at the speed of light, zooming out of the Sun. It's traveling locally at the speed of light, but it's bouncing around every which way. Gradually, it finds itself at the surface of the Sun, where it can then zoom away, but that takes 100,000 years.

It's like a random walk. You have a drunk whose looking for his car keys, which are somewhere underneath a lamppost. The lamppost is shining there, but the drunk is kind of wandering around in every direction, and he gradually makes his way to the lit-up part under the lamppost. But it takes a while to get there; it's not a direct path. In a similar way, the photons take a very indirect path in coming out. So, if you shut down the fusion reactions in the Sun right now, we wouldn't know it for 100,000 years because a bunch of photons are already on the way, and it'll take them 100,000 years to get out. So, you could seriously question whether, in fact, fusion is occurring in the middle of the Sun. It was a legitimate question to ask. Do we have direct evidence for such a physical process? Photons don't provide it.

Remember, in the last lecture, I said that a ghostly little particle called a neutrino is produced during the fusion reactions. Neutrinos are very strange particles. They're very light. For a long time, people thought they were massless. I will give you evidence for a very slight mass today. People thought that they also travel at the speed of light, or nearly at the speed of light, and that they hardly interact at all. That last fact is true; they're very antisocial particles. You could have a bunch of neutrinos going through a light year of lead, and most of them would make it through unscathed. That is, most of them would not encounter anything along the way. They wouldn't bounce off of anything; they wouldn't reflect off of anything. They go through a light year of lead as though it were a vacuum. That's how non-interactive they are. In fact, they don't care that there's this opaque cloud of gas around them. They're not affected by the Sun's gases. They just go zooming right out. It takes them two seconds to go from the core of the Sun out to the surface, and another eight minutes to reach us here on Earth. So, they are a direct probe of what's going on in the core of the Sun. If we could detect neutrinos coming in great numbers from the Sun—that would prove that nuclear fusion is going on right now.

Where do these neutrinos come from? Recall the first step of the proton-proton chain, where two protons combined to form a deuteron. In the process, one of the protons turned into a neutron, a positron—that is, an antielectron—and a neutrino. The positron annihilates with the electron and produces some photons, some energy. That's part of the way in which the Sun shines. But the neutrino just goes zooming out of the Sun. More precisely, this is a kind of neutrino called an electron neutrino because it is associated with the process that forms electrons or—in this case—antielectrons. This is one flavor of neutrinos. By the way, if you want to learn more about particle physics, there's a whole Teaching Company course on particle physics. I'm going over things very briefly here, just to give you the idea of what's going on.

John Bahcall calculated how many neutrinos should be coming out from the Sun and reaching the Earth. It turns out that every square centimeter of the Earth, every second, is bathed by about 60 billion neutrinos. That's how many are passing through your body every second from the Sun if the theory is right, 60 billion per square centimeter of your body per second. That's an awful lot of neutrinos. Some might interact. When you've got that many, some will interact, and maybe you can detect them. John Bahcall inspired experimental physicists and chemists to search for neutrinos, the rare neutrinos that might interact with something here on Earth.

How do you do it? How do they interact? A neutrino can combine with a neutron in an atomic nucleus, turning it into a proton and an electron. Specifically, a nucleus of chlorine can occasionally absorb an electron neutrino, turning it into a radioactive form of argon. That radioactive argon can, in principle, be detected with a Geiger counter or some other device that detects radioactivity. There were experiments designed to detect these radioactive argon atoms and variations on this theme. The most famous and long-lasting of such experiments was conducted by Ray Davis in the Homestake Gold Mine in South Dakota, about a mile underneath Earth's surface. He did it way down there to shield the experiment from cosmic rays, from charged particles coming from outer space and from the Sun.

Here's a photograph of the tank that he had holding 100,000 gallons of dry-cleaning solution, perchloroethylene, C_2Cl_4. Here are John Bahcall and Ray Davis in 1964 next to his contraption, looking it over, trying to perfect the method by which the detections of the argon atoms are made. Indeed, the experiment was amazingly precise. Out of a 100,000-gallon tank of cleaning solution, Davis could detect one radioactive argon nucleus. He told his graduate students, "Put a couple of them in there. Don't tell me how many you put in, and I'll try to find them." By golly, he found them. He found the right number. That's amazing, one atom. It was a very highly sophisticated experiment. Solar neutrinos, in a nutshell, were detected. It was a fantastic discovery; they're there. A few of them occasionally interact with the chlorine to produce argon.

Some time later in Japan, in a similar experiment, Masatoshi Koshiba made a similar discovery. He could actually tell from where the neutrinos are coming. He could tell that they're definitively coming from the Sun. Davis was only detecting neutrino interactions with chlorine. He couldn't definitively say that they're coming from the Sun, although it's sort of like, where else would they be coming from, if not the Sun? It's shining brightly in neutrinos, and nothing else comes close because all the stars are much farther away. Davis presumed they're from the Sun; Koshiba showed that they are from the Sun. They shared the Nobel Prize in Physics in 2002 for their discovery.

I think John Bahcall deserved part of the Nobel Prize as well, because it was his calculations that inspired the experimentalists, and he calculated ever more correctly and precisely just how many neutrinos of different kinds of energy should be coming out from the Sun. Without his calculations, the experimentalists would not have been able to interpret or evaluate their data nearly as completely and thoroughly, and in such a physically meaningful way. I really believe that my friend, the late John Bahcall, who died in 2005, deserved part of the Nobel Prize, and it's a shame that he wasn't given that prize. Many of my colleagues in physics and astronomy feel the way I do.

The detection of neutrinos was a great breakthrough, but there was another puzzle. Davis was only detecting roughly one-third of the expected number of neutrinos. Let's look at a graph of the number of neutrinos plotted on the vertical axis, versus the year in which the data were gathered on the horizontal axis. The units are called solar neutrino units. You don't have to worry about what those are, but it's just a unit of measurement for the detection of solar neutrinos. Theory predicted that eight solar neutrino units worth of neutrinos should be detected every year. Davis found that only two or so were. But there's uncertainty in the measurements, and there's uncertainty in the theory—and two is consistent with one third of six, the lower limit for the theoretical value. Basically, after a while, everyone started saying that we're only detecting about one-third of the expected number of neutrinos, one-third to one-quarter.

This was a big puzzle. It suggested that something is wrong, perhaps in our theory of what's going on in the middle of the Sun. For example, perhaps the temperature in the core isn't quite as high as Bahcall thought. That would mean that an energetic rare form of neutrino produced by a rare variation of the proton-proton chain—I won't go into the details, but there are variations on the theme that I discussed in the previous lecture. Only certain reactions produce the very most energetic neutrinos that the Davis experiment was able to detect. Maybe the temperature in the Sun isn't high enough to allow that rare branch of the proton-proton chain to occur. Or maybe the Sun simply isn't fusing much at all, period, right now. Maybe it's taking a break. It's on vacation from fusion, and the light we're seeing is those 100,000 years worth of photons still leaking out.

Those are astrophysical solutions, or possible solutions, to the solar neutrino problem. There were also solutions proposed based on fundamental physics. Maybe the electron neutrinos, which are the ones that Davis could detect, are turning into some other flavor of neutrinos. They come in different kinds. There are muon neutrinos, and there are tao neutrinos as well—so maybe they're turning into other kinds of neutrinos. What we really need are other experiments that might detect those other forms of neutrinos, and also experiments to detect lower-energy neutrinos. A possible solution was found in 1998, or at least hints of a solution, when a Japanese group made a measurement of a type of neutrino called the muon neutrino using an underground lab—kind of like Ray Davis's, but more sophisticated—where they have a giant swimming pool of water surrounded by a whole bunch of photomultiplier tubes that are able to detect flashes of light that occur when muon neutrinos interact in various ways with the water. Here you can see them cleaning the photomultiplier tubes.

In 2001, by the way, one of the tubes broke, and it set off a chain reaction that sort of popped three-quarters of their photomultiplier tubes—a huge disaster for them. Fortunately, they had gathered some fantastic data before that, data that turned out to be revolutionary. What they did was they measured muon neutrinos produced by the interaction of cosmic rays from space with molecules in our atmosphere. Here, a cosmic ray goes crashing into a molecule and produces some muon neutrinos, denoted by these red arrows. Then this experiment, the Super-Kamiokande Experiment, could detect these muons neutrinos—both those coming from the direction just above the detector, close to the surface of the Earth, and also those coming from the back side of the Earth, and also those coming from other random directions.

I won't go into details, but suffice it to say that the number of neutrinos coming from the back side of the Earth was smaller than the number coming from the part of the Earth close to where the detector was—suggesting that the muon neutrinos going through the Earth turn into some other kind of neutrino or some other kind of particle on the way because their detector was sensitive only to muon neutrinos. If the ones that traversed the whole diameter of the Earth turn into something else along the way, you'll find fewer of them coming from that direction than from the direction right above you. That's what they found. This was the first real evidence that neutrinos can change flavor. There was additional evidence announced through other experiments of a similar nature around that time. It was very exciting because it hinted that at least muon neutrinos can change flavor, and so perhaps the electron neutrinos from the Sun can change flavor as well.

That's good that muon neutrinos can change flavor, but the Sun produces electron neutrinos. Maybe the electron neutrinos don't really change flavor the way the muon neutrinos appeared to do. What we'd really like is an experiment of an analogous nature that attempts to measure the electron neutrinos from the Sun and whether they change with time into other kinds of neutrinos. That would be the real test of what's going on in the Sun. In the last two years, there's been a brilliant experimental test of this sort done in Canada, at the Sudbury Neutrino Observatory. Once again, it is one of these underground tanks. They have them very deep underground to shield them from natural cosmic rays coming in all over the place in our atmosphere, and surface radioactivity from various radioactive elements in surface rocks and stuff.

Here's the chamber at the Sudbury Neutrino Observatory, or SNO for short, that has a whole bunch of water inside, and it is surrounded by lots of these photomultiplier tubes, as in the case of the Super-Kamiokande Experiment in Japan. The difference in the case of SNO is that they used a heavy form of water. Water whose hydrogen atoms, or at least one hydrogen atom, consists of a deuteron being circled by an electron rather than a proton being circled by an electron. This is called heavy water, deuterated water. Remember, deuterons are protons plus neutrons together, one each. That's still hydrogen. There's one proton, so it's still hydrogen, but it's a heavy form of hydrogen because there's a neutron attached as well.

In this heavy water, two types of reactions can occur. In one type of reaction, a neutron in the deuteron absorbs an electron neutrino, transforming itself into a proton and an electron. The deuteron gets destroyed. It gets turned into two protons, which repel each other. But along the way, an energetic electron is formed, and that electron is so energetic that it can travel faster than the speed of light in water. You might say wait a minute, didn't I tell you earlier that nothing travels faster than the speed of light? I did, and that's one of the fundamental assumptions of Einstein's special theory of relativity. But light slows down in water and other substances. Indeed, that's what leads to the bending of light, the refraction of light. In fact, relativity says that you can have particles going through water faster than the local depressed speed of light, but still slower than the speed of light in a vacuum—so you're not violating anything in relativity.

When you have such a particle traveling faster than the speed of light in water, it forms electromagnetic radiation called Cerenkov light. This is analogous to an airplane moving through air at faster than the speed of sound. It produces a sonic boom. Cerenkov light is an electromagnetic boom. It's light emitted as a result of the particle traveling faster than the local speed of light. That particle has to be charged for this to happen. The charged particle moving through water produces this light, and it leads to the detection of the neutrino that had combined with the neutron to form the proton and the rapidly moving electron. When they did this kind of measurement, they found that only one-third of the expected number of neutrinos was detected—so that confirmed Ray Davis's result.

But there was another type of interaction that's sensitive to all kinds of neutrinos. That interaction is shown here. A neutrino of any flavor—electron, muon, tau, it doesn't matter—can scatter or reflect off of an electron in the water. That kick makes the electron that's in the water move away with a rapid speed, faster than the speed of light. That charged particle—the electron—traveling faster than the speed of light produces, once again, Cerenkov radiation. The neutrino slammed into it, kicked it, made the electron move fast, and then you detect the Cerenkov light. This reaction is insensitive to the type of neutrino that slammed into the electron. All of them do it about equally well. This type of reaction potentially finds all of the neutrinos. When they counted up all the neutrino events from this reaction, guess what? They found Bahcall's expected total number of neutrinos. Is that great or what? They're all there, but the electron neutrinos are only one-third of them.

How can that be? The nuclear reactions produce only electron neutrinos; they don't produce the muon and tau kind. The electron neutrinos must have changed flavor along the way. That's the conclusion. Electron, muon, and tau neutrinos are different flavors—which are, in fact, manifestations of different combinations of a more fundamental type, or set of types, of neutrinos, boringly called type 1, type 2, and type 3. The 1, 2, and 3 are the fundamental neutrinos. The electron, muon, and tau neutrinos are combinations of those fundamental neutrinos that can actually be observed. You observe the electrons, muons, and taus, but the true fundamental kind are the 1's, 2's, and 3's. What can happen is that, in the Sun, an electron neutrino might be produced, but it's a combination of, say, a type 1 and a type 2, just for illustrative purposes.

The type 1's and 2's have a different quantum mechanical wave function associated with them. They can change. They travel along like this, and you see they're in phase initially—that's an electron neutrino—but after a while, they're out of phase. The type 2 is sort of oscillating up here, and the type 1 is oscillating down. That may be a different kind of neutrino, a muon or a tau neutrino. Indeed, here where they cross, they're going in opposite directions than what they were initially—so this might be the muon or tau neutrino. I'm just showing this for illustrative purposes. Then, later on, they become an electron neutrino again. Maybe later on they become a muon or a tau neutrino. In other words, they change their flavor. What happens is that they change their flavor on the way, and that's why the Davis detector didn't see enough electron neutrinos, because only one-third of them, on average, are electron neutrinos at any one time. The other two-thirds are muon and tau neutrinos.

This solved not only a huge problem in astrophysics, but also raised a lot of eyebrows in fundamental particle physics—because in fundamental particle physics, neutrinos are supposed to be massless, and yet it turns out they cannot change flavors if they're massless. The reason for that is that if they're massless, they must travel at the speed of light; otherwise, they wouldn't exist, it turns out. But at the speed of light, time stops. Recall that we discussed this gamma factor, the time dilation factor, when we studied relativity in Lecture 42. At the speed of light, gamma is 1/0 or infinity. So, time progresses in the neutrino's frame of reference at a rate that is equal to that in the lab divided by gamma. Gamma is infinity; anything divided by infinity is zero—so time doesn't pass in the neutrino's frame of reference. If no time passes, it can't change flavors. It's as simple as that. If it changes flavors, it has to move at a speed less than that of light. In that case, time passes and there's a way for it to change flavors.

If it travels slower than the speed of light, then, in fact, it must have mass. That's the only way in which a neutrino, or anything else, can travel at a speed slower than light. It must have mass. A massless particle has to travel at the speed of light; otherwise, it doesn't exist. So, the great discovery here is that neutrinos have mass. That is sending the world of fundamental particle physics—the physicists studying that world—into a tizzy because the standard model of fundamental particles predicts that neutrinos should be massless. So, this cool result is affecting not just astrophysics and our understanding of the Sun, but also our understanding of the ways in which particles fundamentally behave in their microscopic and submicroscopic scales.

The experiment that's now being done at the Sudbury Observatory is to try to detect a difference in the rate of neutrino oscillations, or changes, when traveling through the Earth versus when traveling through the vacuum of space. Some theories suggest that neutrinos traveling through Earth change flavors at a different rate than neutrinos traveling through the vacuum. The results aren't out yet. I think the data has been gathered. The rumors I've been hearing are that indeed the rate of neutrino oscillations is different inside the Earth than when they are traveling through a vacuum. This is a fantastic story. The detection of neutrinos and neutrino oscillations is one of the greatest achievements of the past few decades. It affects not just astrophysics, confirming our understanding of the core of the Sun—it also affects fundamental particle physics, throwing a giant wrench into the theory, and that's really exciting.

Brown Dwarfs and Free-Floating Planets

Lecture 50

"As we've seen, true stars are defined by the fact that they produce nuclear fusion in their cores."

Deep in the core of the Sun, temperatures are so high that protons can fuse together through the proton-proton chain, forming helium nuclei. This nuclear fusion provides a long-term source of energy for the Sun and, presumably, for other stars. Pre-main-sequence stars release energy through gravitational contraction, replenishing the energy supply that is lost to surrounding space and heating up the gas. As contraction continues and the center of the star gets hotter, nuclear fusion begins to take place, providing a stable new source of energy and halting contraction. Thus, a star is born, having reached hydrostatic and thermal equilibrium. Some astronomical bodies cannot reach sufficiently high temperatures (at least 3 million K for the least massive stars; higher for more massive ones) to allow fusion to begin. Thus, contraction remains the only source of energy (except deuterium fusion; see below). Such bodies are called **brown dwarfs**, which some astronomers think of as failed stars. Brown dwarfs are cool, dim, and small, and they continue contracting until a new form of pressure takes over—*electron degeneracy pressure*.

Electron degeneracy pressure is a strange quantum-mechanical pressure arising from the fact that electrons repel one another (not electrical repulsion but, rather, quantum mechanical). Electrons are a type of fundamental particle called *fermions*, which cannot occupy the same quantum state. Another fundamental particle is a *boson*. Unlike fermions, two or more bosons (e.g., photons) can be in the same quantum state. Eventually, the density becomes so high in a brown dwarf that the electrons start overlapping spatially. To be in different quantum states, their momenta and, hence, their energies must be different. Some of the electrons must have very high energies and momenta because all of the lower-energy and lower-momentum quantum states are already fully occupied. These high-energy electrons exert an extra pressure—degeneracy pressure—which helps support the contracting brown dwarf. Before a brown dwarf becomes **degenerate**, temperatures are sufficiently high and densities sufficiently low that electrons are spread more randomly and exert normal thermal pressure. Brown dwarfs are about the same size as Jupiter. Because they are cool and small, they are faint and difficult to notice. What little light they do emit tends to be in the infrared wavelengths.

Brown dwarfs were predicted many decades ago but weren't found until the 1990s. Now, we know of at least 1000, the discoveries of which coincided with the explosive growth in studies of exoplanets. The spectrum of one of the first brown dwarfs showed absorption bands of methane, which is similar to the spectra taken of Jupiter. This confirmed that the object was a very cool brown dwarf; the presence of methane means that the atmosphere is much colder than that of the least massive stars. Brown dwarfs with methane in their atmospheres are called *T dwarfs*, while hotter ones are called *L dwarfs* (but some L dwarfs are genuine stars). After the first few discoveries, many more brown dwarfs were found. Sky surveys taken at infrared wavelengths reveal many brown dwarfs. Some of the brown dwarfs orbit nearby stars, but others appear to have formed in solitude. In order to truly know whether a star is a brown dwarf or an L-type main-sequence star, we need to know its mass. This tells us whether it's capable of high enough temperatures for nuclear fusion to occur.

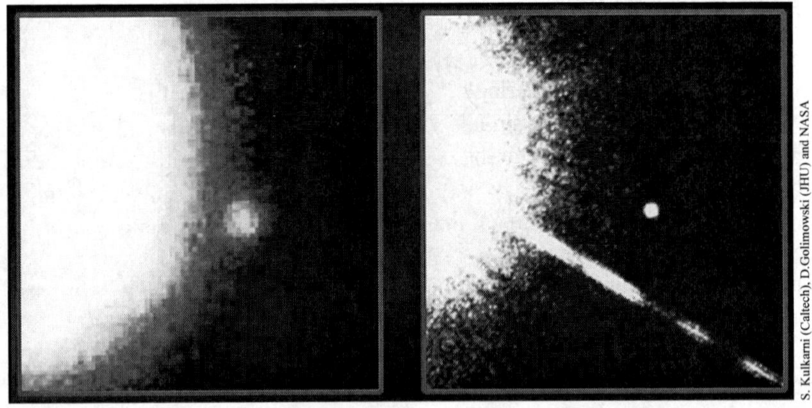
S. Kulkarni (Caltech), D.Golimowski (JHU) and NASA

Brown dwarf Gliese 229B as observed by Palomar Observatory (left) and the Hubble Space Telescope (right). (Note: The spike at right is an artifact of the telescope.)

Some brown dwarfs are actually in binary systems, allowing us to measure their masses using Newton's laws of motion and the law of universal gravitation. Just as there are exoplanets orbiting normal main-sequence stars, it is possible that exoplanets orbit some brown dwarfs. Debris discs have also been found around some brown dwarfs, which could potentially form small planets.

Brown dwarfs do experience fusion, though not through the proton-proton chain, which requires temperatures over 3 million K. At lower temperatures, *deuterium fusion* can occur, essentially bypassing the difficult first step in the proton-proton chain. Brown dwarfs begin fusion with deuterium (normally formed in the proton-proton chain), which collides with protons to fuse into light helium nuclei. This type of fusion is short-lived and occurs in brown dwarfs between 13 and 75 Jupiter masses. Above 75 Jupiter masses, the normal proton-proton chain occurs. Astronomers disagree about whether or not brown dwarfs should be called true stars or failed stars. One solution is to simply recognize that brown dwarfs and normal stars are "fusers." Normal stars undergo nuclear fusion of protons, whereas brown dwarfs fuse deuterium. Below 13 Jupiter masses, even deuterium fusion doesn't occur. Thus, we call these bodies planets.

Suppose we plot the distribution of the number of bodies discovered through the Doppler wobble technique (Lecture 38) against their derived masses. As we discussed when considering exoplanets, the Doppler wobble technique actually leads us to infer $M \sin i$—rather than the actual mass, M—where i is the inclination angle between the orbital plane and our line of sight and *sin* denotes "sine." Only the radial component of the total velocity is measured with the Doppler effect. Thus, if $\sin i$ is less than 1, the true mass (M) must be greater than the measured quantity, $M \sin i$. Objects that are 12 or 13 Jupiter masses, minimum, are almost certainly brown dwarfs. Some with a minimum mass below 12 Jupiter masses are probably also brown dwarfs. Even a few objects having a rather low minimum mass (say, 4 Jupiter masses) might actually be brown dwarfs if their true mass exceeds 13 Jupiter masses. Nevertheless, most objects with an inferred minimum mass below about 6 Jupiter masses are probably planets, not brown dwarfs; they don't experience fusion of any kind, even deuterium fusion.

What would we call an object of less than 13 Jupiter masses, not fusing deuterium, and not orbiting a star? Most astronomers would call it a free-floating planet because it has a planetary mass and it is not undergoing fusion. Some free-floating planets were ejected from their planetary systems, while others simply formed on their own through gravitational contraction out of a cloud of gas—like normal stars. Some astronomers would like to call the result of the first scenario a planet and the second scenario a brown dwarf, regardless of whether the object is massive enough for deuterium fusion to ever occur. Others don't want to apply the term *planet* to objects that are not orbiting another star.

"All this is telling us is that, in the process of star and planet formation, there's a range of masses that can occur. It's all part of one continuum, one process that leaves the same sorts of objects, but having a continuum of masses."

Gibor Basri, of the University of California, Berkeley, has proposed that we call objects *planemos* ("planetary mass objects") if they are at least massive enough to be spherical but not massive enough to be deuterium-fusing brown dwarfs. If a planemo happens to orbit a star, we call it a planet. Because we are finding more and more of these relatively low-mass objects through infrared studies of the skies, at some point, we will have to agree on the terminology and classification. Clearly, a range of masses can occur in the process of star and planet formation, from the most massive O-type stars to the least massive normal hydrogen-fusing stars, down to deuterium-fusing stars (brown dwarfs), and to planetary mass objects that don't fuse at all. ■

Important Terms

brown dwarf: A gravitationally bound object that is insufficiently massive to ever be a main-sequence star but too massive for a planet. Generally, the mass range is taken to be 13–75 Jupiter masses.

degenerate gas: A peculiar state of matter at high densities in which, according to the laws of quantum physics, the particles move very rapidly in well-defined energy levels and exert tremendous pressure.

Suggested Reading

California and Carnegie Planet Search, www.exoplanets.org.

Pasachoff, *Astronomy: From the Earth to the Universe*, 6th ed.

Pasachoff and Filippenko, *The Cosmos: Astronomy in the New Millennium*, 3rd ed.

Questions to Consider

1. What are the problems associated with requiring knowledge of an object's formation history to classify it as a brown dwarf, a planet, or something else?

2. Under what circumstances can the true mass of a brown-dwarf candidate identified through the Doppler wobble technique be determined?

3. Do you consider brown dwarfs to be failed or genuine stars?

Brown Dwarfs and Free-Floating Planets

Lecture 50—Transcript

The solar neutrino experiments discussed in the previous lecture prove conclusively that the source of energy inside the Sun is nuclear fusion. Deep in the core of the Sun, the temperatures are so high that protons can fuse together through the proton-proton chain, forming helium nuclei. This provides a long-term source of energy for the Sun, and presumably for other stars, which are simply distant suns. The fusion that occurs in the core is occurring at a temperature of at least 3 million degrees, and typically more like 10 million degrees. Indeed, in our Sun, the temperature of the core is about 15 million degrees. There's this fusion in the core, which then produces energy that leaks out through the rest of the star over the course of something like 50,000 to 100,000 years, and then is radiated out into space. That's how the Sun and other stars shine.

Prior to the onset of fusion, however, the Sun was a pre-main-sequence star. It was releasing energy through gravitational contraction, and that's what replenished the supply of energy in the star that was lost to the surrounding space. As the contraction proceeded, and the center of the star heated up, eventually this contraction stopped because nuclear fusion turned on, providing a stable new source of energy other than gravitational contraction. This resulted in the halting of the contraction because now there was no reason for the star to continue contracting, because it had a new supply of energy, nuclear fusion. At the point when nuclear reaction starts in a contracting pre-main-sequence star, that's when we say that the star is born. It achieves stardom once nuclear reactions start. The star achieves hydrostatic and thermal equilibrium. It's not contracting anymore; it's just staying put. It's producing energy at exactly the same rate that it's losing energy to the surroundings. It is in perfect equilibrium, and that is what we call a star.

In some cases, if the thing that's contracting has less than, say, about 1/12 of the solar mass—that's roughly 75 Jupiter masses, or 0.075 solar masses—then in that case, the temperature never gets high enough for fusion of ordinary hydrogen to begin through the proton-proton chain. You just don't get up to that temperature of a minimum of 3 million degrees. So, fusion doesn't begin, and contraction remains the only source of energy being leaked out into space. You've got to replenish the energy that's leaking out into space from a star. If you don't, the star will just contract—or collapse if it completely loses all of its energy. It replenishes the energy by contracting slowly, and that contraction process has to continue unless there is some other source of energy, such as nuclear fusion.

In a sufficiently low-mass contracting object, you can't get that nuclear fusion, and so the object continues slowly contracting and does not achieve stardom. It does not achieve this nuclear fusion. Such an object is called a brown dwarf. It's often also called a failed star, a star that was on its way to being formed, but nuclear fusion simply never started up. The term "brown" is a little bit misleading. It was chosen because these objects are very cool and very dim, so you can't see them very well. People didn't want to call them black dwarfs because black dwarfs will, as we see, be a term that's used for extremely cool white dwarfs, which is another class of objects that I'll talk about in two lectures. A lot of the names that you could potentially use had already been used up. Brown dwarf was chosen because they're cool, and dim, and small; they're dwarfs. They're sort of an extension of what the main sequence would have been in the regular temperature-luminosity diagram. You've got smaller masses, and you have smaller and smaller dwarfs. So, this is a dim, cool dwarf, and people simply called it a brown dwarf. I think Jill Tarter of SETI fame first came up with that term.

This brown dwarf continues contracting until a new form of pressure starts becoming important, in addition to the normal thermal pressure associated with random motions of particles. The new pressure is called electron degeneracy pressure—not because it's morally reprehensible or anything like that. It's simply the name that quantum physicists give to this weird quantum-mechanical pressure, having to do with electrons basically not liking each other. I'll explain more carefully what I mean in a few minutes. This sort of pressure also occurs in white dwarfs and neutron stars, so we're going to encounter it several times in this course. Why don't we discuss it here in some detail right now?

Electron degeneracy pressure results from the fact that electrons are a type of fundamental particle called a fermion. Fermions don't like each other. They don't like to be close to each other. More precisely, they don't like to be in the same quantum state. The quantum state of a particle, broadly speaking, consists of its position and its momentum, or mass times velocity, and its spin. Particles can spin in different directions. You can have two particles with the same position and momentum, but opposite directions of spin, say clockwise and counterclockwise, and they can occupy the same quantum state, other than the spin, because the spin makes them sufficiently different from one another. They're not completely identical.

Fermions, of which electrons are an example, don't like each other, and more than one fermion cannot be in a given quantum state. The other fundamental kind of particle is called a boson, and bosons are just the opposite. They love being in the same quantum state. Photons are bosons, and, in fact, the operation of lasers is based on this principle that photons like all being together in the same quantum state. What happens is that in this contracting cloud of gas, this brown dwarf, eventually the density becomes so high that the electrons start overlapping spatially quite a lot—they're in about the same place. To be in different quantum states, their momenta, and hence their energies—because the energy is basically proportional to the momentum squared—have to be quite different. Indeed, some of the electrons must have very high energies and momenta because all of the lower-energy and momentum quantum states are already fully occupied.

These electrons that have high energy and momentum exert an extra pressure, over and above the random thermal motion pressure that particles exert. Had they not had this constraint that they can't all go in the same state, they would have gone in the same state, a low state, and they wouldn't have exerted this quantum-mechanical pressure. Because they can't all go into the same or the same few lowest-energy states, some of them have to be in rather high-energy states, which exert this extra degeneracy pressure. Let me illustrate what I mean right here. Here is a schematic of what the energy states might look like in a brown dwarf. Energy goes up vertically along this axis. Energy = 0 is at the very bottom, and here we have all these energy states. Each energy state can have two electrons of opposite spin, but only two. You see that they all line up in these various energy states. By the way, they can't have zero energy because there's a quantum-mechanical jitter that they can't get rid of. No electron is in the zero energy state, but some are nearly at zero, and others are pretty high. It's these that are moving around rather quickly, and they exert a pressure that helps support the contracting brown dwarf.

Let me give you an analogy. It's like an apartment building with lots of apartments, but there's the restriction that only two people can be on a given floor of this apartment building. There's two, two, two, two, and two. The people on the highest floors are energetic people. To get up there, they had to have been energetic. You could say that by climbing those flights of stairs, they lost all their energy. But let's say they took the elevator, and they're living up there, and they're just energetic sorts of people, and they're just bouncing off the walls and stuff. That's what those electrons are doing. They're moving around very quickly because they're forced to do so through this quantum mechanics. They are the ones that exert this pressure. Now they can't go jumping down to lower energy levels because those lower energy levels are already filled. All these guys together, actually, work in a quantum-mechanical way to exert this extra degeneracy pressure. Here we illustrate it for a fully degenerate electron gas—that is, all the electrons are in the lowest possible state that they can be in.

Long before a brown dwarf becomes degenerate, the temperatures are sufficiently high, and the densities are sufficiently low, that the electrons are not degenerate. In that case, this apartment building looks like this. Here are those same levels, and you have one or two electrons per level, sometimes zero electrons per level, but they're all more spread out randomly. They exert the normal thermal pressures of a normal gas, a non-degenerate gas. But as the thing contracts and the densities rise, gradually the electrons start overlapping more and more, feeling each other more, and the degeneracy pressure starts setting in. Here at the right, we have a non-degenerate electron gas. As it contracts and the electrons start overlapping, some of them are in their lowest possible energy states; others are not. Now there's some extra degeneracy pressure.

Then finally, in the fully degenerate electron gas, all of them are in their lowest possible energy states, and all of the pressure is this electron degeneracy pressure. None of it is the random thermal motions that you had in a non-degenerate gas. It's a very weird sort of a pressure, and it's a very weird sort of an object, this brown dwarf. They're very small and very faint.

They have about the same size as Jupiter. They don't get much smaller than Jupiter because the electron degeneracy pressure doesn't let them. Being cool and so small, they don't emit much light, and so they're very difficult to notice. They especially don't emit much visible light because they're so cool. What electromagnetic radiation they do emit tends to be in the infrared.

Well, do they exist? For a long time, astronomers weren't sure. They were looking for brown dwarfs for literally decades. They were predicted many decades ago. In 1994 and 1995, the first few convincing cases of brown dwarfs were found. Now, a little over a decade later, there's over 600 known. This decade—12 years, let's say—has coincided with the explosive growth in studies of exoplanets. Exoplanets were discovered in the mid-1990s, and now we know of lots of them. Brown dwarfs were discovered in the mid-1990s, and now we know lots of them as well. Here's the first clear photograph of a brown dwarf. This one happens to be gravitationally bound to another star. In the Hubble Space Telescope image, it looks like there's a spike sticking out of the star. Maybe the star is firing a laser beam out at aliens near it or something. Don't pay attention to that spike; that's an artifact of the photographic process with the Hubble Space Telescope. Here's this normal star, and next to it—at the same distance from us—is this diminutive, faint, little star, which, in fact, is so dim that it can't be a normal main-sequence star. It must be a brown dwarf.

When the astronomers who found it took a spectrum of this object, they found in the spectrum absorption bands of methane. That's what you see when you take a spectrum of Jupiter or Saturn; you see absorption bands of methane, and ammonia, and things like that, but mostly methane. This really solidified the case that this is a brown dwarf. Indeed, such cool brown dwarfs that have methane in their atmospheres are called T dwarfs. The ones that are hotter are L dwarfs. Some of the L's are actually genuine stars. Remember: Oh, Be A Fine Girl; Kiss Me Lovingly? Well, some of the L's are main-sequence stars; other L's are sort of somewhat more luminous brown dwarfs than the T dwarfs. The T's are sort of an extension of this main sequence. You could say, "Oh, Be A Fine Girl; Kiss Me Lovingly, Tenderly, if you wish." It's sort of an extension of the main sequence, but fusion of the normal sort isn't going on.

After that first discovery, quite a few other brown dwarfs were found orbiting nearby stars. With the technique of adaptive optics, you can see them next to, for example, this bright star and this one here. Actually, these are the same star observed with the Gemini 8-meter telescope in Hawaii and the Keck 10-meter telescope in Hawaii. You can see that the adaptive optics process has made the main bright star look kind of serrated and weird, but that's just an artifact of the optical process. What's important is you can clearly see this faint brown dwarf right next to it. They're so faint and cool that the best way to look for them is actually in the infrared. If you take a look at several photographs, you'll see what I mean. Here's an optical photograph of a star field, where this is a negative image—bright things look black, and the black background sky looks white or gray.

Here's a faint little guy marked with an arrow. In the near infrared, somewhat longer wavelengths than the eye can see, you can see that the object is much brighter than it was in the optical relative to the other stars, which have stayed about the same. In the true infrared, the object is even brighter than it was in the near infrared or the optical relative to the other stars. These things emit a lot of light in the infrared—well, "a lot" is a relative term. They're very dim all around, but if you're going to find them glowing, you might as well look in the infrared. There have been some surveys recently of the whole sky, or much of the sky, at infrared wavelengths that have found lots of these things. Here's an example, where we zoom down into the Orion nebula, first at optical wavelengths. Then we will look at the same field in the infrared, and you'll see a whole bunch of faint, little stars and star-like objects pop up.

Some of those are simply faint L main-sequence stars; others are L-type brown dwarfs, and some are even cooler T-type brown dwarfs. The very faintest, little dots here may be brown dwarfs, and the brighter dots are L stars, which may or may not be brown dwarfs. It's a bit hard to tell sometimes. You have to know the mass in order to really know whether it's above or below this fusion dividing line. These surveys have found a truly remarkable number of brown dwarfs. They're all over the place, but they have low mass, so their overall contribution to the mass of our galaxy is relatively low. Main-sequence stars, though not as numerous, are much more massive, and they outweigh the brown dwarfs.

You can sometimes find binary brown dwarfs. Those are really good because, just as in the case of binary stars, you can use them to measure the masses of the two components using Newton's laws of motion and the law of universal gravitation. Here's an example of a binary brown dwarf. The images are a little bit elongated here because of the photographic process. Here's an artist's conception of what one might look like. If we zoom down into the Orion nebula, we can find one where, again, the artist is sort of simulating what the orbit might look like. But from the spectra of the brown dwarf, taken over the course of time, you can tell that there are two brown dwarfs orbiting each other there. That's great because we can figure out their masses. Indeed, the masses are below this 0.075 solar-mass dividing line, roughly 1/12 of a solar mass.

Just as there are exoplanets around normal main-sequence stars, you might expect there to be exoplanets around at least some brown dwarfs. Indeed, with adaptive optics, those have been found, at least in a few cases. Here is one particularly good case where the image of the brown dwarf shows next to it an even fainter dot. That dot seems to be associated with the brighter object, the brown dwarf, and it is thought that this is an exoplanet orbiting the brown dwarf. That's still a little bit of a controversial conclusion, but I think it's pretty solid. As more of these things are found, I'm sure that people will confirm that, "Yes, brown dwarfs do occasionally have planets going around them." Just as the first few exoplanets that were discovered—people thought, "Well, maybe those are brown dwarfs, interestingly enough." Only a few of the exoplanets had been discovered, and they were pretty massive—at least as massive as Jupiter, or maybe even more massive. People said, "Well, those aren't exoplanets; those are brown dwarfs." Now we know that they are exoplanets, at least in most cases.

In a similar way, I think that, as more of these guys are found, we will see that many of them are exoplanets. The more massive ones would be brown dwarfs in their own right. Debris disks have also been found around some brown dwarfs. Here's one where you can see the artist's simulation of the disk rotating, and planetesimals forming in that disk. Presumably this disk, corresponding to the artist's conception, will form a planet in the future, but it'll be a small planet, very close in to the brown dwarf because the disk that was found is a very physically small disk. So, it'll form a diminutive planetary system.

An interesting fact that even many astronomers don't know is that in brown dwarfs, there actually is some fusion. Some fusion does occur. Many astronomers think that brown dwarfs are completely devoid of fusion. That's not true. They're devoid of fusion via the normal proton-proton chain, where you start out with two protons that have to approach each other, and then the strong nuclear force grabs them and binds them together into a deuteron, and so on. That's hard to do. That requires temperatures of at least 3 million degrees. But at lower temperatures, another form of fusion can occur, and that's deuterium fusion. Let's go back to the first few steps of the proton-proton chain. This first step doesn't occur in a brown dwarf. That's where two protons came together, formed a deuteron, a positron, and a neutrino. I explained how that has to happen twice, and then two deuterons form helium nuclei, which then themselves fuse into a larger helium nucleus, and so on. That was the standard proton-proton chain.

But suppose you bypass this first step, and you already have some deuterium. It turns out deuterium is produced shortly after the Big Bang—I'll talk about that later. If you start with deuterium and bang a proton toward it, it can, in certain circumstances, fuse into a light helium nucleus, a He_3 nucleus. As long as there's some deuterium in the contracting object to begin with, that source of fusion will occur, that form of fusion. It occurs for a while. It doesn't occur for very long because the amount of deuterium in the universe, the abundance of deuterium, is only 1/100,000 of the abundance of normal hydrogen, of normal protons. The deuterium fusion doesn't last long in a brown dwarf, but it is there, and so the brown dwarf does fuse for a while.

This is true for brown dwarfs between 13 and 75 Jupiter masses. You get this fusion of deuterium for a while. So, some astronomers say, "Okay, they're fusers, but they're still not really stars because they're not fusing normal hydrogen in a normal way for a long time." Well, they're not doing it for a long time, but they are fusing deuterium for a while—so they're undergoing nuclear fusion. So, I (and others) think that they're not really failed stars. They're just not the normal sort of star, but they undergo fusion. A compromise would be, perhaps, not to call it a star, but to call it a fuser.

Some astronomers—like my colleague Gibor Basri—say that they're really stars, pretty normal stars. They just have a different fuel supply. Okay, they don't last long, but O-type main-sequence stars don't last very long either, just a few million years, and yet we call them stars. So, why don't we call brown dwarfs stars as well?

Then other astronomers say, "Yeah, O-type stars don't last very long, but at least when they're done with their hydrogen fusion, they turn into something vastly different." We will see later that they turn into a red supergiant, and finally they blow up. Once they're done fusing normal hydrogen, they become distinctly different. Whereas once a brown dwarf is done fusing deuterium, it doesn't become distinctly different. Those astronomers say that's why you shouldn't call brown dwarfs true stars. Isn't this reminiscent of the Kuiper-belt object Pluto—the discussion that we had before, where we don't really have a good, self-consistent definition of what is a planet? We don't really have a good, self-consistent definition of what a star is either, in that gray area between stars and planets. So, I think they're fusers. I think they're genuine stars, but other people don't.

Below 13 Jupiter masses, even deuterium fusion doesn't occur. Those, we would say, are planets. Some of the objects found by Marcy and Butler, and the other planet hunters of the world, are surely brown dwarfs. If you look at the distribution of the number of planets versus their mass, or more correctly, their minimum mass—the measured quantity is M sin i, so M might be bigger than the indicated quantity if sin i is less than one. Anyway, these guys that are already near 12 or 13 Jupiter masses at a minimum, those are almost certainly brown dwarfs because they're more massive than 13 Jupiter masses. The ones that are below 12 Jupiter masses, some of them might be above 13, and so they're probably truly brown dwarfs as well. But the ones that have low mass, those are planets because they're orbiting a star, and they don't have any fusion going on. The fact that there are more and more of them as you go to smaller and smaller masses, yet the smaller masses, if anything, are harder to find, really convinced astronomers that most of the objects that the exoplanet hunters are finding really are planets and not brown dwarfs. But some of them near the dividing line may well be brown dwarfs.

Now suppose you have an object below 13 Jupiter masses. It's not fusing deuterium, but it is not orbiting a star; it's just flying around on its own. What do you call that? Do you call it a planet? Most people think that a planet is an object that orbits another star. If this thing is the mass of a planet, but doesn't orbit another star, what do you call it? Do you call it a brown dwarf? Do you call it a free-floating planet? Most astronomers would say it's a free-floating planet because, though it's not orbiting a star and can't be seen from the light that it reflects from that star, nevertheless it has a planetary mass, and it is not undergoing fusion of any sort—so it's not a brown dwarf; therefore, it's a free-floating planet, some astronomers say. But then you might say, "Well, was the free-floating planet produced through an ejection from another planetary system?"

For example, here's a planet going around a star near another planet. If they get close to one another, a gravitational interaction between the two can eject one of them and leave the other one in a slightly perturbed orbit. That can be one way of forming free-floating planets. Indeed, a few years ago, there was a well-publicized case where it looked like a planet was being shot out from a star, and this was hailed as the first ejected planet ever found. Later, it was found out that, in fact, this was just a faint background star and has nothing to do with the foreground star. In any case, certainly some of the free-floating, planet-like things out there may well have been ejected from planetary systems. If they were once planets, why not continue to call them planets?

On the other hand, if they formed not in a planetary system like this, but rather if they formed on their own through gravitational contraction out of a cloud of gas—just like normal stars form, and just like brown dwarfs form—some astronomers say the formation process, being so similar to that of stars, means that these are really star-like objects. They're an extension of the brown dwarf sequence, and so we might as well call them brown dwarfs. They're not escaped planets, and they're not really planets at all. They formed through gravitational collapse—not in the disk of gas around a star the way normal planets form. People don't really know what to call these things.

My colleague, Gibor Basri, has a solution. He says, "Let's call objects that are at least massive enough to be spherical, but not massive enough to be deuterium-fusing brown dwarfs—let's call them planemos, planetary mass objects. If a planetary mass object, a planemo, happens to be orbiting a star, then we call it a planet. If it's not orbiting a star, then we call it just a planemo, or a free-floating planet. If you don't like the word "planet," just call it a planemo, a planetary mass object. He has recently proposed that terminology. I'm not sure that the astronomical community will accept it, but we will see in the future whether they do or not.

Quite a few of these free-floating planets, or planemos, have been found in the recent past through, once again, infrared studies of the sky. If you look at clouds of gas at infrared wavelengths, you find not only low-mass, main-sequence stars, and not only faint, small brown dwarfs, but you find some objects that are so faint, so dim, that the calculated masses are in this planemo range; they're below 13 Jupiter masses. So, more and more of these things are popping up in various surveys. They appear not at all to be rare, and we have to deal with them. I think what all this is telling us is that, in the process of star and planet formation, there's a range of masses that can occur—from the most massive O stars to the least-massive normal hydrogen-fusing stars, then down to the deuterium fusing stars, and then down to the planetary mass objects that don't fuse anything at all. It's all part of one continuum, one process that leaves the same sorts of objects, but having a continuum of masses. That has to teach us something about star and planetary formation.

Our Sun's Brilliant Future

Lecture 51

"Through observational studies of stars at different stages of their lives and using the physics of gases held together by gravity, astronomers can predict with accuracy how stars are likely to evolve."

The laws of physics help us understand how stars will evolve based on the nuclear reactions that occur, pressures and temperatures in the core, and other factors. In this way, we can predict what will happen to our Sun as it evolves. For about 9 or 10 billion years, the Sun will remain on the main sequence—that is, it will fuse hydrogen into helium in its core through the proton-proton chain, releasing energy. At 4.6 billion years old, our Sun is about halfway through its normal main-sequence lifetime. As the Sun evolves, it will gradually brighten, because as it fuses hydrogen into helium, its core temperature necessarily has to rise, and the fusion rate will increase. For every four initial protons, only one helium nucleus will be produced. Therefore, the number of particles per unit volume in the middle of the Sun—that is, the number density—will gradually decrease. Pressure is proportional to the product of number density and temperature. If the number density decreases (because hydrogen is fusing to helium), the temperature has to rise to compensate and maintain the same pressure. In a few hundred million years, Earth will be substantially warmer because the Sun's luminosity will have increased somewhat. In about half a billion to a billion years, the oceans will be gone, and the Earth will be like a scorching desert.

It's possible that some compensating effect will keep Earth's temperatures lower than anticipated. We know that an inverse compensating effect must have occurred billions of years ago when the Sun's luminosity was about 30% lower than it is now. Without such an effect, the Earth's oceans would have been frozen, yet fossil evidence proves that certain life forms existed in relatively warm conditions. It's possible that a more pronounced greenhouse effect was occurring on Earth at that time.

What will happen in about 5 billion years when the Sun reaches the end of its main-sequence life? A similar future awaits other sun-like stars. Its helium core will increase to about 10% to 15% of the total mass. Remember, the core is the only part of the star where temperatures are high enough for nuclear fusion to be sustained. However, nuclear reactions in the core will stop because helium nuclei repel each other much more than protons do, and the temperature will not be high enough for helium nuclei to fuse into heavier elements. Heat will gradually diffuse out of the helium core, causing it to gravitationally contract in order to replenish the supply of energy lost to its surroundings. The energy released by the contracting helium core will heat the surrounding hydrogen-fusing shell. This will increase the fusion rate in that hydrogen shell, causing the star to become much more luminous than before. The energy liberated by the hydrogen-burning shell causes the surrounding layers (all the way out to the surface) to expand. Thus, contraction of the helium core leads to an expansion of much of the rest of the star.

"Our Sun will probably form a spherically symmetrical planetary nebula. We don't know exactly what it'll look like, but I'm hoping that it'll be really pretty, so that future aliens, looking at our dying Sun, will say, 'Wow, that's a real beauty.' "

The expanding atmosphere of the star cools and its color shifts to more yellow, then orange (like that of a K-type star), and eventually, perhaps even somewhat red, because the peak of the spectrum shifts to longer and longer wavelengths. The Sun will become a red giant—perhaps 100 times more powerful than it is now—and much more luminous because of the vigorously burning hydrogen shell around the small helium core. The Sun will bloat to such a large size in its red-giant stage that it might extend to the orbit of Mercury. Contraction of the helium core will heat it up. Eventually, this slowly contracting helium core will reach temperatures of about 100 million K, sufficient for three helium nuclei to fuse, forming a carbon nucleus. The carbon nucleus can pick up another helium nucleus and turn into oxygen. Both of those reactions, the formation of carbon and the formation of oxygen, liberate still more energy. At that point, the Sun will have two sources of energy: the helium-fusing core surrounded by an inert helium shell and a hydrogen-burning shell around that.

The fusion of helium into carbon and oxygen lasts only about 1 million years because the individual fusion reactions don't produce much energy compared to the original proton fusion reactions. Yet because the star is very luminous, it produces a prodigious amount of energy quickly; therefore, helium is used up rather quickly, and a carbon-oxygen core forms. The carbon-oxygen core does not have high enough temperatures for fusion; thus, it begins to contract, just as its predecessor helium core had done. That contraction heats the carbon-oxygen core and liberates energy to a helium-fusing shell, which burns even faster, making the star's luminosity rise. The hydrogen-burning shell also fuses at a faster rate, releasing even more energy. This extra energy pushes out the outer envelope of the star even more, causing it to become a still larger red giant, which in the case of the Sun, might encompass Earth's orbit (or at least Venus's orbit). Different stars have different red giant time phases, ranging from about 100 million years to a few billion years. The Sun's red-giant stage will last about half a billion years.

During the red-giant stage, our Sun will become unstable, and its outer layers will be ejected in a series of relatively nonviolent outbursts. At such a huge size, a star's outer atmosphere is barely bound to the star; gravity is weak there because the gases are so far from the core. The star begins losing its outer atmosphere through a steady *stellar wind* (analogous to the current solar wind) as radiation pushes out the gases. The star also becomes unstable, oscillating in size, and some of these pulsations actually eject the outer parts of the star—10% to 20% of its mass—in a relatively nonviolent way, like a cosmic burp. When such material is ejected, the star becomes an expanding, glowing shell of ionized gas. The ejected shell can appear in the shape of a disk or a ring, called a **planetary nebula**, though it has nothing to do with planets. The term was derived in the 19th century before scientists knew that these nebulae were actually dying stars. Over a few tens of thousands of years, the nebula's light spreads so much that it fades.

The gas in a planetary nebula is ionized because many ultraviolet photons are emitted from the hot central star, whose surface used to be the core of the star (prior to the ejection of the outer atmosphere). The gas glows as the electrons recombine with positive ions. Also, electrons flying around in the gas hit other electrons bound in atoms, kicking them up to higher energy levels. These excited electrons subsequently jump back down to lower energy levels, thereby emitting light. Many interesting photographs have been taken of such light emitted from planetary nebulae, showing not only brilliant colors but some fascinating structures. Deep photographs can capture layers ejected long ago.

Dying stars can also produce *bipolar ejections*—that is, ejections that occur along an axis, forming planetary nebulae that are not spherically symmetric but, rather, shaped somewhat like an hourglass. We think that bipolar ejections are formed in a binary system in which one star expands into a red giant and begins transferring material to its companion. The transferred material envelops the companion and the red giant. As the common envelope contracts, a disk forms, forcing the ejecta of the expanding nebula to be expelled predominantly along the plane of the rotating disk. ■

Important Term

planetary nebula: A shell of gas, expelled by a red-giant star near the end of its life (but before the white-dwarf stage), that glows because it is ionized by ultraviolet radiation from the star's remaining core.

Suggested Reading

Kippenhahn, *100 Billion Suns: The Birth, Life, and Death of Stars*.

Pasachoff, *Astronomy: From the Earth to the Universe*, 6th ed.

Pasachoff and Filippenko, *The Cosmos: Astronomy in the New Millennium,* 3rd ed.

Questions to Consider

1. Do planets form directly from a planetary nebula?
2. How can we be so confident in our theory of the Sun's future evolution?
3. It is often said that we on Earth have about 5 billion years before we need to worry about the Sun's death. Why is this incorrect?
4. Why doesn't the helium core of a red giant immediately start fusing to heavier elements?

Our Sun's Brilliant Future

Lecture 51—Transcript

In the previous lecture, I discussed stars of such low mass that the temperatures in the cores are not high enough to allow normal hydrogen fusion to occur. These are the so-called "brown dwarfs," failed stars. But they're not completely failed because deuterium fusion does occur for a short time. Now let's get back to the Sun and consider its evolution. What will happen to it in the future as it ages? For about 9 billion or 10 billion years, the Sun sits on the main sequence—that is, it fuses hydrogen into helium through the proton-proton chain, releasing energy along the way—and it's about halfway through its normal main-sequence lifetime. It's about 4.6 billion years old, so it's a middle-aged star. The hydrogen that fuses into helium releases energy because the helium product is slightly less massive than the original four protons, or hydrogen nuclei, that went into making it, and that mass deficit gets converted into energy. It's just $E = mc^2$, and it's a huge amount of energy, despite the fact that only 0.7% of the mass of the original hydrogen got lost, or used up, in binding energy to form the helium nucleus.

What will happen when the Sun starts running out of hydrogen fuel in its center? What the Sun will do in its future will be typical of what other low-mass stars do in their future. By studying the Sun as a representative star, we can understand other stars as well. Conversely, if we look at other stars at various stages of their lives, we can come to an understanding of what our Sun will do. It's sort of like, if you want to study how a typical person evolves during their lives, you can study one person and assume they're typical—and from studies of that one person, you deduce how the rest of them work. But it takes less time, and you get a more representative sample, if you study a bunch of people at different stages of their lives: babies, adolescents, adults, and older people. You say, "Okay, the babies of today will transform themselves into adolescents, and then adults, and then older people." This will be representative of how humans as a class, as a species, evolve during their lives. You can make that kind of a study in a short amount of time by looking at groups of people, and you get a much more representative group than if you just choose one person and study his or her life for 70 or 100 years.

We can learn about our Sun as a representative star by looking at other stars in different stages of their evolution. We can also use the laws of physics and try to understand how the stars will evolve, and how the Sun will evolve, based on the nuclear reactions that occur in gases, and the pressures and temperatures associated with them, and so on. The study of the physics of gases is a pretty advanced state. We know how gases behave pretty well, so we're pretty confident about the broad overview that I'm going to give you in this lecture. The details might still be a bit shaky, but we're pretty certain of the overall concepts. Here's a temperature-luminosity diagram, where you see the main sequence, and the Sun is sitting on the main sequence. Stars having different initial masses are at different places on the main sequence. The initial mass with which the star is born dictates where it lands on the main sequence. Here are the low-mass stars, intermediate-mass stars, and then the massive stars, the most luminous ones of all.

As the Sun evolves, it'll gradually brighten, ever so slowly. The reason for that is it'll be gradually using up the hydrogen in its core, turning it into helium. For every four initial protons, there will end up only one helium nucleus. Therefore, the number of particles per unit volume in the middle of the Sun—that is, the number density—will gradually decrease. Pressure is proportional to the product of number density, number of particles per unit volume, and temperature. If the number density is decreasing because hydrogen is forming into helium, to compensate and maintain the same pressure, the temperature has to rise. The temperature in the core of the Sun will gradually rise, and that'll mean that the protons will come closer together as they're all zooming around. There'll be more chances for the strong nuclear force to grab hold of them and fuse them. Therefore, the fusion rate will increase, and the luminosity of the Sun will slightly rise.

Let me show you how that works, or at least how much it will rise, if I look at this animation here. Right now, the Sun has a central temperature of about 15 million degrees, and it has a luminosity of 1 solar luminosity. If the temperature rises by a few million degrees, the luminosity will rise as well—not a lot; I've exaggerated the effect a little bit right here. So, in a few hundred million years, the Earth will be substantially warmer because the Sun's luminosity, or power, will have increased a little bit. The oceans will start boiling away. Indeed, in about a billion years, the oceans will be gone, and the Earth will be bone dry. Worldwide, it will be like a scorching desert. We'll have to do something about life on Earth if we want to survive. We'll have to somehow find ways to make water and store it deep underground. I don't know; I don't have the imagination right now to know what we should do. Maybe we should move off the planet entirely.

In any case, these are all very long time scales, and if we don't do something about the threat of a devastating comet or asteroid collision with Earth, that'll actually kill us first. There are a number of cosmic catastrophes we need to worry about. The oceans boiling away, I think, is on a longer time scale than the comet and asteroid threat. Maybe there will be a compensating effect that'll keep the temperatures lower than anticipated. We know that a compensating effect of an inverse type must have occurred when the Earth was a lot younger—because when the Earth was a lot younger, billions of years ago, the Sun's luminosity was about 30% lower than it is now. Remember, the luminosity is gradually increasing, ever so slightly. Billions of years ago, it was 30% lower.

That means that the Earth's temperatures would have been significantly lower than they are right now—and, in fact, the Earth should have been in a deep freeze because, with significantly lower temperatures, the oceans basically freeze over after a while. If the Earth were in a deep freeze, we wouldn't have the sorts of life that we see back then: 3 billion, 2 billion, 1 billion years ago. There are certain forms of life that we can tell existed in relatively warm conditions. Something was keeping the Earth significantly warmer billions of years ago when the Sun's luminosity was lower. We think that may have been a more pronounced greenhouse effect on the Earth. That's what kept it warmer; that's the speculation. Maybe in the future when the Sun will be too hot, and the luminosity will be too high, then maybe there will be some other compensating effect that'll keep the Earth's temperature low. But without that compensating effect, the Earth will become hot, the oceans will boil away, and it'll be a rather inhospitable place.

What will happen in about 5 billion years when the Sun reaches the end of its main-sequence life? It will have built up a helium core of about 10% to 15% of its total mass. That's the part of the star where the temperatures are high enough for nuclear fusion to be sustained. You have such high temperatures only in the central 10% to 15% of the star's mass, not in the whole thing. In that core, where a helium core will have been built up, nuclear reactions will stop because the helium nuclei repel each other much more than protons do. Remember, helium nuclei have two protons each, so the electric repulsion of helium nuclei is a lot greater than the electric repulsion of protons. You need higher temperatures to get the helium nuclei sufficiently close together to fuse through action of the strong nuclear force.

For a while, at least, there will be no nuclear reactions in the core, and yet light will be coming out from the helium core. It goes from hot regions to cold regions; heat diffuses out. Therefore, the helium core will have no choice but to slowly gravitationally contract to replenish the supply of energy lost to its surroundings. That contracting helium core will also heat up because gravity is doing work on the gas during the contraction. It's pulling on it, compressing it. A good analogy is a pump and a bicycle tire or a basketball. If you compress the air in the basketball or the bicycle tire, that air heats up. You can feel it if you just put your hand on the rubber. The heat from the air is slowly diffusing through the rubber, and you can actually feel it. You've been compressing the air and doing work on it, so you heat it.

In a similar way, gravity compressing the core heats that core, and you end up with a structure that looks like this: a contracting helium core that's getting ever hotter due to the gravitational contraction, and a hydrogen-burning shell around it. This is a region where the temperatures are sufficiently high to fuse hydrogen into helium, so that's adding to the mass of this contracting helium core. But meanwhile, the core itself is getting hotter and heating the region surrounding it. So, what happens is that the hydrogen-burning shell feels the heat of the core and starts fusing hydrogen more quickly than it would otherwise have done. It's being, in a sense, propelled by extra energy from this contracting helium core. The hydrogen-burning shell kind of goes wild. It starts producing energy at a much greater rate than it normally would have, and that energy gets soaked up by the surrounding layers—which respond by expanding. So, the contracting helium core leads to an expansion of much of the rest of the star.

The expanding atmosphere of the star cools. An expanding gas cools for the same reason that a gas that's contracting heats up. As the star expands and cools, its color shifts from sort of whitish, in the case of the Sun, to more yellowish, to orange, to red, because the peak of the spectrum shifts to longer and longer wavelengths. As the Sun bloats out like this, its color will change. It'll become a giant star, and it'll become more orange or red. We will then call it a red giant. It'll be much more luminous than it is right now because of this very actively fusing, hydrogen-burning shell. It's just kind of going wild. The Sun might become 100 times more powerful than it is right now, and Earth will be fried. We will be completely devastated by then. With a 100-times-more-powerful Sun, there's not going to be any moisture left on Earth.

Indeed, the Sun will become so big in its red giant stage that it might come out even to the orbit of Mercury. So, it won't quite reach the Earth, but it'll be a very big star at that point. The structure of the Sun will then be like this: a small helium core with a hydrogen-burning shell around it, vigorously burning—and an expanding, bloated envelope of hydrogen. Plotting it on the temperature-luminosity diagram, we see that here's the main sequence. Stars sit on the main sequence, relatively unchanging for a long time. They change, as I said before, by a few tens of a percent in luminosity. But then, when the hydrogen runs out in the core, and the core turns into helium, and you have this hydrogen-burning shell, that's when the star's structure really starts changing fast, and it evolves into a red giant. Its surface temperature goes down, so it becomes more like a K star, and its luminosity rises because its radius is rising. So, the thing will be a big, huge star—more orange or red than it is right now. That's what we call a red giant.

Eventually, this slowly contracting helium core will reach temperatures of around 100 million degrees. At those temperatures, three helium nuclei can come together and form a carbon nucleus, through this process of nuclear fusion. Once you've done that, it's relatively easy for the carbon nucleus to pick up another helium nucleus and turn into oxygen. Both of those reactions—the formation of carbon and the formation of oxygen—liberate still more energy. So, at that point, the Sun will have two sources of energy: the helium-fusing core and a hydrogen-burning shell around it. This fusion of helium into carbon and oxygen doesn't last very long; it lasts only about a million years or so. That's because each fusion reaction doesn't give all that much energy compared to the original proton-fusion reactions. Per reaction, they give a lot more energy than when you fuse helium together.

Yet the star is a very luminous star at this point—it's a red-giant star—so it's producing energy at a prodigious amount and a prodigious rate, and so it has to do a very large number of reactions per second in order to produce that luminosity. Therefore, the helium gets used up rather quickly, in only about a million years, and a carbon-oxygen core forms. Then you have this carbon-oxygen core. Its temperatures aren't high enough for fusion, so it begins to contract—just as the helium core, its predecessor, had done. That contraction heats the carbon-oxygen core, liberating energy to a helium-burning shell around it, causing the helium-burning shell to burn even faster—that is, to fuse even faster than before.

That causes the hydrogen-burning shell to fuse even faster than before, and this means that the star produces really a large amount of energy per unit time. That means its luminosity rises really a lot. That energy pushes out the outer envelope of the star even more, causing it to become an even bigger red giant. At this point, the star becomes a really huge red giant, and might even encompass Earth's orbit. That's how big the Sun might become. The calculations are still a little bit uncertain as to just how big the Sun will become in about half a billion years when it becomes this red giant, and then an even-bigger red giant. Different stars spend different amounts of time reaching and staying in the red-giant phase. It goes from something like 100 million years to a few billion years. For the Sun, that stage, the red-giant stage, will last about half a billion years or so.

At this point, with the star being so big, its outer atmosphere is barely bound to the star—that is, gravity is weak way out there because the outer atmosphere is so far from the star—and the star becomes unstable. It can start losing that outer atmosphere through a steady wind. The radiation is pushing out on the gases and making them move away from the star. We call this a stellar wind, kind of like the solar wind. But also, the star can become unstable sort of pulsationally. It starts oscillating, and some of these pulsations actually eject the outer parts of the star in a relatively non-violent way that I call a "cosmic burp." The outer 10% or 20% of the star's mass can be ejected. There can be many such ejection episodes, which—together with the steadier winds—gradually erode away at the star's atmosphere, exposing more and more of the inner core.

What you get after one of these burps, one of these ejections, is an expanding shell of ionized gas, of gas that's being ionized by the photons from the hot central core, which is a very small object at this point. Those expanding shells of gas can glow and form a disk, or a ring, as pictured here. This is an example of what's known as a "planetary nebula." There is the dying central star. It's ejecting these shells of gas, and they're glowing. Here's another one; look at the beautiful structure in this one. There's the dying star, and here are these shells of ejected gas. The gas can clump up, as was the case in this particular planetary nebula. The term "planetary nebula" came from the 19th century, when astronomers using small telescopes looked at these little disks of gas, didn't know physically what they are, and said, "They kind of look like planets." But they're clearly nebulae.

You can tell from the spectrum that there are ionized gases there—so, these are not planets, but astronomers called them "planetary nebulae." Even in cases where the nebulae look like rings, such as in this one, the famous Ring Nebula in the constellation Lyra, some of the early telescope astronomers used didn't distinguish the ring very well, and so even the ring-like objects were called planetary nebulae. Here's another example. This one actually is about as big as the full Moon is in the sky. These things have nothing to do with planets. This is not the disk from which planets form. Again, this is just the term that early astronomers used.

Why do these objects glow so much? If you look at this diagram, you see the hot central star. It's hot because this is the inner region of the star, and that's where there's a lot of hydrogen and helium fusing in these shells surrounding the core, and a lot of ultraviolet photons get emitted. Those ultraviolet photons can ionize the cloud that's being ejected away from the star. I'm going to illustrate that here by saying that I'm an electron, bound to the nucleus of an atom. There's the nucleus there, that ball. I'm bound, but then I absorb an ultraviolet photon from the central star, the central core. I absorb that energy, and I become ionized. I'm no longer bound to that nucleus; I'm just sort of freely wandering around.

Eventually, I'll find a nucleus that has a positive charge, and I'll be attracted to it, and I can jump into a bound state, releasing a photon in that process. I went from a free state to a bound state, one of lower energy, so I had to get rid of a photon. I emitted a photon. Then I can cascade down to progressively lower levels by emitting more and more photons. In this case, I emitted first a red photon, and then a green photon because the levels that I jump between corresponded to energies, in this hypothetical case, of a red photon and a green photon. So, the energy of the ultraviolet photon that ionized me got transformed into a bunch of photons that come out from this nebula, causing the nebula to glow.

The other way in which the nebula can glow is if an electron flying around in this nebula hits me. Through that collision, it can knock me into a higher level. Someone in the audience can kick me up to a higher level like this, up to a higher electronic energy level, and then I come back down and emit a photon in the process. So, the kinetic energy of electrons flying around and hitting other electrons bound in atoms can transform that energy to emitted light. That's why the nebula glows. You can illustrate this process here through this diagram. The photon comes in, ionizes the atom—the electron is just wandering around on its own now—and then it meets up with another proton, say, releases some energy, and then cascades down to the lower energy levels, emitting photons along the way. Or, as I said, a collision can knock an electron from one bound state to another bound state—and then in coming back down, that electron can emit some light.

The spectrum of the gas might look like this. Here are spectra of many different gases: hydrogen, helium, sodium, and mercury. But in this case, the important point is that these are emission lines, not absorption lines, because the gas is emitting the light. It's not absorbing light from within; it's not like a normal star. It's a glowing nebula now, not the photosphere of a normal star. By taking advantage of the fact that the nebula is emitting mostly in a few strong emission lines, amateur astronomers have taken some stunningly beautiful photographs of planetary nebulae. Here are a few of them, in this case by Richard Crisp. You can take some very nice photographs if you don't take into account these emission lines that I'm talking about, but Richard Crisp and some of his colleagues have taken beautiful photographs by knowing that the nebula emits in certain emission lines, and using that fact to their advantage.

What they've done is they've taken photographs by using filters that pass only the light of a particular emission line. The light from that line goes through the filter, and everything else gets blocked; all other forms of light get blocked. By taking several different photographs through filters that allow light from different emission lines through, they get the structure of the nebula, as seen in these different emission lines. If you then combine all the photographs, you can get a truly stunning picture like the ones that I just showed. Here's a photograph not of a planetary nebula, but of the remnant of an exploding star, the Crab Nebula; it was made by a similar process. To illustrate that, here is a spectrum of the Crab nebula, where we plot brightness versus wavelength. You can see these emission lines by isolating the light in the emission lines. Here's one of doubly ionized oxygen. Here's an image in hydrogen light. Here's an image in singly ionized nitrogen, and here's an image in sulfur that's singly ionized. You combine all those images, and then you get these stunning photographs. I think they're truly marvelous.

As the nebula spreads out, it fades. Over a few tens of thousands of years, maybe 50,000 years, it spreads so much that it fades out, and you can't see it very easily. But there can be many such ejection episodes over the course of a million years or so spent in the planetary nebula, or in this stage where the star is able to emit a planetary nebula. Indeed, you can see in photographs evidence for multiple ejections. Here's one ejection near the central star, and here's an earlier ejection, which has spread out more. Here's another star, where you can see a very complicated ejection. But if you take a deeper picture—the central part of this picture corresponded to the picture I just showed you—you can see that farther out, there's this complex structure, indicating previous ejections earlier in time. Again, amateur astronomers can take really stunning photographs of these planetary nebulae—in some cases, even better than photographs that I've seen taken with space telescopes. Here's the Ring Nebula in Lyra, photographed by Geoff Collins. This is a normal photograph through broad filters, allowing lots of different light to pass. But if you then supplemented this with a photograph taken through a filter that only allows hydrogen emission lines to pass, you get this amazing outer structure, which can be further enhanced through photographic processing, digital processing, to show this beautiful, flower-like arrangement of gases, corresponding to previous sets of ejections, or planetary nebular formation stages, in the slow death of this single aging star down here. Here's the corresponding Spitzer infrared photograph, and I actually think that the amateur astronomer has a more stunning photograph than that of Spitzer.

Of course, Hubble does provide really great resolution, or clarity. If we look at some of the Hubble photographs, we can see really amazing structure in some of these planetary nebulae: all sorts of little knots and things, where clumps of gas apparently collected and were rammed into by other gases moving more quickly. Here, in fact, is an animation of a planetary nebula forming, and it's ejecting gases. But then later on, a more rapid wind, a more rapid ejection, occurs, and it compresses the gases of the previous ejection, forming the overall shape and structure of this nebula, and giving rise to these comet-like clumps, or streaks, that look really cool. They look like little teardrops or little comets. With Hubble, you get this kind of resolution that allows you to see those kinds of things.

The other thing that's fun to see is evidence for bipolar ejection in the deaths of stars. Some planetary nebulae have not a spherically symmetric appearance, but more an appearance that suggests an axis of symmetry, like this one or that one. Look at that; there's an axis of symmetry, and the ejection occurs along that axis, not in a spherically symmetric way. What we think is going on is that there's a disk of gas here, which is preventing the gas from expanding very much along the direction of the disk—but perpendicular to it, it can expand more easily. Here in this photograph, you can almost convince yourself that you see the disk that's blocking expansion in this direction and allowing it in the perpendicular direction. That's one of my favorites of these planetary nebulae that show this bipolar ejection.

We think that the disk and the bipolar ejection are formed when you have a binary system, and one star gets older and expands into a red giant, starts transferring material to its companion, and actually envelops the companion. Now both stars are inside a common envelope. As that common envelope contracts, a disk forms, then when that star decides to become a planetary nebula, the ejection occurs predominantly along the axis of the disk. So, that's how you get these bipolar ejections. Our own Sun probably will not form a bipolar planetary nebula; rather, it'll probably form a more or less spherically symmetrical one. We don't know exactly what it'll look like, but I'm hoping that it'll be a really pretty one, so that future aliens, looking at our dying Sun, will say, "Wow, that's a real beauty."

White Dwarfs and Nova Eruptions

Lecture 52

"By looking at more advanced stages of evolution of other Sun-like stars, and by using the laws of physics—in particular, the physics of hot gases—we have deduced that, in about 5 or 6 billion years, our Sun will expand greatly and then eject its outer envelope of gases, becoming a beautiful, glowing planetary nebula."

In this lecture, we will examine what happens when relatively low-mass solitary stars (such as the Sun) die at the end of their lives. Stars with initial sizes of between 0.08 and 8 solar masses and, perhaps, up to 10 solar masses eventually expel their outer atmosphere of gases during the planetary nebula stage and cease all nuclear fusion to become *white dwarfs*. The Sun is a representative star in this range. After the Sun becomes a red giant and then a planetary nebula, the remaining star (what used to be the core of the red giant) consists of carbon and oxygen. The nuclei repel each other so strongly that fusion cannot take place. Even in the helium shell around the carbon and oxygen core, temperatures are too low for fusion to be maintained. The core continues to contract as the star loses energy, increasing its density. Electron degeneracy pressure keeps the star from contracting indefinitely; it eventually reaches an equilibrium size. Heat is still liberated from the dying star by electrons moving to the lowest energy levels and from positively charged atomic nuclei. Yet no new energy is created because nuclear reactions and gravitational contraction have stopped. White dwarfs are about the size of Earth. The radius of a white dwarf is proportional to its mass raised to the −1/3 power: $R \propto M^{-1/3}$. As mass increases, radius decreases because of the high compression of electrons by gravity. A tablespoon of white dwarf material would weigh several tons.

"At the ends of their lives, stars can do a variety of interesting things: from the surface explosions of white dwarfs, to instabilities in more-massive stars that cause them to brighten and fade occasionally."

The atomic nuclei of a white dwarf are not degenerate; they can still lose energy, because although the electrons are all in their lowest energy states, the nuclei are not. Thus, white dwarfs can be thought of as "retired stars." Their light comes from the supply of thermal energy built up in the nuclei over the star's lifetime. A white dwarf gradually becomes dimmer and dimmer as the atomic nuclei cool down. After tens of billions of years, white dwarfs are no longer visible, and they become *black dwarfs*—though there is no sharp dividing line between white dwarfs and black dwarfs, and some astronomers avoid this term. Black dwarfs are still supported by electron degeneracy pressure, but the positive ions inside have cooled to low temperatures. If we could touch a very cold white dwarf (black dwarf), it wouldn't burn you, despite the fact that many electrons are moving at very high speeds. All of the electrons are already in the lowest energy states possible. In other words, the electrons can't transfer energy to a touching hand because they can't move to lower energy levels and give off excess energy.

Now let's review what we have learned about the post-main sequence, or after-main sequence, evolution of a Sun-like star. We'll also add some details and examine the physical properties of white dwarfs. The position of a star on the temperature-luminosity diagram is dependent on its mass. A main-sequence star remains nearly unchanged in luminosity and surface temperature for a long time. (It grows somewhat brighter with time, but to a first approximation, we can ignore this.) Once a star's core hydrogen is used up, the helium core contracts and the star transforms into a red giant. The helium then fuses to carbon and oxygen. Contraction of the carbon-oxygen core turns the star into an even larger red giant. The red giant's outer atmosphere becomes unstable and is dispersed through winds and gentle ejections. The temperature of the central star's surface increases as its outer layers are peeled away. This pattern of degeneration to a white dwarf is what happens in general to all stars between about 0.08 and 8 solar masses and, perhaps, up to 10 solar masses, following their long period of stability as main-sequence stars.

It turns out that what happens in detail to a star as it dies depends on its mass at birth. Regardless of its initial mass (up to 8 solar masses), those stars that become white dwarfs will always be less than 1.1 solar masses in the white dwarf stage. Our Sun will end up as a roughly 0.6-solar-mass white dwarf. Stars with initial mass below 0.45 solar masses will consist of helium in the white dwarf stage; those stars never achieve a sufficiently high temperature for helium to undergo fusion into carbon and oxygen.

A star with an initial mass between 8 and 10 solar masses fuses carbon and oxygen to form oxygen, neon, and magnesium in its core. Though we don't know for sure, such a star could turn into an oxygen-neon-magnesium white dwarf having a mass perhaps somewhat larger than 1.1 solar masses. Other calculations say that such stars explode at the end of their lives. We are not yet certain what will happen. Stars initially greater than 10 solar masses eventually explode, which we'll discuss in Lecture 53.

White dwarfs have a theoretical maximum mass of 1.4 solar masses. This is known as the **Chandrasekhar limit**, named for a great Indian astrophysicist who derived it. The limit occurs because as a white dwarf accumulates material, its radius shrinks. Eventually, the radius is so small and the electrons are squeezed into such a tiny volume, that their speeds approach the speed of light, and their ability to exert more pressure diminishes.

A star in a gravitationally bound binary system can change its mass by accreting material from its companion. Thus, stars whose initial masses were low can increase in mass—and, hence, core pressure—causing them to behave like more-massive stars, which in turn, speeds up their evolution. In a binary system in which one star is already a white dwarf and its companion star begins "feeding" it material through an accretion disk, a sudden brightening can occur, called a *nova*. The eruption is caused when accreting material forms clumps that fall onto the white dwarf's surface, releasing gravitational energy. The white dwarf can brighten by a factor of 100, sometimes even more, for a few weeks. Or the material can accumulate on the surface and undergo an uncontrolled chain of nuclear reactions, releasing even larger amounts of energy and making the white dwarf up to a million times brighter for a short time.

In addition to white dwarfs, other stars can undergo such rapid eruptions, though the physical mechanisms may differ. One example is Eta Carina, in the southern hemisphere, a massive star that sometimes brightens considerably. Clearly, at the end of their lives, stars exhibit a variety of interesting phenomena, from the surface explosions of white dwarfs to instabilities in more massive stars that cause them to occasionally brighten and fade. ■

Name to Know

Chandrasekhar, Subrahmanyan (1910–1995). Indian-born American astrophysicist. Awarded the Nobel Prize in Physics in 1983 for his work on the physical understanding of stars, especially the upper mass limit of white dwarfs.

Important Term

Chandrasekhar limit: The maximum stable mass of a white dwarf or the iron core of a massive star, above which degeneracy pressure is unable to provide sufficient support; about 1.4 solar masses.

Suggested Reading

Kippenhahn, *100 Billion Suns: The Birth, Life, and Death of Stars*.

Pasachoff, *Astronomy: From the Earth to the Universe*, 6th ed.

Pasachoff and Filippenko, *The Cosmos: Astronomy in the New Millennium*, 3rd ed.

Questions to Consider

1. If you compare a photograph of a nearby planetary nebula taken 100 years ago with one taken now, how would you expect them to differ?

2. Why is the surface of a star hotter after the star sheds a planetary nebula?

3. What forces balance to make a white dwarf?

4. If you wanted to prove that a nova must be a binary star system, what kinds of observations might you make?

White Dwarfs and Nova Eruptions

Lecture 52—Transcript

As discussed in the previous lecture, by looking at more advanced stages of evolution of other Sun-like stars, and by using the laws of physics—in particular, the physics of hot gases—we have deduced that, in about 5 or 6 billion years, our Sun will expand greatly and then eject its outer envelope of gases, the outer atmosphere that it contains right now—becoming a beautiful, glowing planetary nebula. Here's an example of one. There's the dying central star. There is this rich structure of gases around it. We don't know that our Sun will look like this one. Almost certainly it won't because we see such a great variety of planetary nebulae, but it'll look something like this. This basic structure is what we think all stars having an initial mass between about 1/12 of a solar mass and eight solar masses will eventually achieve. Their central star is dying, and it's getting rid of the outer atmosphere of gases.

This central star is basically the denuded core of a once-vibrant star, one that had lots of nuclear reactions going on. But now there are no nuclear reactions anymore. The core consists mostly of carbon and oxygen, and they repel each other so strongly, through electrical forces, that at the temperature of the carbon and oxygen core, you can't get those carbon and oxygen nuclei close enough together for fusion to begin. Even in the helium shell around the carbon and oxygen core, the temperatures are too low for fusion to be maintained. Basically, this is a star that no longer has nuclear reactions going on. It's only a small fraction of the mass that it initially had. Our Sun will end up in this stage with about half of its present mass. Even stars up to eight solar masses lose most of their material through this planetary nebula stage, and through winds, in such a way that they end up with less than about 1.1 solar masses of material in the core, regardless of the mass with which they started, all the way up to about eight solar masses.

This core begins to contract because there are no nuclear reactions going on, yet it's losing energy to the outside world. That energy has to come from somewhere; it comes from the gravitational contraction; therefore, its density will increase. Eventually, the electrons will start feeling each other. They'll say, "Hey, I don't like you. I don't want you to invade my space. Get out of here." Remember, electrons are fermions, and they don't like to be in the same quantum state—yet the contraction of the star is forcing them to be on top of one another spatially. To compensate, they have to achieve a range of energies, a range of momenta. You get, then, this electron-degeneracy pressure where, for quantum-mechanical reasons—the Pauli Exclusion Principle, basically, is what says that these electrons don't like each other. It was named after Wolfgang Pauli, a great quantum physicist of the early 20th century. His exclusion principle says that electrons don't like each other, so the electrons take on a range of energies and momenta, and hence exert a pressure outward in the core, keeping it from collapsing entirely.

At this point, the core will become a white dwarf. It'll be supported by this electron-degeneracy pressure—the pressure of fermions that Wolfgang Pauli talked about long ago. It'll be this weird retired star, essentially, with no nuclear reactions going on. Heat is still liberated from electrons that are coming down to the lowest energy levels and from positively charged atomic nuclei, which are not degenerate, and they can jump down to lower energy levels. But, basically, no new energy is being created in the star because nuclear reactions have stopped, and gravitational contraction has stopped. The only energy coming from the star is whatever thermal energy the particles had at the time of formation of this white dwarf. It's held up by these counterintuitive quantum-mechanical forces. It's very small, about the size of the Earth, and it's very dim.

As you can see here, the companion of the bright star Sirius is a white dwarf, and it's much, much fainter than Sirius itself because the star has only about the size of the Earth and about the temperature of the Sun. It's pretty hot, and hence it looks white, but it's so small that there's almost no area radiating energy, and so the whole thing is of very low luminosity. These things are very, very dim. Being so small and dense, you have about half of a solar mass compressed into the size, the volume, of the Earth. That means that a tablespoon of material would weigh several tons. This is incredible stuff. You can't hold it because it would be so dense, incredibly dense. The degenerate electrons in a white dwarf are moving around very, very fast. You might say, "Well, why can't they lose energy?"

If you go back to the diagram that I showed for brown dwarfs, it's essentially the same reason. All the electrons are piled up in the lowest energy states that they can achieve already. There are no empty states here. There are two oppositely spinning electrons per energy level. All of the apartment-building stories, to use my analogy from two lectures ago, are used up. There are no places to which the electrons can jump down, and therefore they cannot lose any energy because there are no available empty quantum states. It's really a weird thing. The other weird thing about it is that, for a normal solid, as you increase the mass, you increase the radius. If I pile more and more bricks on top of one another, there is a greater mass of bricks, but the radius of the structure is bigger too. Indeed, for a sphere, the radius is proportional to mass$^{1/3}$ because mass is basically proportional to volume, and volume is proportional to the cube of the radius for a sphere.

Radius is proportional to mass$^{1/3}$ for normal material. But for a fully degenerate white dwarf like this, radius actually is proportional to mass$^{-1/3}$. That means as you increase the mass, the radius decreases. That's because these electrons are getting more and more compressed by gravity with the higher mass, but they're resisting that compression, yet they can't quite resist it as much as gravity is pulling in. So, gravity slightly wins, and the radius is smaller for the more-massive objects. That fact will end up being critical when we consider the stability of white dwarfs that are more massive than the Sun.

The atomic nuclei in white dwarfs can lose energy; they're not degenerate. Let's take a look at the energy level diagrams for both the electrons and the nuclei. The electrons are all in their lowest energy states; the nuclei are not. There's a bunch of lower energy states into which the nuclei could jump. Gradually, over time, as the white dwarf cools, the nuclei are jumping down to progressively lower energy levels, releasing radiation. That's the light that we see coming from a white dwarf. It is not light coming from nuclear reactions right now. It is, rather, light coming from the store, the supply, of energy that the nuclei built up over the normal course of the star's life. So, we can think of white dwarfs as retired stars. They're spending their life savings of energy. They're not generating any new income, any new energy. They're just liberating, radiating, the energy that they stored up during their normal lives. They get cooler, and dimmer, and cooler, and dimmer.

After tens of billions of years, there's almost no light to be seen anymore. At some point, they become so cool and dim that we essentially lose sight of them. At that point, some astronomers call them black dwarfs because you no longer can see them. Don't confuse them with brown dwarfs, which are a different sort of object that we discussed previously, and don't confuse them with black holes, which are yet again different. We will consider them in the future. They're still supported by electron-degeneracy pressure at this point. The electrons are holding the whole structure up, but the positive ions inside have been gradually cooling down.

If you find yourself a cold white dwarf, you could actually put your hand on it and not get burned, despite the fact that some of the electrons are moving at speeds close to the speed of light. You might think, "Whoa, won't they hit my hand and burn me?" After all, when I put my hand on a hot stove, an electric burner, I say, "ouch"; it stings. It stings because the electrons and atoms in the stove are moving very quickly. That's what a high temperature means, lots of random motions. When they hit my finger—and specifically the nerve endings on my finger—they transfer some of their energy to my finger, and I say "ouch". In transferring some of their energy to my finger, they have to lose energy, and they drop down to lower energy states.

In a cold white dwarf, the electrons are already in the lowest energy states possible, so they can't lose any energy and transfer it to my finger or hand—because to transfer that energy to my hand would mean that they'd have to jump to lower energy levels, and all of those are already occupied. You have this gas where some of the particles are moving at close to the speed of light, exerting tremendous pressure, but your hand won't get burned by it—pretty cool. Be sure if you do this to find a cold one, though. If you find a warm white dwarf, remember the positively charged ions are not degenerate, and they could transfer their heat to you and jump down to lower energy levels. Only try this experiment with a very cold white dwarf, where the ions are not in any high-energy states. You want one in which the ions also are in low-energy states.

Let's take a look at what's called the post-main-sequence (or after-main-sequence) evolution of a Sun-like star. Here's the temperature-luminosity diagram. As I've emphasized before, where a star lands on the main sequence is dictated by its mass. Then, on the main sequence, it basically sits there—almost unchanged in luminosity and surface temperature. I mentioned last time that the luminosity does actually gradually grow a little bit during the star's life, but not much. Then, once the hydrogen is used up in the core, and it becomes helium, then you get the ascent to the red giant stage, followed by helium burning to carbon and oxygen, and then a further ascent to an even bigger red giant. Then the outer atmosphere of the star is unstable and starts getting blown off through winds and these cosmic burps that produce the beautiful glow of ionized planetary nebulae. Then the central star becomes hotter, and hotter, and hotter. That's because you're exposing deeper and deeper layers of this star as the planetary nebulae ejections progress further and further.

That is, with each ejection of the outer atmosphere, you're exposing what was previously a more interior, and hence hotter, part of the star. It zooms over to the left on the temperature-luminosity diagram, and then the central star is on its way to becoming a white dwarf. It contracts for a while and dims down and cools. Then finally, it can contract no further because it's supported entirely by electron-degeneracy pressure. At that point, it continues to cool, but not contract. It cools because the ions are losing energy, but it remains roughly the same size because the electrons are holding it up. Then eventually, at some point, some astronomers call it a black dwarf, though the distinction between a white dwarf and a black dwarf is sort of arbitrary. So, this general pattern is what all stars between about 0.08 and 8 solar masses do. The details may differ from star to star, but they all do this basic sequence—this movement through the Hertzsprung-Russell diagram—after a period of long stability on the main sequence. The main sequence is by far the longest active stage of a star's life. The white dwarf stage is longer than the main-sequence stage, but at that point, the star is a retired star. It no longer has nuclear reactions going on.

What happens in detail depends on the amount of mass with which the star was born. Our Sun will lose roughly half of its mass on its way to becoming a white dwarf. We think it'll actually end up having about 6/10 of its present solar mass. That's the mass it will have when it becomes a white dwarf. More-massive stars become white dwarfs that are more massive, but always less than 1.1 solar masses. Even an 8-solar mass star manages somehow to eject most of its material and end up as a carbon-oxygen white dwarf, no more massive than 1.1 solar masses. So, that's what happens to the more-massive stars than the Sun. What about less massive? Below about 0.45 solar masses, the white dwarf that remains consists of helium, not of carbon and oxygen because below 0.45 solar masses, initial mass, the helium never achieves a sufficiently high temperature to undergo fusion into carbon and oxygen. Those stars, during their active lives, stop during the helium stage and don't become carbon-oxygen cores.

On the other hand, if a star is between 8 and 10 solar masses initially, we don't know exactly what happens to it in all cases. But in some cases, what could happen is that the carbon and oxygen will fuse to form oxygen, neon, and magnesium before the thing turns into a white dwarf. Then you get this planetary nebula stage, and you end up with an oxygen-neon-magnesium white dwarf, at least according to some theoretical calculations. Other calculations say that no, stars between 8 and 10 solar masses explode at the end of their lives. We don't really know what happens in that range, 8 to 10. Above 10 solar masses, as I'll explain later, stars don't do this kind of thing at all. They don't become white dwarfs; rather, they explode. I'll discuss that in a few lectures.

Let's get back to the white dwarfs. It turns out that white dwarfs have a limiting mass, 1.4 times the mass of the Sun. This is known as the Chandrasekhar limit, named after Subrahmanyan Chandrasekhar, a great Indian astrophysicist who derived this maximum possible mass for a white dwarf in 1930, when he was not quite 20 years old, on a voyage from India to England, where he was to be educated. He had finished college at the age of 19, and he was going on to graduate school in England. The voyage took two and a half weeks. He said, "Well, why don't I combine the new fields of quantum mechanics and relativity, and apply them to stars?" He figured out that white dwarfs have this fundamental maximum limit—which will be related, as we will see, to the explosions of stars. He derived this during this two-and-a-half-week journey, and 53 years later, he won the Nobel Prize in physics, largely for this work. It's kind of a cool thing. You could say, "Wow, here's what I did on my trip to graduate school." That's reminiscent of Newton saying, "Well, here's how I spent my vacation away from college during the plague. I developed calculus, and the laws of gravitation, and all that." Some people are just really smart and really creative.

The Chandra X-ray Observatory is named after Chandra because he went on to do many, many great things in astrophysics. Interestingly, when he arrived in England and presented his results to the astronomers there, the most famous and most respected one, Sir Arthur Eddington, publicly ridiculed the result, and sort of personally ridiculed Chandra, apparently, at a scientific meeting. Chandra was devastated. By the way, I call him Chandra because that's the way he's known throughout the world. Chandrasekhar is kind of a long name, and he liked going by the name Chandra. Chandra fortunately didn't get so affected by this public humiliation that he just quit astrophysics and went on to do something else. He stuck with it, and he was eventually proven right. Indeed, he won the Nobel Prize. He did so many very, very important things. He was really one of the great astrophysicists of the 20th century.

This limit occurs because, as you pile up more and more mass on a white dwarf—say you've got a white dwarf, and it's accumulating material from a companion star, for example—the radius shrinks. Remember, the radius is proportional to the mass$^{-1/3}$. Eventually, the radius is so small, and the electrons are squeezed into such a tiny volume, that their speeds approach the speed of light. But they can't exceed the speed of light, and it's even hard for them to get up to the speed of light. In fact, it's impossible to get completely up to the speed of light. Their ability to exert pressure on their surroundings diminishes—or more correctly, it keeps growing, but it doesn't grow as quickly as for electrons that are not moving quite that close to the speed of light.

They're approaching this limiting speed, and so their ability to continue providing more and more pressure diminishes. Eventually, at 1.4 solar masses, the electrons just say, "We give up; we can't do it." In fact, Chandra predicted that such a star should be unstable and would collapse. We now know that another solution for the star's dilemma is to explode, and I'll discuss that in a couple of lectures. But Chandra saw that there's this limit combining relativity and quantum mechanics to whole new fields of physics.

He came up with a very fundamental result, and it's really amazing that he was able to do this at the tender age of not quite 20.

The evolution that I've described up until now is applicable to single stars—stars that were born with a certain amount of mass and didn't gain any mass from their surroundings, or from any companion, or anything like that. The evolution is dictated largely by the initial mass with which the star was born. A little bit depends on the initial chemical composition, but most of them are mostly hydrogen and helium to begin with. So, we can say with considerable certainty that the basic idea that I just spelled out is correct, and is about the same no matter which group of 4-solar mass stars you're looking at, or 3-solar mass, or 8-solar mass. For a given mass, they all will behave in about the same way.

However, if the star is bound in a system with another star—gravitationally bound as most stars are—it can in principle, if it's close enough to the other star, gain material from that other star, and its mass will change. Here's a schematic of two stars, gravitationally bound to one another. This figure-8 here shows the Roche lobes of the stars. The Roche lobes are those regions within which the gravity of a particular star—either that one or that one—dominates. If you were to drop a particle here, it would move toward this green star. If you were to drop a particle there, it would move toward the blue star. These two guys orbit each other. Eventually, the more-massive one, which evolves more quickly than the less-massive one, will bloat and fill its Roche lobe. It can then start transferring material to the other star, increasing that other star's mass.

But if the star's mass increases, then the pressure in the core increases. The reaction rate has to be faster to counteract the gravity trying to pull in on the star, and so the star ends up behaving like a more-massive star, and it starts evolving more quickly. So, an initially less-massive star can gain mass and evolve move quickly as a result of this process. An accretion disk forms in the transfer of material from one star to the other because the whole thing is a rotating system, and the transfer of matter from one star to another cannot occur just along a straight line between these two stars. A disk has to form. Indeed, we saw this formation process when we were considering the bipolar outflows of some planetary nebulae. Here's a star expanding to a red giant and spilling its material onto the other star. You can see that a disk formed around that other star.

If one of the stars is highly evolved—if it's already a white dwarf—and the other star starts spilling material onto it, then you can get brightenings of stars called novas. *Nova stella* means "new star" in Latin, but really this is an old star that's accreting material from its companion. That material can fall in clumps onto the surface of the white dwarf, releasing gravitational energy, or it can accumulate on the surface of the white dwarf and then undergo an uncontrolled chain of nuclear reactions, releasing a lot of energy. So, in a number of ways, the white dwarf that's gaining material from its companion star can quite suddenly release large amounts of energy, becoming a factor of a hundred to a million times brighter than it was before, for a very short amount of time.

Here's this accretion disk formed from the spilling of material from the more normal star onto the white dwarf. The white dwarf is stealing this material. As this material falls onto the white dwarf and accumulates, nuclear reactions can occur, or gravitational energy can be released, and so the star can brighten. Here's an example, in 1934, of a star becoming a nova, suddenly brightening by, in this case, a factor of a few hundred thousand, I think it was. After this surface explosion ends and dies down, the star fades. But then it continues gaining material from its companion star. Once there's enough material on the surface, it can once again explode, producing a nova once again, a new star. But really, this is an old star that's undergoing surface nuclear reactions—so novae can recur.

Amateur astronomers often find novae because they scan very large parts of the sky—whereas professional astronomers usually look at only small parts of the sky in great detail. Amateurs look at the whole sky, and they can find bright novae, like this one that occurred in the constellation Cygnus in 1975. Here are Hubble Space Telescope pictures of another nova; it went off in 1992, also in Cygnus, but unrelated to the one in 1975. You can see the ejected material came out in a ring, not in a spherically symmetric way. Moreover the ring expanded in the time interval between the two Hubble exposures, and you can also see the dramatic improvement in clarity after Hubble's optics were fixed by the space shuttle astronauts.

Some stars are not white dwarfs that undergo outbursts, but rather are pretty massive, big stars that undergo outbursts unrelated to the nova phenomenon, the white dwarf phenomenon. But they still look like novae. They brighten considerably, and you have to study them in detail to see that they are not, in fact, white dwarfs with surface explosions. One such star is Eta Carina, in the Southern Hemisphere, where you have a nebula, which through binoculars looks something like this. In the middle of that nebula, there's a very massive star that's undergoing a bipolar ejection like this, kind of like the planetary nebula stage that I showed you earlier. Eta Carina sometimes brightens considerably. In fact, in the 19th century, there was a while when it was the brightest star in the sky. Though it could be confused with this white dwarf surface-explosion process, the process of a nova, in this particular case, that's not what's happening.

Massive stars can sometimes masquerade as white dwarfs that are undergoing the nova process. One really cool example of this was a variable star that occurred in the constellation Monoceros. It brightened dramatically in 2002, becoming an incredibly bright star. What we think happened was that this star was in a binary system, and for some reason, the main star suddenly grew bigger. We're not sure whether it was influenced by its companion or what. But by becoming really big, I mean almost as big as Jupiter's orbit is around our Sun. Sort of between Mars and Jupiter—that's the kind of size we're talking about here. By becoming really big, it became correspondingly brighter, but we don't know why it became that big.

In this particular eruption, no matter was ejected from the star. We can tell from spectra that the gas was moving out pretty slowly. But, apparently, previously during this star's life, there were ejections of gas. This particular brightening lit up the gases surrounding the star, the gases that were emitted in previous ejections having been done by this star. You can see the echo, the light reflecting off of the gases that had been previously ejected from this star. Though ejection is not happening in this particular outburst, the light from this outburst illuminated that surrounding gas, producing a beautiful nebula like this. In fact, you can understand how these things happen. If you have a flash of light that hits clouds of gas and then gets reflected in many directions, including the direction toward our eyes—when the light reaches us, we see these glowing shells. The glowing shells should change with time, and indeed they did.

As the Hubble Space Telescope took more and more photographs of this object over the course of a couple of years, it could see progressively more and more extended shells being lit up by the flash of light from the current outburst of this very strange star. Clearly, at the ends of their lives, stars can do a variety of interesting things, explosive things: from the surface explosions of white dwarfs, to instabilities in more-massive stars that cause them to brighten and fade occasionally. As we will see in the next lecture, some stars literally explode at the ends of their lives, catastrophically destroying themselves. Those are the supernovae, the exploding stars.

Exploding Stars—Celestial Fireworks!

Lecture 53

In the last lecture, we saw that most stars die rather quietly, becoming red giants, planetary nebulae (gentle ejections of matter), and white dwarfs. A small minority of stars, however, ends their lives with catastrophic explosions: *supernovae.*

A supernova can increase a star's luminosity to as much as 10 billion Suns. At its peak, a supernova can rival the brightness of an entire galaxy of stars. The star's gases may be ejected at speeds greater than 10,000 kilometers per second, which we can determine by examining the spectra of supernovae, as we'll see in the next lecture. Supernovae heat the **interstellar medium**—the gases between the stars—causing winds to blow out of entire galaxies. They also give rise to compact remnants in some cases, such as neutron stars and bizarre black holes. Supernovae accelerate charged particles to very high speeds, creating cosmic rays that cause at least some of the mutations that led to the evolution of life.

From the human perspective, the most important aspect of supernovae is that they create and disperse into the cosmos the very elements of which life is made. Though the hot, dense, early Universe (the so-called Big Bang, to be studied in future lectures) produced hydrogen and helium, all the heavier elements were created inside stars. If some stars didn't explode, those heavy elements would remain forever locked up inside white dwarfs, unavailable as the raw material from which new stars, new planets, and even life could form. Indeed, all of the elements in the upper part of the periodic table, such as silver and gold, were produced from such "stardust." In addition, iron, calcium, carbon, and oxygen, ejected by exploding stars, are essential for life on Earth.

We can look at the spectra of **supernova remnants** and see those elements, which weren't present when the star first formed. After tens of thousands of years, nebulae expand even more, gradually merging with other existing clouds of gas and dust within galaxies to form new stars, new planets, and even life. DNA, the basis of life itself, owes its existence to previous generations of stars. The ejection of the heavy elements into the cosmos, and the production of the elements themselves, is the most centrally important aspect of supernovae. Is our Sun, then, a second-generation star? From what supernova did we come? In reality, our Solar System was formed from a mixture of exploding stars in which debris from many explosions coalesced. In other words, many different generations of stars gave rise to the cloud of gas from which our Solar System formed over a vast time scale.

Let's discuss some specific supernovae. The most famous supernova remnant is the Crab Nebula, an expanding set of gases created by a supernova that occurred in A.D. 1054, first seen on July 4 by Chinese astronomers. The supernova was alleged to be visible during the day for 23 days. In A.D. 1006, another supernova left a bright remnant recently photographed with the Chandra X-ray Observatory and other telescopes. A clear supernova was last seen in 1604 in our own Milky Way Galaxy by Johannes Kepler, of which we can see the remnant. Kepler's mentor, Tycho Brahe, also witnessed a supernova in 1572 that has produced an expanding remnant. Supernovae are rare; in a big galaxy like our own, they might occur a few times per century. The reason we may not have seen a bright supernova since 1604 is that some may be hidden by the extensive gas and dust in the plane of our Galaxy. One supernova occurred in the 1670s in the constellation Cassiopeia, but only one person possibly noticed it. However, today, we clearly see its remnant and can even detect a neutron star in the middle. Ironically, supernovae are easier to find than other galaxies. Because they are so rare, we have a better chance of finding one by looking at many galaxies over time to observe changes.

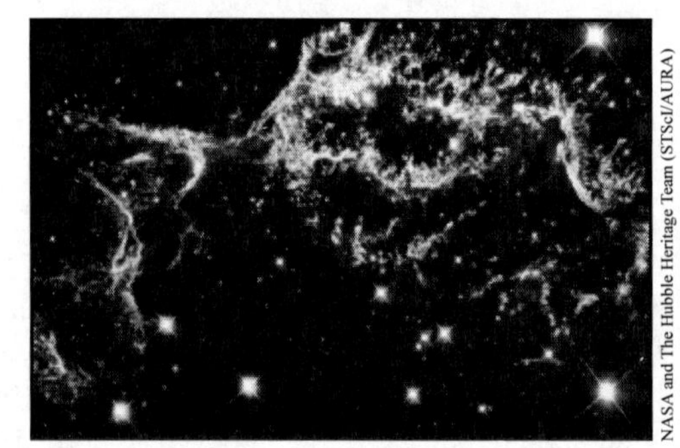

NASA and The Hubble Heritage Team (STScI/AURA) using data collected by Principal Astronomer Rob Fesen

Supernova remnant, Cassiopeia A.

The Lick Observatory's Katzman Automatic Imaging Telescope (KAIT), owned and operated by my research group, is programmed to take pictures

(CCD images) of more than 1000 galaxies over the course of a single night in search of supernovae. Each week, new images of 7000 or 8000 galaxies can be compared with previous images to see if anything new appears. The computer software automatically makes the comparisons and identifies supernova candidates; then, undergraduate students examine them to determine which ones are likely to be genuine supernovae. My group has discovered more than 600 relatively nearby supernovae during the past decade, about half of all the bright supernovae that have occurred during this interval. We are the world's leaders in finding such objects. Supernovae are named in the order of discovery in any given calendar year. For example, the first supernova of 2000 is named SN 2000A, the second is named SN 2000B, and so on, up through SN 2000Z. The next two after that are SN 2000aa and SN 2000ab, and so on. Though this is not scientifically important, my group even found both SN 2000A and SN 2001A— the first supernova of the new millennium, regardless of one's definition of the new millennium (Jan. 1, 2000, or Jan. 1, 2001). We now study nearby supernovae in great detail for a better understanding of how stars explode.

"You want to spend your time with the Keck telescopes and others studying the supernovae, not searching for them. It's really a great form of cooperation between professionals and amateurs."

Amateur astronomers have discovered many supernovae by making similar observations. Rev. Robert Evans, who lives in Australia, was the first amateur astronomer to systematically find bright supernovae. He conducted visual observations of galaxies through his telescope and found about 40 supernovae over the course of a few decades. Inspired in part by the success of Evans and in part by our KAIT search at the Lick Observatory, several amateur astronomers have now found more than 100 supernovae each by repeatedly taking CCD images of galaxies and comparing them to look for new objects. For this reason, amateurs are important in our studies of the heavens. Amateur astronomers in general contribute to our study of the stars, helping to increase our chances of discovering interesting celestial phenomena. ■

Important Terms

interstellar medium: The space between the stars, filled to some extent with gas and dust.

supernova remnant: The cloud of chemically enriched gases ejected into space by a supernova.

Suggested Reading

Filippenko, "Stellar Explosions, Neutron Stars, and Black Holes," in *The Origin and Evolution of the Universe*.

Harvard-Smithsonian Center for Astrophysics, *Supernova*, www-cfa.harvard.edu/oir/Research/supernova/SNlinks.html.

Kirshner, *The Extravagant Universe: Exploding Stars, Dark Energy, and the Accelerating Cosmos*.

Marschall, *The Supernova Story*.

Pasachoff, *Astronomy: From the Earth to the Universe*, 6th ed.

Pasachoff and Filippenko, *The Cosmos: Astronomy in the New Millennium*, 3rd ed.

Questions to Consider

1. Does the fact that you are made of stardust give you a sense of unity with the cosmos?

2. Why do we think that the Crab Nebula is a supernova remnant?

3. If one or two supernovae occur in a typical galaxy every century, how many galaxies would you need to monitor to find 20 supernovae each year?

Exploding Stars—Celestial Fireworks!

Lecture 53—Transcript

I discussed last time how most stars die rather quietly. They expand to form red giants, then gently eject their outer atmospheres—becoming beautiful, glowing planetary nebulae. Meanwhile, the central star contracts, grows dimmer and dimmer, and finally becomes a white dwarf that gets dimmer all the way down to the black-dwarf region, where it can no longer be seen at all. Some of those white dwarfs are bound sufficiently closely to a companion star that they can steal material from that companion and re-brighten in a series of nova episodes. But that's only some stars; most other stars, even those in binary systems, have the other star too far away to contribute substantially to its mass, and so they just die quietly like the solitary stars do.

But a small minority of stars end their lives with a catastrophic explosion, literally blowing themselves to smithereens and becoming as luminous as millions or billions of normal suns. This is basically a previously normal star, which suddenly, in a day or over the course of a few weeks, grows in brightness by factors of up to 1 billion to 10 billion, dwarfing the previous brightness that it had during its normal life. Here is the star at the end of its life. It has a normal luminosity, and then bam, a day later, it was millions of times brighter than it was before—a colossal explosion of a star. Looking at a series of frames of this process, you can see the thing brightening and fading. Here, I've sped up the process. For this type of supernova, the brightening takes a few weeks, and the fading takes many months. But at its peak, the supernova can rival the brightness of an entire galaxy of stars. It can be as bright as a few billion stars, and although our own galaxy is hundreds of billions of stars, some galaxies are only a few billion stars, and so a single star can rival the brightness of an entire galaxy. That's just an incredible explosion.

Here's an example, again showing the power of a supernova. There's the galaxy before the supernova went off, and here it is after it went off. That is a single star in this galaxy of stars that is now easily visible through a telescope. This galaxy is 100 million light years away, yet a single exploding star within it is comparable in apparent brightness to a star in our own galaxy that is only 1,000 light years away. Here's one 1,000 light years away; there's another one 100 million light years away. You can do the math using the inverse-square law. The supernova was, in this case, about 10 billion times the brightness of the star.

Wow, that is incredible. If you were near one of these things, and you experienced the explosion from a close distance, you'd need sunblock of a billion or 10 billion, if the Sun were to do this. We know about sunblocks of 40 or 50 when we go out to the beach and want to block the ultraviolet rays from our skin. But if our Sun were to do this, which it won't—don't worry, it won't—you would need sunblock of a billion or 10 billion to give you adequate protection. Needless to say, we don't make sunblock that's that powerful. In fact, it's impossible to make. The gases get ejected with speeds greater than 10,000 kilometers per second, in some cases. We can tell that by examining the spectra of supernovae, as I'll show you in the next lecture.

Supernovae are important for a number of reasons. First and foremost, they are just these amazing explosions. Seen from afar, from a safe distance, who can resist watching them? Look at how much we like Fourth of July fireworks shows. Everyone goes to these things. I experienced one this past Fourth of July. It was great; I loved watching the fireworks. In a sense, a cosmic explosion is the biggest Fourth of July fireworks show you're ever going to see—but again, viewed from a safe distance. My own fascination with explosions actually goes back quite some time. In high school, I lit up all sorts of powders and stuff that went off in flashes. Here I am, a senior in high school, doing a science club demonstration where I lit up a flash powder, and it fortunately didn't harm me. It was a bit more of an explosion than I reckoned it would be.

Interestingly, the smoke went through the school ventilation system, and it started pouring into the administrative offices. They were about to pull the fire alarm, the principal was, and then they said, "Wait, let's look at the weekly calendar of activities. Ah, there's a science club demonstration at noon today by Alex Filippenko. That's probably what's going on." They came over, rushed into the room, saw that it was just me doing my explosions, and said, "Okay, that's all right." That's me in high school, playing with explosives. Fortunately, I don't do it to be a terrorist or anything like that. I just like the process of the explosion and what a bang it produces. I'm naturally drawn to the biggest bangs in the Universe besides the birth of the Universe itself, cosmology, which I'll discuss later. These exploding stars are among the biggest bangs in the Universe.

They're not just fun to watch; they're actually important. They heat the interstellar medium, the gases between the stars. These ejected stars slam into the gases between the stars and heat them up, causing winds to come out of entire galaxies. You can see galaxies ejecting material that's been heated by supernova explosions—fountains of gas squirting out of galaxies. It's really cool. Supernovae also give rise to compact remnants in some cases, neutron stars and bizarre black holes that I'll be discussing shortly. That's cool. They accelerate charged particles to very high speeds; the cosmic rays that come in and cause at least some of the mutations that lead to the evolution in life. But most important from the human perspective is that supernovae, exploding stars, create and disperse into the cosmos the very elements of which life is made.

I'll discuss later how the Big Bang produced only hydrogen and helium, basically. All the heavier elements were produced inside stars. We've already seen some of this in our discussion of nuclear fusion. Hydrogen goes to helium; helium goes to carbon and oxygen. In some cases, I even mentioned that carbon and oxygen can go on to neon and magnesium. These nuclear furnaces, stars, produce these elements. But if some stars were not to explode, those heavy elements would remain forever locked up inside the white dwarfs and wouldn't become available as the raw material from which new stars, new planets, and even life, could form. So, you want some stars to explode to disperse the heavy elements that were created during the normal lives of stars.

These explosions also produce many of the heavy elements. Indeed, all of the elements in the lower part of the periodic table, the really heavy ones, were produced during the explosions themselves; silver, and gold, and things like that. Even the iron and other elements were produced during the explosions. Iron is in our red blood cells. Calcium is in our bones. Oxygen, we breathe. Carbon is in most of our cells. All of these elements were produced by stars, and in some cases, the explosions themselves. Certainly they were dispersed into the interstellar medium, becoming available as the material from which new stars, planets, and life were made. We can tell that this happens because we can look at the remnants of stars, supernova remnants, of which a schematic is shown here, and real-life photographs are shown in the next few pictures.

Here's the remnant of a supernova. You can look at the spectrum of the gases and see that they are enriched in heavy elements that weren't there prior to the explosion. Here's another beautiful supernova remnant. All of these elements that we see, and that are so critical for life as we know it, were produced in stars and in supernovae. The iron, the nickel, the other heavy elements; everything except the hydrogen and helium was produced by stars. Looking at an explosion, simulating what it might look like, there's a bang. Then the ejected material goes streaming away from the center of explosion, becoming ever more dispersed, laden with heavy elements. Those supernova remnants expand over thousands of years. Here's one, the Crab nebula, that's only 1,000 years old. But after tens of thousands of years, the nebulae expand even more and gradually merge with other existing clouds of gas and dust within galaxies—eventually becoming gravitationally unstable, as in the Eagle nebula here, and collapsing to form new stars, new planets, we think in some cases, and even life.

Here are the protoplanetary disks that form around young stars. Some of them, we think, coalesce into planets. Indeed, searches for extrasolar planets have found many of them—not yet Earth-like ones, but getting close. We're down to five Earth masses for some of the techniques, as I've already discussed. Some of the planets around stars surely will be rocky, Earth-like planets where the material for life collected and was able—at least in the case of the Earth—to form separate self-replicating, evolving molecules. DNA, the basis of life itself, owes its existence to previous generations of stars. This ejection of the heavy elements into the cosmos, and the production of the elements themselves, is the most centrally important aspect of supernovae, in my opinion. In the course of the next few lectures, I will show you the evidence for the creation of heavy chemical elements in exploding stars.

This process was first understood by two groups of astronomers. One consisted of Margaret Burbidge, Geoff Burbidge, Willie Fowler, and Fred Hoyle. They wrote a paper: Burbidge, Burbidge, Fowler, and Hoyle, now known as simply B^2FH—it's such an important paper—where they outlined the series of nuclear reactions that probably produces the heavy elements of the periodic table. Independently, Al Cameron came up with very similar nuclear reactions and deduced that the heavy elements come from stars—as did some other physicists dabbling with, or studying in great detail, these processes. Many of them realized that stars are nuclear furnaces that have the potential to produce heavy elements—that's what I mean by dabbling with the subject—but some explored the subject in great detail and showed the specific reactions that probably occur in stars and in exploding stars.

People often ask, "Is our Sun, then, a second-generation star? Is it a third-generation star? From what supernova did we come?" Well, it's not that way. It's a mixture of exploding stars that produce all this debris in galaxies. The debris from many explosions mixes together, coalesces, and finally forms a contracting cloud of gas and dust that gives rise to stars and planets. Some of the stars that contributed to that cloud had short lives. The massive stars have short lives and explode soon after their birth; other stars have long lives before they explode. So really, it was many different generations of stars that gave rise to the cloud of gas from which our Solar System formed. Typically, maybe it would be 100 generations, but some of the stars exploded shortly before the pre-solar nebula formed, and other stars may have exploded 6 billion years before it formed. It's really a vast time scale.

The most famous supernova remnant is the Crab nebula. It's an expanding set of gases that's a favorite photographic object of amateur and professional astronomers. Here are a bunch of photographs, including one that my team took with a telescope that we run at Lick Observatory. This photograph I've already shown. It was taken by Richard Crisp through this tri-color process of isolating light that comes out in the form of various emission lines. It's a beautiful picture. Then the Hubble picture is just astonishing. Look at the detail of the filaments of these expanding gases, rich with heavy elements.

The supernova that gave rise to that remnant occurred in the year 1054, first seen on July 4 by Chinese astronomers. I don't know what significance July 4 had to the people of the Sung Dynasty. I doubt that they celebrated the American Revolution back then. Anyway, here are some writings from the Sung Dynasty, where it is said that "...in the first year of the period Chiho, the fifth moon, the day chi-ch'ou, a guest star appeared several inches south-east of Tien-Kuan. After more than a year, it gradually became invisible." I can't read Chinese, by the way. This is what my colleagues tell me it says. Elsewhere it says that the star became four times brighter than Venus, and it was visible in daytime for 23 days. It only became invisible at night after about 653 days. This was an amazing guest star that Chinese astronomers studied.

Remarkably, it was not noted in Europe, and so it may have simply been a casualty of the Dark Ages. Either people didn't look, or they were afraid to write about what they actually saw in nature. It was possibly seen by the Anasazi Indians in Chaco Canyon, New Mexico. There is a drawing made by the Anasazi Indians that looks like a bright star next to the crescent moon. Dating that drawing brings you to a time close to July 4, 1054. However, this might have been simply Venus close to the crescent moon. We don't know the exact date when that drawing was made, and there are many times when Venus happens to be close to the crescent moon, or some other star is close to the crescent moon—so it's not clear whether the American Indians saw this explosion.

There have been other colossal explosions in our Milky Way. In the year 1006, there was a very bright one, whose remnant is shown here, imaged with the Chandra X-ray Observatory. A clear supernova was last seen in 1604 in our own galaxy by Kepler. He studied this supernova in some detail. Now we can take pictures of its remnant. Here's one that is a combination from the Chandra X-ray Observatory, the Hubble Space Telescope looking at visible wavelengths, and the Spitzer Space Telescope looking at the infrared. Prior to Kepler, his mentor, Tycho, saw one in 1572. Indeed, Tycho became quite famous because of the book he wrote about the supernova of 1572. Now we can see its remnant expanding as well. Here is a radio image—taken with the Very Large Array in New Mexico—of the expanding remnant of Tycho's supernova. Here's a Chandra X-ray Observatory image of the hot gas associated with that expanding remnant.

Supernovae are rare. In a big galaxy like our own, they might occur a few times per century. But in smaller galaxies, they might occur only once per century, or only maybe once every two or three centuries. It depends on the galaxy that you're looking at. It's probable that our own Milky Way Galaxy is overdue for a good supernova. As I said, there was one in 1572, and there was one in 1604. You might think, "Hey, it's been several hundred years." But some of them get hidden by the extensive gas and dust in the plane of our galaxy. Here's the Milky Way, and you can see these dark blotches, which are clouds of gas and dust that extinguish or dim the light from behind them. So, it's possible that some supernovae have occurred in the past few centuries, and we simply haven't seen them.

There was one that occurred in the 1670s in the constellation Cassiopeia, right about here, but no one saw it, or perhaps one astronomer saw a very faint, new star there. We don't know why it was so faint. Possibly it was intrinsically not very luminous, and possibly it was hidden by a dark cloud of gas and dust. But now, today, we clearly see its remnant, and even a neutron star in the middle. There have been a few supernovae in our galaxy in the past 400 years, but I wouldn't be surprised if a bright one occurred during my lifetime. I hope one occurs because I love these things.

Supernovae are easier, ironically, to find than other galaxies. First of all, there are a lot more galaxies out there than just our Milky Way—so if you confine your search to just the Milky Way, you're probably not going to find one. But if you look at other galaxies, you improve your odds. It's still hard; supernovae are rare. They're like two-headed snakes; they're rare. You occasionally find them, but you're unlikely to find one in your own backyard. You have to look over a wide distance to find a rare animal like a two-headed snake. Similarly, you have to look at distant galaxies, and many of them, to find an appreciable number of supernovae. They can be subtle. For example, here's a galaxy where there are some foreground stars in our own Milky Way, and then there's one of these things, which is a supernova, in that galaxy. It turns out to be this one. But just looking at that galaxy once, you wouldn't be able to tell that that's a supernova and that it's not just a star that belongs there because it's in our own galaxy and happens to be along the same line of sight.

You can't just look at a galaxy once. You have to look at a galaxy repeatedly to see if anything has changed. I could force each of my students to look through a telescope. Here I'm looking through a telescope. I suppose I could do this myself. But I could force each of my students to look through a telescope at one—and only one—galaxy each night until they find something new, an exploding star. Then they would graduate and move on to greener pastures, and I could write about the supernova. But there are some crimes that are so egregious that even a tenured professor can get fired. Subjecting students to decades of searching through the eyepiece of a telescope for an exploding star would be one such abuse of one's power as a professor. That wouldn't be a good thing to do.

It's better, of course, to not just look at one galaxy, but rather to look at many galaxies, because if there's one supernova per century per galaxy, that's equivalent to their being one supernova per year per 100 galaxies. Each of these 100 galaxies is only going to produce a supernova once in a century, but they're going to do it randomly. So, on average, one of those galaxies will do it in any given year. If you look at 1,000 galaxies, 10 of them will produce a supernova in any given year. So, I could have my students looking through the eyepiece at many, many different galaxies, but that would be kind of cruel and unusual punishment as well.

Nevertheless, there are some amateur astronomers that love doing this. The Reverend Robert Evans has committed to memory the star fields around about 1,500 galaxies, and he looks through the eyepiece of his telescope at this set of galaxies every once in a while, and he sees what's new. He notices it by comparing with his memory of what the galaxy and the star field around it should look like. Indeed, in the past two and a half decades, he has found, visually, 40 supernovae, which professional astronomers then studied in great detail. I really like Reverend Evans for having found lots of bright supernovae. He tells me that occasionally he feels that a supernova is going to occur in a particular galaxy on a particular night. Maybe he has some sort of inside communication from God or something, since he's Reverend Evans.

Here's an example of one of his discoveries, taken with a telescope that I run. Here's the galaxy before the supernova occurred, and there's the galaxy after the supernova occurred. Most things look the same, but here you can see something looks different. This image gives us a clue to how we might look for supernovae and find lots of them. You just take pictures of galaxies, and you look for arrows. Where you see arrows, you find supernovae. There's one, there's one, there's one. It happened twice in this galaxy. By the process of mathematical induction, this technique must work every time. Just take photographs of galaxies and look for arrows, and the arrows will show you where the supernova is. Well, of course, it's not that easy. You have to identify the supernova, and then you use Adobe Photoshop or something to put in the arrow.

Our team runs a telescope at Lick Observatory, which is about a two-hour drive southeast of San Francisco. This telescope, the Katzman Automatic Imaging Telescope, is not a large telescope by today's standards. It's only three-quarters of a meter in diameter. But our team, and in particular, Dr. Weidong Li, a member of my team, has programmed this telescope to take pictures of lots of galaxies over the course of a single night. We can take images of over 1,000 galaxies on an average night. So, over the course of a week, we can get images of 7,000 or 8,000 galaxies, and then the next week we take new pictures of those same galaxies. We can compare the new pictures with the old pictures and see if anything new has shown up. Usually, there's nothing new, but here's an example where something new showed up.

The new image of the galaxy looked different from the old image, and that difference is this dot here, which is a candidate supernova. We don't know from this one image alone whether it's a supernova because it might have been an asteroid flying through the field of view at the moment that we took the image. But we can take additional images to try to confirm it. In general, the students that I have looking at the images verify that the computer program that identified the object as being a new star, it really is a star and not some sort of a cosmic ray or some other imperfection, or some asteroid flying through the field of view. These undergraduate students have a chance to get their hands dirty with research at an early age. This gives them training with which they go on to graduate school and do other important work.

I've had many, many students in my group over the years helping the automatic software discover supernovae by confirming what the software thought it had found, and getting rid of all the fake supernovae that sometimes occur when an asteroid flies through, or a cosmic ray hits the

CCD, or two stars are blended together and look like they're a supernova or something like that. I'm proud of the record that my students have achieved. We've found hundreds of supernovae over the past decade or so. Indeed, we are the world's leaders in finding supernovae. The first one we found, Supernova 1997bs, you might think was of questionable integrity, given its name: 1997bs. But it turns out that the naming scheme for supernovae is that the first one that you discover in any one year is given the designation year plus an a, the second one b, all the way up through z for the 26th one. And then you go aa, ab, ac, and so on through az, and then ba, bb, bc, and so on. So, I will leave it as an exercise to the listener to figure out what number sequentially that supernova happened to be that year.

Finding one in one year was not a world record, but we then found many in the subsequent years, setting a world record several times. We even found the first supernova of the new millennium (regardless of your definition of the new millennium): Supernovae 2000a and 2001a. We found a lot of these things. We now study them in great detail and try to come to a better understanding of how it is that stars explode. We've found about half of all the bright supernovae that have been found over the past decade, but I should say that amateur astronomers have found most of the others. Most of the ones we don't find, they find. This is really a great thing. They look with their telescopes, taking CCD images in general—or as Reverend Evans does, he looks just visually through the eyepiece—and they do the same sorts of comparisons that we have done, although we have software that does it relatively automatically, and they painstakingly look at their images night after night, trying to see what's new. They've also developed some software that helps them out as well.

I'd like to pay a tribute to some of the amateurs who have done such a great job in finding supernovae and in developing software that is similar to the software we have developed, with which they—using their system—can find supernova candidates. Tim Puckett, living in Georgia, has found over 150 supernovae. That is an amazing accomplishment. Tom Boles, in England, has also found over 100 supernovae. Here's his 100th discovery, Supernova 2006bk, where there the markers show a star next to a galaxy, or perhaps embedded in the galaxy, and a previous image of the same field did not reveal that star in the field. This was a new star. Boles discovered it, and it was his

100th discovery. But really, it wasn't a new star—it was the death of an old star. Here's Mark Armstrong, who also has found many supernovae—several dozen, I think. There are other amateur astronomers, Michael Schwartz and others, who have also found dozens of exploding stars.

It's really great that amateur astronomers can contribute to our field in such a valuable way. By finding lots of supernovae that professional astronomers can then study, they increase our chances that we will come to a complete understanding of the phenomenon. They also free up more of our time using the bigger telescopes, so that we can examine in detail the known supernovae, rather than spending most of our own telescope time searching for supernovae. You want to spend your time with the Keck telescopes and others studying the supernovae, not searching for them. It's really a great form of cooperation between professionals and amateurs. Through studies of supernovae over the past few decades, we've now come to a pretty good understanding of how they explode. In the next few lectures, I'll explain the different ways in which stars having different masses and different configurations—like single stars versus binary stars—can sometimes, very rarely, explode at the ends of their lives.

White Dwarf Supernovae—Stealing to Explode

Lecture 54

In the previous lecture, we looked at supernovae, catastrophic explosions of a small minority of stars at the ends of their lives. We begin this lecture by discussing the two main kinds of exploding stars.

As we have seen, spectra of stars provide us with information about the stars' chemical compositions. Similarly, spectra of stellar explosions can tell us a tremendous amount about the stars from which the supernovae arise and about the explosions themselves.

There are two main types of exploding stars. Those that show hydrogen in their spectra are called *Type II*; those that do *not* show hydrogen in their spectra are called *Type I*. The Type I class is further divided into subclasses Type Ia, Ib, and Ic, based on the details of the optical spectra. Type Ia supernovae were formerly known simply as Type I before the other two subclasses were recognized. Type Ia supernovae reach their peak brightness over the course of about 3 weeks, then decline for many months or years. Type II supernovae reach their peak brightness in just a day or two, then maintain that brightness for up to 3 months before quickly declining. Type Ia supernovae occur in all kinds of galaxies, including elliptical ones that consist only of old stars. Type II supernovae, as well as Types Ib and Ic, tend to occur in the arms of spiral galaxies and those galaxies where young stars are forming. This suggests that Types II, Ib, and Ic are somehow related to the deaths of massive stars.

The spectra of supernovae show evidence for the rapid ejection of gases at speeds sometimes exceeding 10,000 kilometers per second. We can deduce this by plotting their brightness against wavelength and seeing how the absorption lines are blueshifted relative to the emission line. The invention of robotic technology and computer software allows us to examine supernovae in great detail. For example, we can obtain the spectra and compare the light curves of many such events, teaching us about the physics of explosions.

We can also study the rates at which different kinds of supernovae occur (that is, how many per century per galaxy, on average) and in what types of galaxies. This information is important because different types of supernovae produce different kinds of chemical elements, and they infuse their galaxies with that material. If certain kinds of galaxies produce more Type Ia supernovae than Type II, the chemical evolution of those galaxies will be different from the galaxies that produce more Type II supernovae, for example. The data can also help us determine the rate of formation of **neutron stars** and **pulsars**, as well as how quickly the gas between the stars is heated by these explosions.

What produces a Type Ia supernova? Traditional Type Ia supernovae don't show hydrogen in their spectra, which means that there's very little or no hydrogen present in their ejecta. This is significant because hydrogen is by far the most common element in the Universe. In addition, as we said earlier, Type Ia supernovae tend to occur in galaxies that have only old stars. Further, all supernovae of this type have nearly the same observed properties—similar light curves and similar peak power. These characteristics suggest that Type Ia supernovae arise from carbon-oxygen white dwarfs, perhaps surrounded by a thin helium layer, but possibly with little or no hydrogen.

"An understanding of Type Ia supernovae and how they occur will be very important in our studies of cosmology, the overall structure and evolution of the Universe."

Such a white dwarf in a binary system can sometimes accrete hydrogen from a companion star that is on its way to becoming a red giant. The accreted material increases the white dwarf's mass. If the accretion rate is just right, the star can avoid nova-like surface explosions that prevent the mass of the white dwarf from growing substantially. Instead, its mass increases. Once the white dwarf reaches its limiting mass, the Chandrasekhar limit of about 1.4 solar masses, it becomes unstable, setting off a runaway chain of nuclear reactions (starting with the fusion of carbon and oxygen) that releases tremendous energy and completely obliterates the star.

About half of the white dwarf's mass fuses to a radioactive isotope of nickel. This nickel-56 decays into radioactive cobalt-56, then into stable iron-56. The decay process emits gamma rays, which are extremely energetic photons. Those gamma rays bounce around inside the exploding star, gradually being converted into optical light. The optical radiation escapes from the expanding gases, giving rise to the optical display of light that we see for a few months or a few years. If radioactive nuclei had not been produced, the explosion would not be visible because all of the released energy would have been used up in the star's expansion. We are not certain about the details of such explosions. For example, what happens if we incorporate rotation and magnetic fields? How, exactly, is the thermonuclear runaway initiated, and how does it proceed?

Although we think that only carbon-oxygen white dwarfs in binary systems can accrete enough material to reach the Chandrasekhar limit and explode in the observed manner, we don't exactly know which kinds of binary systems are suitable and how the white dwarfs reach this limit. Main-sequence stars are generally much smaller than their Roche lobes—the region within which a star's gravity dominates—and, therefore, are not capable of spilling material onto their companion white dwarfs. On the other hand, if a red giant is spilling material onto a white dwarf, we would expect the explosion to rip some of the gas away from the envelope of the red giant, causing the hydrogen to glow and show up in the spectrum. Yet it doesn't.

Some physicists have proposed the existence of sub-Chandrasekhar white dwarfs—with masses of less than 1.4 solar masses yet still explosive. For example, an explosion can be initiated at the boundary between the helium envelope and the carbon-oxygen core. The problem is that the spectra and light curves from such models don't agree with what is observed. Some physicists have proposed that two white dwarfs in a binary system will gradually spiral together and merge, causing an explosion. However, we know of too few binary white dwarfs in our own Galaxy to account for the observed number of Type Ia supernovae in a typical galaxy. We don't really know how a star reaches the Chandrasekhar limit, but this question offers a great opportunity for future astronomers to discover the fundamental mechanism by which white dwarfs reach their explosive stage. ■

Important Terms

neutron star: The compact endpoint in stellar evolution in which typically 1.4 solar masses of material is compressed into a small (diameter = 20–30 km) sphere supported by neutron degeneracy pressure.

pulsar: An astronomical object detected through pulses of radiation (usually radio waves) having a short, extremely well-defined period; thought to be a rotating neutron star with a very strong magnetic field.

Suggested Reading

Filippenko, "Stellar Explosions, Neutron Stars, and Black Holes," in *The Origin and Evolution of the Universe.*

Harvard-Smithsonian Center for Astrophysics, *Supernova*, www-cfa.harvard. edu/oir/Research/supernova/SNlinks.html.

Kirshner, *The Extravagant Universe: Exploding Stars, Dark Energy, and the Accelerating Cosmos*.

Marschall, *The Supernova Story*.

Pasachoff, *Astronomy: From the Earth to the Universe*, 6th ed.

Pasachoff and Filippenko, *The Cosmos: Astronomy in the New Millennium*, 3rd ed.

Questions to Consider

1. Distinguish between what goes on in novae and Type Ia supernovae.

2. Do you think the observed widths of emission and absorption lines in the spectra of supernovae give some indication of the speed of the ejected gas?

3. Most white dwarfs have a mass of about 0.5 to 0.6 times the mass of the Sun. Does this pose a problem for the hypothesis that most Type Ia supernovae arise from the explosion of a merged white dwarf binary whose mass reaches the Chandrasekhar limit?

4. Given that the nuclear reactions at the surface of a white dwarf undergoing a nova explosion release enough energy to eject the material accreted from the companion star, why do you think it is important for a white dwarf to avoid the nova process if it is to eventually become a supernova?

White Dwarf Supernovae—Stealing to Explode

Lecture 54—Transcript

In the previous lecture, I introduced supernovae: catastrophic explosions of a small minority of stars at the ends of their lives. They just go kazam, like that, and blow themselves to smithereens! They're absolutely spectacular. For a few days or weeks, they can be as luminous or powerful as millions, or even billions, of normal stars like our Sun. They're really important because they produce and eject into the cosmos heavy chemical elements that are the raw material for the formation of new stars, rocky Earth-like planets, and ultimately even life itself. We came from stars, from supernovae, and I'll give you that evidence in the next few lectures. They're amazingly brilliant, luminous objects. What kinds of stars actually explode, and how do they do it? I mentioned last time that the Sun won't explode; it'll die a rather gentle, quiet death, becoming a planetary nebula and then a white dwarf. You can get major clues about stellar explosions from the spectra of those explosions.

Again, just as a review, you pass starlight through a prism, disperse that light into the rainbow or spectrum of colors, and you can measure various lines—absorption lines and emission lines—produced by the chemical elements in the ejected gases. If you plot the intensity or brightness of the spectrum as a function of wavelength or color—going from the ultraviolet at the left to the red and the near infrared colors on the right—you find that the spectra of supernovae look like roller coasters. They have all these up and down undulations, and there are different kinds of spectral signatures. Some seem to have hydrogen in their spectra; those are observationally called the Type II's. They have other things as well—iron, and calcium, and oxygen—but the notable feature is hydrogen.

Then there are some that have calcium, magnesium, iron, sulfur, silicon, and oxygen, but notably no hydrogen. Those are observationally called the Type I's. Type I has no hydrogen; Type II has hydrogen. It's a purely observational distinction. It says nothing yet about the physical mechanism for the explosion. They have all these undulations. You look at the spectrum, and you see silicon, and oxygen, and calcium. These are the sorts of elements that were synthesized during and before the explosion and expelled into the cosmos, making them raw material for the formation of new stars, planets, and life. If you look at the Type I's and the Type II's, those that don't have hydrogen are generically the Type I's, but it turns out that, among the Type I's, there are several different subclasses—now called Ia, Ib, and Ic. All of them don't have hydrogen, but there are some other differences among those subclasses. The classical, most historically long known Type I's are now called the Type Ia's; they're the generic Type I's. We now know about these Type Ib's and Type Ic's, and I'll talk about them in the next couple of lectures. They actually have a very different explosion mechanism than the Type Ia's.

If you look at other differences among Type I's and Type II's, you can find them. For example, if you plot the brightness on the vertical axis versus time on the horizontal axis—here the units are days since peak brightness, so here's the peak right there—you find that Type Ia's rise for about three weeks, reach their peak, and then decline for many months and even years. Initially, the decline is pretty fast, and then it slows down. The Type II's rise in a day or less, in some cases; maybe in some cases two days, but anyway a very short rise. Then they sit around at about the same brightness for up to three months, and then they decline steeply, followed by a more gradual decline. Those are representative light curves, or plots of brightness versus time, for these two main classes: the classical Type Ia's and the Type II's. Other differences are the environments in which these supernovae are found. Type Ia's occur in all kinds of galaxies, including galaxies called elliptical galaxies like this one, which consist only of old stars—no young, massive stars. There's one that went off in an elliptical galaxy. It looks like Type Ia's probably come from an old population of stars.

Type II's, on the other hand, and also Type Ib's and Type Ic's, tend to come from the arms, or from near the arms, of spiral galaxies like this one. You never see them in elliptical galaxies, which consist only of old stars. You always see them in spiral and other irregular-type galaxies that tend to have a lot of star formation going on right now. There are massive stars being born. There are young, massive stars—and massive stars don't live very long before they die. The fact that Type II and Ib and Ic supernovae occur in spiral galaxies in or near regions of massive star formation also suggests that they are somehow related to the deaths of massive stars. If you look at the spectra of these supernovae, you find that, in all cases, there's evidence for very rapid ejection of gases—speeds of thousands of kilometers per second, sometimes exceeding 10,000 kilometers per second.

The way we deduce that is by looking at the spectrum—again, brightness versus wavelength. Here's an emission line of hydrogen. Then the absorption line here is blueshifted relative to the emission line. From the amount of this blueshift, you can tell how rapidly the ejecta are expanding. Also, the general breadth or width of this emission line tells you that the gases are expanding very quickly. Some of the gases are moving toward you; those would be blueshifted. Some are moving away from you; those would be redshifted. The overall ensemble of atoms moving toward you and away from you at different speeds leads to this broad emission line and a broad absorption line. Let me show you in a little bit more detail how that occurs.

If you have a star with only a photosphere—the surface, so to speak, from which the photons come—you get a plot of brightness versus wavelength that's a continuum; sort of a continuous spectrum with no lines. If you have a stationary absorbing shell of material outside the photosphere, that absorbing shell can produce an absorption line, and that absorption line will be at the rest wavelength if that shell is stationary. By rest wavelength, I mean the wavelength you would have measured for that line in a laboratory gas at rest. But now, if that shell is expanding away from the photosphere, the line that you see is blueshifted because this gas is moving towards you, and so absorption occurs of photons that are blueshifted relative to the rest wavelength. If you have a bunch of shells of gas, they're moving at different speeds away from the photosphere, and you get a bunch of absorption lines, all of which are blueshifted relative to the rest wavelength, and which together form a broad absorption line, as seen by us.

Then you have to remember that all these shells are also producing some emission as well. If you add that in as well, you find an emission line centered at the rest wavelength because most of the emission is coming from shells of gas that are moving perpendicular to your line of sight. That's where the emission is coming from. It's coming from these regions here—whereas the absorption is coming from those regions there. You get a blueshifted absorption line and a broad emission line centered at the rest wavelength of the line coming from all these shells over here. That's a bit of a complex thing. I just want you to get the basic idea of what I'm saying. The idea is that, from the width of the lines, and specifically from the blueshift of the absorption, you can tell how rapidly this gas is being ejected away from the star.

For many years, there were only a few bright supernovae to study, and a lot of those had been found long after they went off. Therefore, we couldn't study how they were exploding in the first few minutes or the first few days or weeks of the explosion. But now with robotic searches, like that conducted by my group at Berkeley—and also with the help of many amateur astronomers—many bright supernovae are found each year and can be studied in great detail. With our search at Lick Observatory—conducted by myself, and Weidong Li, and our whole team of students—we find lots and lots of supernovae: many dozens per year, up to 100 per year, and here are five examples. We can study them in great detail. We can follow how quickly they brighten and fade, and we can get lots of spectra of them. Here's one, Supernova 2000cx, for which we got really great measurements of the brightness, or magnitude, versus time. You can see that it brightened, and then it declined. It had slightly different light curves as viewed through a filter that transmits blue light, or visual light, or red light, or infrared light. The light curves differ in each of those bands. We can learn about the physics of the explosion through studies like this.

Measuring those actual light curves used to be a really tedious, labor-intensive process. But thanks to the great programming skills of Weidong Li and a student of mine, Mohan Ganeshalingam, we now have a largely automated program. Here's Mo, who helped Weidong write this program. We can now measure the light curves, or brightness versus time, of many supernovae just overnight. Once we get the data, you let the program run through its course, and you get the light curve. Here's a bunch of light curves of Type II supernovae, all showing this characteristic feature where they stay at about a constant brightness for a few months, and then they decline. Here's a bunch of Type Ia supernovae, where they brighten over the course of, say, three weeks and then decline. These are really great light curves. Data like this are of unprecedented quality. We can do that now with all these robotic telescopes and the automatic programs with which we measure these light curves.

We can also study the rates at which different kinds of supernovae occur. With our Lick Observatory sample, we've actually found over 700 supernovae, and we've determined what types they are and what kinds of galaxies they're in. Therefore, we can figure out what kinds of supernovae occur at what rate, how many per century on average, in different types of galaxies. This is important because different types of supernovae produce different kinds of chemical elements, and they pollute those galaxies in which they occur. If certain types of galaxies produce more Type Ia's than Type II's, their chemical evolution will be different from that of galaxies that produce more Type II's, for example. The rates are also important for determining the rate of formation of neutron stars and pulsars, and how quickly the gas between the stars is heated by these titanic explosions. Therefore, astronomers really want to know the rate at which supernovae occur.

I have a student, Jesse Leaman, who is a quadriplegic, in fact. He can hardly move anything, but he's got a tremendous brain. He's got a great mind. He talks to his computer, and he can program the computer, and he can tell it to analyze the data. He's been doing a marvelous job at analyzing the data set that we have of over 700 supernovae, sitting in his wheelchair, talking to his computer, programming the computer, asking it to write papers. I mean it doesn't write them automatically. You don't just say, "Hey, write the paper." He has to dictate it, of course, but it does it. He's been doing tremendous science, and I think he'll be a real inspiration to many people who are also in disabled or disadvantaged positions like this.

Let's now talk about what produces these supernovae. The traditional kind, the Type Ia's, don't have any hydrogen in their spectrum. You might say, "Okay, well, maybe that's because the gas is not at the right temperature to produce hydrogen. Maybe it's too cold or too hot to produce absorption lines in the visible part of the spectrum." But when you measure the spectrum, you find that it corresponds to, roughly speaking, a black body having a temperature of around 10,000 degrees Kelvin. At such a temperature, hydrogen optical absorption and emission lines should be very prominent if there were hydrogen present in the ejecta. But there's no hydrogen in the spectrum, so we conclude that there's no hydrogen present in the ejecta, or very little. That's very significant because hydrogen is by far the most common element in the Universe. Any star that doesn't have hydrogen is a peculiar star; it's not a normal sun or anything like that.

The other interesting properties of the Type Ia's is that they sometimes, as I said, occur in galaxies having only very old stars. Also, all of the Type Ia's look very similar in their observed properties. They have the same light curve, the same peak power, and they're almost identical twins. All these characteristics sort of shout out "white dwarf"—white dwarf at some limiting mass, the Chandrasekhar limit that I mentioned when I discussed white dwarfs. A carbon-oxygen white dwarf surrounded by a thin helium layer might not have any hydrogen at all. I mean it's stealing some hydrogen from a companion star, perhaps, but that hydrogen can quickly turn into helium through fusion. There might not be any hydrogen, and the carbon-oxygen white dwarf might achieve always the same configuration when it explodes. I'll explain that in a minute.

Here's what we think is happening. There's a white dwarf made predominantly of carbon and oxygen—perhaps with a helium shell around it—stealing material from a companion star. The way this happens is that the companion star can turn into a red giant near the end of its life and spill material over to the white dwarf. Let me show that. Here are the Roche lobes, the regions within which the gravity of each star dominates. Here's the star as it was during its normal main-sequence life. Then near the end of its life, it starts expanding. It's on its way to becoming a red giant, becoming bigger and bigger. You can see that eventually it reaches the size of its Roche lobe, or the region within which its gravity dominates. Then material can spill across this point right here, become dominated by the gravity of the white dwarf, and accrete gradually onto the white dwarf, increasing its mass.

What we have here, then, is this picture where the material from a relatively normal star expanding to a red giant is spilling onto the white dwarf. That's very similar to the picture that we showed when we were discussing novae in Lecture 52, where I was saying that a white dwarf collects material from the companion star, and then undergoes a series of surface explosions that make the thing look somewhat brighter than before—maybe a factor of 100 or a million times brighter than before. In the case of the Type Ia supernova, what we think actually happens is that the mass of the white dwarf is able to grow without undergoing nova explosions, which tend to eject material.

If the mass of the white dwarf grows enough, it reaches this limiting mass, known as the Chandrasekhar limit—after Subrahmanyan Chandrasekhar, who first figured it out when he was only about 20 years old—and that limit, 1.4 solar masses, is a mass at which the white dwarf has to do something catastrophic. It's unstable at that mass, and that's what Chandra knew. He didn't know exactly what would happen to the white dwarf, but he knew that it would become unstable. What we think happens is that there's a runaway chain of nuclear reactions that completely obliterates the star. It just goes kabloom, like that, through an uncontrolled chain of nuclear reactions.

The Chandrasekhar limit is given on this T-shirt that I made. It depends on the number of protons per number of protons plus neutrons in a star. Then there's a bunch of fundamental constants: Planck's constant, the speed of light, Newton's constant of gravity, and the mass of the proton. All these things together give you this fundamental limit at which a star has to become unstable, and Chandrasekhar figured that out.

It's a runaway nuclear explosion because the matter is degenerate. That means that all of the electrons are in their lowest-possible energy states, and they're holding the star up just barely, trying to keep it from collapsing. The nuclei, the carbon and oxygen nuclei, hit each other, they fuse, and they release some energy. That energy is not able to expand the star, as would have been the case in our Sun if you were to suddenly heat up the center of our Sun, because degenerate matter doesn't really care what temperature it's at. It just keeps the same pressure because the pressure is dictated largely by the electrons. Therefore, that energy doesn't go into expanding the star. Instead, it goes into moving the other nuclei a little bit faster, and they slam into each other, creating more nuclear reactions, releasing more energy, which doesn't expand the star, but rather goes into heating up the nuclei even more. They run into each other, creating more nuclear reactions, and you can see it's a nuclear runaway—more specifically, a thermonuclear runaway because it's driven by the heat, the energy, emitted by the first few, or any preceding, nuclear reactions.

It's an amazing explosion, and you can see how bright it becomes here. It just becomes this hugely luminous star, about the same power as a billion normal suns. A Hubble Space Telescope picture of the same supernova shows it much more sharply in this beautiful, dusty spiral galaxy. Now we observe the ejecta of some of these things in our own galaxy, say 1,000 years later, as in the case of this one that occurred in the year 1006. So, a huge number of nuclear reactions occur. They fuse carbon and oxygen into heavy elements, primarily radioactive nickel—nickel with 28 protons and 56 nucleons, or protons plus neutrons. That radioactive nickel decays into radioactive cobalt, and then into stable iron. That decay process emits gamma rays, very energetic photons, because nuclei emit very energetic photons, not just optical and infrared photons like electrons do. Those gamma rays then bounce around inside the exploding star, gradually getting converted into optical light. Eventually, when the stuff gets out, we see it as the optical light. But, roughly, 0.6 of a solar mass of nickel-56 forms in this explosion, driving the optical display of light that we then see for a few months or a few years.

If the radioactive nuclei had not been produced, the explosion would be an optical "dud" because all of the energy that's released through fusion of light elements into heavier elements would have been used up in the expansion of the star. Expansion of material uses up energy. So, if all the energy were used up, and there was no energy associated with radioactive decay, then you wouldn't see anything in this exploding star. It would simply explode, but you wouldn't see the explosion because there'd be no new energy driving the visible light from the supernova. It's good that there's this radioactive decay. It produces bright fireworks for us to see. The details of all this are uncertain. For example, the white dwarf is probably rotating, and it probably has magnetic fields. These are all hard problems to solve. People are now trying to figure out what changes in detail in the exploding star if we incorporate rotation, and magnetic fields, and things like that.

We don't really know yet. We're trying to figure it out. We don't exactly know which kinds of stars will explode, but we think that it's only the white dwarfs in binary systems that are able to accrete enough material to reach the Chandrasekhar limit. Then you might say, "Well, gee, Sirius, 8.7 light years away, has a white dwarf companion. Will it explode one of these days?

Being at a distance of only 8.7 light years, is that going to hurt us?" We don't think that Sirius B is going to explode, at least not anytime soon. First of all, it only has 1.05 solar masses, quite far from the Chandrasekhar limit of 1.4 solar masses. Moreover, it's pretty far from Sirius A, and it's not at all clear that there will be sufficient accretion from Sirius A to produce enough mass to make Sirius B reach the Chandrasekhar limit. If it does, it'll only do so far, far in the future, when the Sun will probably be a lot farther away from Sirius. It could become a nova in between. It might just accrete a little bit, and then blow off the shell that it accretes, and not really grow in mass. We don't think it's going to become a supernova. Nova explosions at a distance of 8.7 light years are nothing to be afraid of.

You might say, "Okay, but what if it were to go off? Let's just suppose. Would it hurt us?" Optically, it wouldn't be a problem because it would only be about 1% of the brightness of the Sun, and it would only be that way for a few weeks. That would be visible in a daytime sky—it would be brilliant—but it wouldn't hurt us. You might worry about the X-rays and the gamma rays that are produced by the explosion because they can not only deplete the ozone layer, but also cause ionization of our atmosphere, which leads to currents in the atmosphere and what's called an electromagnetic pulse. You can get this danger during a nuclear war as well, when a whole bunch of nuclear weapons go off, and they ionize the atmosphere, and there are charge imbalances in the atmosphere that create voltage drops and, essentially, huge bolts of lightning. Electronic circuitry gets short-circuited, so that can be a really bad situation.

A Type Ia at this distance would indeed produce some bad effects, but we don't know exactly what would kill us first. Would it be the X-rays or the gamma rays, or would it be energetic charged particles called cosmic rays, which are produced by supernovae, and which we see in our detectors? We don't really know the effects of radiation on human tissue very well yet, but probably a supernova within 10 light years would be very bad news—between 10 and 100 light years might be bad news. Beyond 100 light years, I think we're safe. It's been estimated that, at most, one of the mass extinctions that has occurred in the history of life on Earth was perhaps caused by a supernova explosion. But even that is controversial, and we don't really know of any direct evidence for an extinction of life on Earth being caused by a supernova explosion.

Let's go back to the Type Ia's and uncertainties in their explosion mechanism. We're pretty sure that a Type Ia results from the explosion of a white dwarf that's reached the Chandrasekhar limit. The spectra that you compute, the light curves that you get from computer models, agree very well with what's actually observed, so that we're pretty sure of. What we're not sure of at all is how a white dwarf actually reaches the Chandrasekhar limit in the first place. How does it become unstable? If there's a main-sequence star trying to spill material to the white dwarf, that's hard because a main-sequence star is much smaller than its Roche lobe, so it doesn't spill material to the white dwarf.

On the other hand, if it's a red giant star spilling material, then shouldn't the explosion rip some of the gas away from the envelope of the red giant, causing the hydrogen to glow? After all, the red giant is made out of hydrogen. Shouldn't we see some hydrogen in the spectrum? We don't, so people say it probably isn't a red giant. If it's not a main-sequence star, and it's not a red giant, what is it? We really don't know what it is. Moreover, if you look at this diagram here of the star spilling over to the white dwarf, as I said before, it looks a lot like the picture we had for novae. In the case of novae, you have the hydrogen accreting onto a surface layer on the white dwarf, and then it explodes. Then the star gathers material for a few years, and then it explodes again, and so on. Each of these nova explosions can get rid of all of the accreted hydrogen, or even more, so the star's mass doesn't really grow up to the Chandrasekhar limit.

You might say, "Well, all right, maybe you can avoid these nova explosions by having a very rapid accretion rate." But in that case, the envelope around the white dwarf swells up, and once again you would get hydrogen in the spectrum because the white dwarf would explode in a swollen hydrogen envelope, and it would look like a Type II. So, it's a real problem. We don't know how the star gets up to the Chandrasekhar limit. Some physicists have seriously considered sub-Chandrasekhar white dwarfs, where the mass is less than 1.4 solar masses, and the thing is somehow able to explode. For example, an explosion can be initiated at the boundary between the helium envelope and the carbon-oxygen core, and maybe that's what happens. The trouble is that the spectra and light curves from such models don't agree with what's actually observed.

So, what do we do? Some physicists have said that maybe what you really have are two white dwarfs in a binary system, which gradually spiral together and merge, causing an explosion. You might have a picture like this, where you have two evolved stars, two white dwarfs, spiraling in toward each other. As they spiral in, they emit gravitational waves. I'll discuss those later on when I cover general relativity and gravitational waves, but here you can see these stars spiraling in toward each other. They're emitting energy in the form of these waves, these ripples in the actual shape of space. As they get closer and closer together, they orbit more and more quickly, until finally they just merge and explode. That might be what produces the actual supernova.

In this animation, they merged and formed a neutron star, and it's still sitting around there, spinning quickly. But in other cases, maybe there's an actual explosion. The animation itself is for a pair of white dwarfs that was recently found that are orbiting each other in just a little bit over five minutes. They're really whizzing around close to one another. They might merge and form a Type Ia supernova. But in general, the problem with this model has been that there are too few binary white dwarfs known in our own galaxy to account for the observed number of Type Ia supernovae each year or each millennium in a typical galaxy. So, binary white dwarfs appear to be really rare. I think it's this interesting situation where we know pretty much how a Ia forms. It's the explosion of a white dwarf star that has reached the Chandrasekhar limit, or in some other way is undergoing a thermonuclear runaway explosion.

We're in this embarrassing situation where we don't really know how the star gets up to this unstable limit. How does it actually get there? That's a huge embarrassment, but it's also a great challenge and an opportunity for astronomers of the future to figure out this fundamental mechanism by which stars reach the explosive stage. As we'll see later on, an understanding of Type Ia supernovae and how they occur will be very important in our studies of cosmology, the overall structure and evolution of the Universe.

Core-Collapse Supernovae—Gravity Wins

Lecture 55

In previous lectures, we've seen how some white dwarfs reach an unstable mass, the Chandrasekhar limit, causing them to explode. In this lecture, we will look at another kind of mechanism for stellar explosions in Type II supernovae and related subclasses.

Type II supernovae appear in spiral galaxies, usually in or near spiral arms, where lots of massive stars are forming or have recently formed. Red supergiants are most likely to become supernovae, and these stars typically have masses initially at least 10 times that of our Sun. A cross-section of such a star would show an iron core with shells of progressively lighter elements surrounding it: silicon and sulfur; oxygen, neon, and magnesium; carbon and oxygen; helium; and finally, hydrogen. Most of the volume, however, is hydrogen. Red supergiants have this onion-like layering because the ashes of one set of nuclear reactions become the fuel for the next set. Hydrogen fuses to helium; helium fuses to carbon and oxygen; and carbon and oxygen fuse to neon, magnesium, and so on, all the way up to iron. Each set of reactions liberates energy because the products are more and more tightly bound compared to the reactants; the binding energy is released during nuclear fusion. Iron and other elements of similar mass—nickel and cobalt, for example—are the most tightly bound elements. For this reason, their fusion does not produce energy; rather, it requires energy.

Very heavy elements (such as uranium) can undergo *fission*, or break up, into lighter elements, releasing energy. Per nucleon (proton or neutron), the binding energy of the products is higher than the binding energy of the very heavy elements. However, this is not what occurs in a red supergiant. We mention fission of very heavy elements simply to stress that iron-group elements are the most tightly bound (that is, have the highest binding energy per nucleon). Thus, as a red supergiant undergoes successive stages of nuclear burning, iron eventually forms in the core, with a silicon-sulfur-fusing shell surrounding it.

As the iron core gains mass, it eventually reaches the Chandrasekhar limit of roughly 1.4 solar masses. At its mass limit, the iron core collapses due to gravity, to a radius of only about 10 kilometers, liberating a tremendous amount of gravitational energy. Microscopically, the electrons and protons combine to form neutrons and neutrinos. The pressure support, previously provided mostly by the electrons, disappears until the star is just a ball of neutrons—a neutron star. During the core's collapse, it rebounds from itself because its constituent particles repel one another when they get too close together. Because material surrounding the core is no longer supported by pressure, it collapses, but then it collides with the rebounding core and is propelled outward at high speeds.

But this *prompt mechanism*, or rebounding effect, is not enough to completely eject the material into space. The gravity of the central neutron star pulls it back; thus, a stronger force is needed to give the material an extra push and create the full visual effect that we see in an exploding star. When the star collapses, its protons and electrons combine to form neutrons and neutrinos. Far more neutrinos are produced simply because of the fact that the young neutron star has an extremely high temperature, about 100 billion K; neutrinos are efficiently produced at such temperatures. Some of the released neutrinos can hit the surrounding layers of gas and eject them into space, creating a successful supernova explosion. During the explosion, elements even heavier than iron can form, generally through the sequential capture of many neutrons (followed by radioactive decay of some of them into protons). In this process, we get the rich periodic table of the elements, of which we and other Earth-like, rocky planets consist.

It is difficult to know precisely when a red supergiant will explode because we can't tell what the core is doing, and there are different time scales associated with the various stages of nuclear burning. For example, because Betelgeuse is a red supergiant, we know that it's at least in the helium-burning stage, which lasts 500,000 years. (For comparison, the main-sequence stage lasted perhaps 7 million years.) It might be in the carbon-burning stage, which lasts only 600 years, or it might even be in the silicon-burning stage, which lasts only a day. Most likely, it'll explode sometime in the next half a million years. In the last few stages of a red supergiant's life—oxygen fusing to silicon and sulfur, and silicon fusing to iron—the temperatures are so high that many neutrinos are formed, which escape from the star immediately at speeds close to the speed of light. If we could develop a highly sensitive neutrino detector, we might be able to predict more accurately when a star is about to explode.

Red supergiants are not the only stars that can explode in this way, as *core-collapse supernovae.* Several other subclasses of stars belong to this category. Most core-collapse supernovae have a hydrogen shell, formally making them Type II supernovae, but some massive stars lose this shell before exploding. Such a **progenitor** star has a helium envelope and other elements (carbon, oxygen, and so on) in its core, ending with iron at the center. These stars explode as Type Ib supernovae. The progenitor stars of Type Ic supernovae have lost both their hydrogen and helium outer layers, leaving an envelope of carbon and oxygen, with shells of successively heavier elements inside. These subclasses of stars are important because we now recognize that not all Type I supernovae obliterate themselves and produce a large amount of iron (first in the form of radioactive nickel), as Type Ia supernovae do. Instead, Type Ib and Ic supernovae form compact neutron stars and eject large quantities of intermediate-mass elements, such as oxygen, calcium, magnesium, and sulfur. Thus, Type Ib and Ic supernovae affect the chemical evolution of galaxies in a way that differs from that of Type Ia supernovae.

"It looks like gravity is ultimately victorious in all these stars, all these massive stars, regardless of how much of an envelope they still retain."

How does the envelope of hydrogen—and, in some cases, helium—get stripped away from a star? Some massive stars experience winds and gentle ejections, where the pressure of photons expels the envelope. Stars in a binary system can also transfer part of their gas atmospheres onto a companion star. In some cases, only partial stripping of the hydrogen envelope occurs, creating a low-mass shell of hydrogen. We call the subsequent explosion a Type IIb supernova. Thus, core-collapse supernovae come in a number of varieties: Type II, the hydrogen-rich and most common kind, and Types Ib and Ic, classified according to whether or not they have helium envelopes. Regardless of how much of a hydrogen envelope they have, their iron cores ultimately collapse to form neutron stars. ■

Important Terms

progenitor: In the case of a supernova, the star that will eventually explode.

stripped massive stars: Stars that have lost their hydrogen and helium envelopes, either through stellar winds or through transfer of gas to a companion star; thought to be the progenitors for gamma-ray bursts.

Suggested Reading

Kirshner, *The Extravagant Universe: Exploding Stars, Dark Energy, and the Accelerating Cosmos*.

Marschall, *The Supernova Story*.

Filippenko, "Stellar Explosions, Neutron Stars, and Black Holes," in *The Origin and Evolution of the Universe*.

———, "A Supernova with an Identity Crisis," *Sky & Telescope*, Dec. 1993.

Pasachoff and Filippenko, *The Cosmos: Astronomy in the New Millennium*, 3rd ed.

Questions to Consider

1. What are the observational and physical differences between Type Ia and Type II supernovae? Which kinds of stars explode, and how do they explode?

2. In a supernova explosion of a 15-solar-mass star, about how much material is ejected (blown away)?

3. Do you expect Type Ib and Ic supernovae to produce a burst of neutrinos, just as a Type II supernova does?

Core-Collapse Supernovae—Gravity Wins

Lecture 55—Transcript

In the previous lecture, I described how some types of white dwarfs may undergo a thermonuclear runaway at the end of their lives, literally blowing themselves to smithereens. These Type Ia supernovae are extremely powerful, luminous objects, amazing to watch, and they generate some of the heaviest elements we know of in the Universe. How they get to this explosive state is, if you'll pardon the pun, a burning question. We don't really know how the white dwarf gets to this unstable Chandrasekhar limit, but once it gets there, it explodes. You need these special circumstances, and most white dwarfs never reach those special circumstances, but occasionally some do, and they explode.

There's another very different type of physical mechanism for stellar explosions, and it's thought to be relevant for Type II supernovae. I mentioned last time that Type II supernovae, like this one here, appear in spiral galaxies, usually in or near spiral arms like this one, where lots of massive stars are forming or have recently formed. We think that these stars that become Type II supernovae are massive stars at the end of their lives. This is also suggested by the light curve of a Type II supernova. You can see brightness versus days here, and it stays at about the same brightness for several months. This suggests that the star that's exploding has a very thick envelope that's able to retain the energy for long periods of time and gradually allow that energy to leak out. If there weren't a thick envelope, the energy would leak out very quickly.

All this points to red supergiants as the most likely progenitor star, or the star that becomes the supernova. Here are these red supergiants that are very, very luminous. They come from massive evolved stars—that is, stars that have left the main sequence. These are stars that are at least 10 times the mass of our Sun initially. Some astronomers think that maybe a star at least eight times the mass of the Sun could explode in this way. It's safer to say that they are at least 10 times the mass of the Sun initially. If you make a cross-section cut of such a star, here's what it looks like. You have an outer envelope of hydrogen, within which there is a helium shell, and then inside that, there is a shell consisting predominately of carbon and oxygen. Then there are shells of even heavier elements, like neon and magnesium and then silicon and sulfur. Finally, at the very center, there's an iron core. You have all these shells of gas. Here's sort of a fancier plot of the cross-sectional view of a red supergiant. It kind of looks like an onion, with all these different layers: the iron core, a layer of silicon and sulfur surrounding it—these are, by the way, just the main elements in each of these layers—then oxygen, neon, and magnesium, and then a layer of carbon and oxygen, then helium, and finally, hydrogen. This whole thing is way down in the middle of an extremely large star, most of whose volume is just hydrogen.

It has this onion-layer structure because the ashes of one set of nuclear reactions became the fuel for the next set. Hydrogen burned, in a nuclear sense, to helium. Helium went to carbon and oxygen; carbon and oxygen went to neon, magnesium, and things like that. This happens because each of these heavier elements is progressively more and more tightly bound than its predecessors. If we look at a plot of the binding energy per nucleon—that is, per proton or neutron in the nucleus of an atom—versus the atomic weight, which is just the number of protons and neutrons—hydrogen has an atomic weight of 1; deuterium has an atomic weight of 2. Then there's helium, beryllium, carbon, oxygen, and so on. The binding energy per nucleon increases along this curve, meaning that the nuclei are progressively more tightly bound. If you take light initial particles and bring them together, and fuse them together, the mass of the end product is less than the mass of the initial reactants. It's that mass difference that gets converted into energy according to Einstein's famous $E = mc^2$. I discussed this process in some detail when I considered the fusion of hydrogen into helium during the main sequence stages of a star's life.

If we look at the red supergiant, we have this hydrogen envelope with the helium shell inside, then carbon and oxygen, and then shells of progressively heavier elements ending all the way with iron in the core. Iron doesn't liberate energy. All the preceding stages liberated energy, keeping the star hot inside, and keeping the pressure high. But iron, if it were to fuse to a heavier element, would absorb energy. That's because iron and the elements surrounding iron are at the top of the binding energy curve. This is an expanded view of the plot I showed you before, where you have the binding energy per nucleon versus the atomic weight. All the light elements can fuse to heavier elements, liberating energy along the way, because the products are more and more tightly bound compared to the reactants.

Similarly, heavy elements can fission, or break up, into lighter elements—releasing energy because the binding energy of the lighter elements per nucleon is higher than the binding energy of the very heavy elements like uranium. Uranium can split into lighter things, and carbon and hydrogen can fuse into heavier things. Both of those processes liberate energy. But right around iron, you have the highest binding energy per nucleon, so iron and its similar elements—cobalt, nickel, and things like that—are the most tightly bound nuclei, so fusion of those elements does not produce energy; it actually would require energy. What you get is a halt, or a stop, to the nuclear fusion process in the core of a star when you reach an iron core. You get this iron core, then, in the middle, surrounded by layers that are still fusing up to heavier elements, liberating energy as they fuse. So, hydrogen fused to helium; it liberated energy. That was the main-sequence stage.

Then helium fused to carbon and oxygen. As you can see, carbon and oxygen fused to heavier elements. All those fusion reactions liberate energy, keeping the star hot inside and preventing it from collapsing due to the inexorable force of gravity trying to pull it in. But once you reach iron, you don't get any extra energy. So, the iron core, then, builds up, and there are these fusion reactions surrounding it. Eventually, when the iron core reaches roughly 1.4 solar masses, roughly the Chandrasekhar limit, it gets to this unstable stage, similar to a white dwarf. But instead of undergoing a thermonuclear runaway, like a white dwarf in a Type Ia supernova, the iron core in a massive star decides to collapse. Gravity is ultimately victorious in this case, and you have this iron core collapsing down to a tiny, little volume, roughly the size of a city, 10 kilometers in radius, a neutron star, and it liberates a tremendous amount of energy in this process.

Microscopically, what happens is that the electrons and protons combine to form neutrons and these funny, little particles: neutrinos, which I discussed when I considered fusion in the Sun. The pressure support that had been provided by the electrons mostly—and to a lesser extent, the protons—disappears. You just get a ball of neutrons. This core collapses to this tiny ball of neutrons, a neutron star essentially, where you have a sphere whose diameter is roughly 20 kilometers, a mass 1.4 times that of the Sun, and an incredible density. I'll talk about neutron stars more later. During this collapse, a tremendous amount of energy is liberated, so this neutron star is collapsing down, or this core is collapsing down, to form a neutron star. But just like when you jump on a trampoline, you don't reach an equilibrium configuration right away—rather, you bounce back up—so, too, this neutron star, collapsing in on itself, reaches supernuclear densities, even greater densities than nuclei. At that point, the material stars to repel, and it bounces—it rebounds, like you are rebounding off of a trampoline.

When you rebound off of a trampoline, you've got a lot of energy coming up. If you have stuff coming down on top of you, you can hit that stuff and propel it upwards. Let me show you a demo. Suppose this little basketball is like the core, the iron core, of a red supergiant collapsing in on itself. It's going to collapse, and it's going to rebound. Here, it rebounds off of the floor, but you can think of it as rebounding off of itself. Here it is; it rebounds, bounces back up, like me jumping off of a trampoline. Meanwhile, the stuff surrounding it has no more pressure support because the core has collapsed, and it too falls. Left on its own, it would just sort of rebound a little bit. But if the stuff surrounding the collapsing iron core hits this outwardly moving rebounding core, it can get an extra amount of energy and get propelled upwards with a huge speed. Let me show you when I put, now, this tennis ball, which are the surrounding layers around the core, on top of the imploding and then rebounding core. Let's see what we get.

Look at that! The thing just takes off. In fact, in lectures, if students aren't paying attention, I can cleverly put the ball on one side here and propel the ball towards them and wake them up. That illustrates to you how the imploding core of a red supergiant can rebound and propel the surrounding layers outwards. That's really kind of cool. That's called the prompt mechanism, and it turns out that it's really not quite enough. Just like the tennis ball eventually came back down because of the Earth's gravity, so, too, the gravity of the neutron star that just got formed is pretty large, and it tends to want to bring this rebounding material back down. You need a bit of an extra push. That extra push is provided by the flood of neutrinos that got emitted by this process. First of all, I said that when the core is collapsing, protons and electrons are combining to form neutrons and neutrinos, and those neutrinos come out, and some of them can hit the surrounding layers and push them out.

Moreover, the young neutron star that has formed has a temperature of about 100 billion degrees. That's because a huge amount of gravitational energy got released as this Earth-sized iron core collapses down to form a neutron star that's only about 10 kilometers in radius. That energy, the gravitational binding energy of a neutron star, is about 1/10 of a solar mass. Using $E = mc^2$, if you convert that into energy, it's a huge amount of energy. At these temperatures of 100 billion degrees that are generated when so much energy is released in such a small volume, it turns out that neutrinos are produced in huge quantities at such temperatures, and so you just get a veritable flood of neutrinos. Even though neutrinos hardly interact at all, some of them will interact with the material surrounding them and help propel that material out. That is what actually allows the star to achieve a successful explosion. The neutrinos actually do most of the work.

The person who first figured this out was Fritz Zwicky at Caltech, along with his colleague, Walter Baade, and they did this in 1933, just a year after James Chadwick had discovered the neutron. They were both brilliant men; Zwicky in particular is one of my heroes. He just was decades ahead of other astronomers in so many ways. Though brilliant and very creative, he was also rather arrogant and abrasive. He actually referred to his Caltech colleagues as "spherical bastards," because "They're bastards, any way you look at them!" This is not a good way to make friends and retain friends, especially since people at Caltech are generally pretty smart. Here's Zwicky, perhaps indicating what he thought of their brain size or something. I'm not sure what he's doing there, but he just didn't get along very well. He was ignored, in part because he was just sort of viewed as a crazy guy. But now, it turns out he was right on a lot of things. He was wrong on certain things too, but he was right as well. He and Walter Baade had figured out this explosion mechanism.

Here's David Arnett at the University of Arizona, formerly in Chicago, showing what the Chicago Sears Tower would look like if compressed to the density of a neutron star. It would be this little gumball here. It would be an incredibly dense material, nothing like what we've created in the laboratory. Only atomic nuclei are of that sort, and we didn't really create them; they already sort of exist. This would be like a ball of neutrons, like a gigantic atomic nucleus, the mass of a building in a gumball. It's just some incredible stuff. This ejected material takes on complex shapes, and there are various instabilities that occur, but the point is that there's so much energy released in this process, that a lot of elements heavier than iron can form due to nuclear fusion. The star has given up; it has exploded. There's plenty of energy floating around. Now that energy can go into colliding neutrons with heavy nuclei, producing even heavier nuclei. That process absorbs energy because you're producing really heavy nuclei—like iron, gold, silver, and things like that—but there's plenty of energy to go around.

The star is no longer trying to hold itself up. It's no longer trying to generate energy to keep itself hot inside; rather, it's already exploded, so there's plenty of energy to go around. Atomic nuclei hit each other, fusing to heavier ones. In particular, neutrons tend to hit things and fuse into even heavier nuclei. In this process, then, you get the rich periodic table of the elements. You get all the elements heavier than iron, one way or another, through Type II supernovae, and also through Type Ia's—but in particular, in Type II's, you get all sorts of nuclei that are otherwise quite rare to produce in any other way. One way or another, then, all of the heavy elements of which we consist, and of which Earth-like, rocky planets consist, were produced by these supernovae. It turns out that, in some cases, you can get an iron nucleus inside the envelope of a red giant, and it can absorb a neutron and produce some heavier elements. So, some nucleosynthesis, or the synthesis or creation of elements, occurs not in supernovae, but rather in red giants. But the seed elements—the iron that was necessary for the nucleosynthesis in those red giants—came from some previous supernova explosion. One way or another, all of the heavy elements came from supernovae.

Given that red supergiants are supposed to explode, you might look in the sky and say, "Well, when is Betelgeuse, the shoulder of Orion, going to explode?" It turns out we don't really know. We don't know because, if you look at the stages of nuclear burning that go on in a red supergiant, and you look at the time scales associated with them, the last few stages—the fusion of silicon to iron—that takes just one day, and it occurs at a temperature of 4 billion degrees. The preceding stage, oxygen to silicon and sulfur, takes just six months. Even the stage where you're fusing carbon to oxygen, neon, and magnesium takes 600 years. These are all very small quantities compared to the stage during which a red supergiant, such as Betelgeuse, did or is fusing helium into carbon and oxygen; that stage is about 500,000 years. These time scales depend a little bit on the mass of the star that you started with, but here's a representative set of time scales for a 25-solar mass star. All these stages are short compared to the main-sequence stage, which, in this case, is about 7 million years.

Since Betelgeuse is a red supergiant, we know that it's at least in the helium-burning stage, so it's at least in the stage that lasts 500,000 years. It might be in the carbon-burning stage, which lasts only 600 years, or it might even be in the silicon-burning stage, which lasts only one day. If the latter, then it's going to explode tomorrow. More likely, however, it'll explode sometime in the next half a million years, and we just don't know when. All these stages are very short. In any case, you look at Betelgeuse, and you say, "Well, we don't really know what its core is doing. Maybe it's in the silicon-burning stage, or maybe it's in the hydrogen- or the helium-burning stages. It can't be in the hydrogen-burning stage because it's a red supergiant at this point. How can we tell at what stage it's at?

It turns out that, in these last couple of stages—in particular, oxygen going to silicon and sulfur, and silicon going to iron—the temperatures are so high, billions of degrees, that a lot of neutrinos are being formed. They escape from the star essentially immediately; they travel at the speed of light, or close to the speed of light, and they just escape from the star; unlike the photons, which bounce around a lot and can't escape. If we had a really sensitive neutrino detector, and we pointed it at Betelgeuse, and we see nothing, nothing, nothing; and then suddenly, we see a large increase in the number of neutrinos, we could say, "Aha, Betelgeuse is about to blow up." We could point to Betelgeuse and actually watch its explosion in the next day or two. That would be kind of fun.

It turns out that it's not just red supergiants that can explode in this way. There has been recently discovered a new set of species of these so-called core-collapse supernovae. In the diagram here, you can see essentially what I'm talking about. True, most core-collapse supernovae have a hydrogen shell.

But suppose the star has gotten rid of its hydrogen shell prior to exploding, in which case it only has a helium envelope, and the normal carbon, oxygen, and so on, core. This could lead to a core-collapse supernova that doesn't have hydrogen in its spectrum, yet is not the explosion of a white dwarf; that we call a Type Ib because it doesn't have hydrogen in its spectrum, so it's a Type I. But it does have helium in its spectrum, unlike a normal Type Ia, so we call it a Type Ib.

If the star has somehow lost not only its hydrogen layer, but also its helium layer prior to exploding, then it has a carbon and oxygen outermost layer, and then successively heavier elements inside—oxygen, neon, magnesium, silicon, sulfur, all the way up to iron. In those cases, the exploding star would have a spectrum devoid of hydrogen, so it would be a Type I, but it wouldn't look like the spectrum of a Type Ia in detail because it's not an exploding white dwarf; rather, it's a core-collapse object with a whole bunch of these intermediate-mass elements—carbon, oxygen, calcium, magnesium, and silicon—surrounding the core and pummeling outwards. That would look spectroscopically different from a Type Ia. These things were suspected in the 1960s, that some Type I supernovae are different from the others, but they were first recognized as a clear, new subclass of supernovae in the mid-1980s.

You might expect this subclass to be pretty important because if some Type I's don't completely obliterate themselves and produce a whole bunch of iron, and nickel, and cobalt—but rather are the core collapse of a massive star that leads to a bunch of oxygen, and sulfur, and stuff being ejected—that would affect the chemical evolution of galaxies because not all Type Ia's and Type Ib's and Ic's produce the same chemical elements. Here is, in fact, a spectrum of a Type Ia long after it has exploded, and you see a bunch of iron in the spectrum; whereas a Type Ic has so-called intermediate-mass elements: oxygen, calcium, magnesium, and things like that. It kind of looks like the spectrum of a Type II—but without the hydrogen.

It turns out that I got interested in supernovae in 1985 because my former thesis advisor, Wallace Sargent, and I were observing at Palomar Observatory, and we were looking at a whole bunch of galaxies, and we stumbled across a supernova of unprecedented observational properties. It turned out to be one of the defining examples of this new subclass. Indeed, it showed that not all Type I's are these Type Ia supernovae. We had a great five-night observing run in February, where all the nights were clear, and we got a lot of great data, and we should have been happy. I was actually kind of depressed at that time because my girlfriend had just dumped me, so I was kind of feeling bummed out, but I was happy that we had gotten good data.

In the last hour of the fifth and final night, we were looking at a bunch of charts in front of us, saying, "Well, which galaxies should we observe?" We had time only to observe two more galaxies. I said, "Well, let's observe NGC 4618 because that galaxy had been classified as a peculiar galaxy in some catalog of galaxies. We pointed the telescope to NGC 4618, and we saw a bright, star-like object in the middle of this peculiar galaxy, where we had expected to see no such bright star, and previous astronomers hadn't seen such a star. We took a spectrum of this star, and the spectrum was truly bizarre. Here's brightness versus wavelength. It had these strong emission lines of oxygen, calcium, magnesium, and things like that, sodium, intermediate-mass elements, and essentially no hydrogen. Yet it didn't look like a Type Ia because it didn't have the spectrum of a Type Ia. We just got really excited. It was so fantastic.

We said, "This is a new type of supernova. This is one whose progenitor star, the star that exploded, somehow had lost, or gotten stripped of, its hydrogen envelope prior to the explosion." We wrote a paper about this, and it got reported in the newspapers. The headline that came out, at least in the *San Francisco Chronicle*, was "Astronomers Peek at a Strip-Teasing Star." We had told the reporter that the envelope had somehow gotten stripped from the star prior to the explosion. We didn't tell him to call it a strip-teasing star. Anyway, this just changed my career. I got really interested in supernovae, the physics of how they explode, and how they can be used to understand the chemical evolution of galaxies and the production of neutron stars and pulsars.

Ultimately, as I'll tell you later in this course, supernovae turn out to be enormously important for cosmology, the study of the structure and evolution of the Universe as a whole. This happened in 1985, and it just changed my career. It just shows how important it is to be observant and to take advantage of opportunities that are thrown your way. Sometimes you get a great new opportunity, and if you don't capitalize on it, it's lost forever. But if you do capitalize on it, all sorts of wonderful things can happen. I've just been having so much fun the past 20 years studying supernovae.

Then the question is: How did the envelope of hydrogen—and in some cases, even helium—get stripped? Well, massive stars go through winds and gentle ejections, as is shown here, where just the pressure of the photons, and these gentle burps that I've talked about before, expel the envelope. In some cases, the envelope just expels outwards in two bulbous, bipolar directions, like this one for Eta Carinae. I've mentioned this star before. In other cases, it's a more or less spherical ejection. That's how a star can lose part of its atmosphere of gas, or it can transfer part of its atmosphere of gas onto a companion star. If it's in a binary system, as most stars are, the companion star can steal some of that material when the primary star expands and starts becoming a red supergiant. You can get this transfer of material.

You might even expect that, in some cases, partial stripping would occur—so, not all of the hydrogen envelope, but just some of it, got stripped away or stolen by a companion star. In that case, you'd have a helium core with all the oxygen, iron, and other stuff surrounded by just a low-mass thin shell of hydrogen. This would look sort of like a Type IIb, you might call it. Initially when it explodes, you'd see the hydrogen, so that would be Type II—but as you start peering down into the expanding ejecta, there wouldn't be any more hydrogen to see, and it would start looking like a Type Ib. This is sort of a hybrid, a Type IIb. I actually observed one of these things, Supernova 1987K, shown here, before the prediction that these things might exist became well known. I said, "Well, here's a star that started out as a Type II." Then later, the spectrum turned into a Type Ib, and so I called it a Type IIb. In fact, there was a theoretical model suggesting that that might even be the case.

Some astronomers thought that I was crazy, that maybe I had observed two different supernovae at different times and gotten confused by their spectra—a Type II on one occasion, and then a few months later, I accidentally observed a different star. They thought that this just can't occur, but it does occur. My results were vindicated in 1993, when there was a bright supernova called 1993J that went off. It was actually discovered by a Spanish amateur astronomer, Francisco Garcia. It went off in a bright galaxy just 11 million light years away. Here it is, going off, and there's the star that exploded right there. That's a before picture, and there's an after picture, in this bright nearby galaxy. It had the same characteristics as Supernova 1987K—that is, it went from Type II to Type Ib in terms of its observational properties.

We think that this star actually lost part of its material through transfer to a companion, and part of its material through a wind. Other astronomers—in particular, in Cambridge, England—have found evidence for the surviving companion. They can see it in the spectrum. In the spectrum of Supernova 1993J now, a decade or two later, you can see evidence for the surviving companion—so it's there. On the other hand, a wind must have been ejected as well from the star that later became a supernova, because we can see the interaction of the ejected material with the supernova. The spectrum now is showing a lot of evidence for the supernova ejecta interacting with gases that had been gently blown off in a wind prior to the explosion. Here, in fact, is a bunch of pictures taken at radio wavelengths, where you can see the expanding ejecta, and they continue to glow so brightly because they're slamming into the surrounding material that had been gently blown off in a wind prior to the explosion.

We think that these core-collapse supernovae come in a number of species or varieties: Type II, the more normal kind; Type IIb; Type Ib; Type Ib/Ic, where the helium envelope is sort of thin; all the way down to Type Ic, where the hydrogen and helium envelopes are gone, but the rest of the core of the star looks about the same. So, it looks like gravity is ultimately victorious in all these stars, all these massive stars, regardless of how much of an envelope they still retain. The iron core ultimately collapses, forming a neutron star and blowing off the rest of the material.

The Brightest Supernova in Nearly 400 Years
Lecture 56

The brightest supernova in nearly 400 years was studied in great detail. It occurred in the Large Magellanic Cloud, a satellite galaxy of the Milky Way, only 170,000 light years away from Earth.

Though astronomers have never directly witnessed the explosion of a visible white dwarf as a Type Ia supernova, we have seen some very massive stars explode as Type II supernovae. One in particular, SN 1987A, was observed in great detail, both before and after the explosion. Supernova 1987A, the first supernova to be witnessed in the year 1987, was the brightest supernova in nearly 400 years. It occurred in the Large Magellanic Cloud, a satellite galaxy of our Milky Way Galaxy. The Large Magellanic Cloud is about 170,000 light years away, which means that the explosion occurred about 170,000 years ago, just about the time of early hominids on Earth.

The supernova was discovered by Ian Shelton, a student working at Las Campanas Observatory in Chile. Using a small telescope, he took numerous photographs of the Large Magellanic Cloud in his study of the variable brightnesses of stars. One of Shelton's photographs indicated an extra point of light that hadn't appeared in previous photographs. The supernova appeared near the Tarantula Nebula and was visible to the naked eye. Other astronomers, both amateur and professional, had also witnessed the same supernova from other parts of the world, though Shelton is credited with its discovery. SN 1987A was a Type II supernova; that is, it had an envelope of hydrogen. But it was a peculiar object. For example, SN 1987A faded much more rapidly than it should have at ultraviolet wavelengths. Moreover, it didn't brighten as much as astronomers had expected.

Being so bright and nearby, SN 1987A was studied with a broad range of telescopes, allowing us to test our theories of Type II supernovae. By examining pre-explosion photographic plates, we discovered that the progenitor star (the star that exploded) was a blue supergiant with an initial mass of about 20 solar masses. Blue supergiants have about the same luminosity as red ones, but blue supergiants have much higher surface temperatures. Previously, astronomers didn't think blue supergiants could explode because we thought they were on their way to becoming red supergiants and would not yet have been able to build up their iron cores. The supernova showed peculiar characteristics, such as the rapid decline of ultraviolet light and a much dimmer than expected appearance. Its apparent magnitude began at about 4.5, then gradually grew to only 2.5 (remember, the lower the magnitude, the brighter the object). Thus, it was visible to the naked eye, but it should have been much brighter than what was actually seen. The deficit of light is consistent with the star being a blue supernova because such stars are smaller than red supergiants. The smaller size translates to less radiating area; thus, the star could not become as bright as a larger star would.

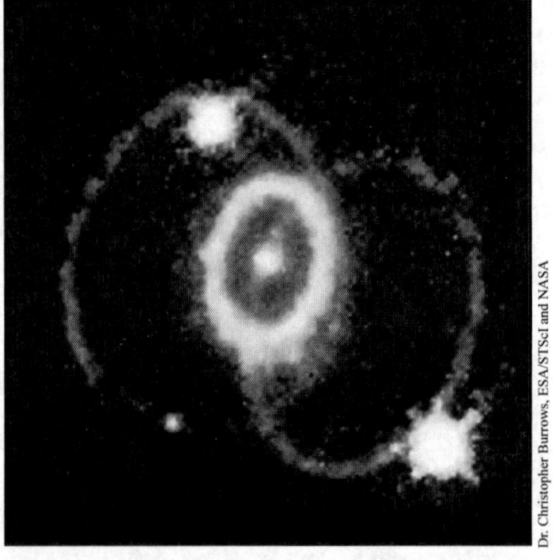

Dr. Christopher Burrows, ESA/STScI and NASA

Supernova 1987A rings, taken by the Hubble Space Telescope.

The discovery of SN 1987A taught us that under some conditions, blue supergiants can develop an iron core and explode. The mechanism is unclear, but it is possible that, because the Large Magellanic Cloud is deficient in heavy elements relative to our own Galaxy, stars deficient in heavy elements have a different structure in their envelopes (outer layers), allowing them to explode as hotter but smaller stars—blue supergiants. Some astronomers think that the star was a blue supergiant when it exploded because it could have been part of a binary system and perhaps swallowed its companion star prior to the explosion, changing its outer structure. Another theory tested was whether multitudes of neutrinos are emitted when a Type II supernova occurs. Indeed, neutrinos were detected by at least two underground tanks of water that had originally been designed to search for the decay of protons. Calculations showed us that the total energy emitted by SN 1987A, in roughly a few seconds, was comparable to the total amount of energy emitted by all the normal stars in the rest of the observable Universe during those few seconds.

Aside from the Big Bang itself, this type of explosion is about the biggest we can get. More than 99% of the explosion energy was in the form of neutrinos. Most of the other 1% was the kinetic energy of the ejected gases. Only about 0.01% of the energy was emitted as optical radiation. Thus, even though SN 1987A was a bright, naked-eye supernova, the visible light constituted only 1/10,000 of the true energy emitted by the explosion.

"I'm sure that in the next decade or two, Supernova 1987A will have additional secrets to tell us."

When neutrinos are emitted by a supernova and hit Earth, they generally don't indicate from which direction they came. In other words, how did we know that the neutrinos detected around the time of SN 1987A came from that star? One obvious factor was that the neutrinos were detected at about the same time as the supernova was discovered. Two reactions can occur when neutrinos released from an exploding star interact with material on Earth. First, neutrinos can scatter off electrons, propelling the electrons forward in roughly the same direction from the supernova. These electrons produce Cerenkov light cones, which can then be detected by a light detector and tell us from which direction the neutrinos came. However, a second reaction—and far more common—is for antineutrinos to combine with protons, forming energetic positrons and neutrons. In such a case, positrons can go in any direction with nearly equal probability (described by the term **isotropic**); thus, they do not indicate where in the sky the supernova occurred.

In general, if a supernova were to occur in our Galaxy (that is, close by), it would produce enough detectable neutrinos (scattering off electrons) to allow us to know in which direction the supernova was likely to become visible within a day. The neutrinos would reach Earth before the light of the exploding star, telling us where to look for the impending supernova light. Although neutrinos don't quite travel at the speed of light, they get a "head start" over the photons from the surface of the exploding star (which are formed only after the shock wave coming from the star's central region reaches the surface, traveling at about 1/10 the speed of light).

Another theory confirmed by SN 1987A was that heavy elements are synthesized by supernova explosions. Electromagnetic radiation from such elements was detected by gamma-ray telescopes and other instruments launched above the Earth's atmosphere after SN 1987A was discovered. Gamma rays were detected arising from specific radioactive elements. The explosion formed radioactive nickel, which quickly decays into cobalt, which—on a longer time scale—decays into iron. These radioactive elements could have been produced only by SN 1987A. Because they are short-lived, they would not have remained in the star at the time of its death had they been present in the material from which the star was first formed, millions of years earlier.

Supernova 1987A is surrounded by rings of gas released before the actual explosion. These rings can be used to study the progenitor star's behavior during the last few thousand years before its death. Ejected gases from the supernova explosion itself are now colliding with the external rings, which are consequently beginning to increase in brightness. During the next few years, the supernova will experience a renaissance, appearing substantially brighter than it is now. ■

Important Term

isotropic: The same in all directions (that is, no preferred alignment).

Suggested Reading

Filippenko, "Stellar Explosions, Neutron Stars, and Black Holes," in *The Origin and Evolution of the Universe*.

Goldsmith, *Supernova! The Exploding Star of 1987*.

Kirshner, *The Extravagant Universe: Exploding Stars, Dark Energy, and the Accelerating Cosmos*. Marschall, *The Supernova Story*.

Pasachoff, *Astronomy: From the Earth to the Universe*, 6th ed.

Pasachoff and Filippenko, *The Cosmos: Astronomy in the New Millennium*, 3rd ed.

Questions to Consider

1. Is it surprising that SN 1987A occurred near a giant cloud of gas (called the Tarantula Nebula) where massive stars have been produced for the past few tens of millions of years?

2. If only 10 neutrinos from SN 1987A were detected by each of two underground tanks containing several thousand tons of water and if a typical human consists of 100 pounds of water, what are the odds that your body directly detected a neutrino from SN 1987A (assuming that you were alive in Feb. 1987)?

3. How compelling do you find the arguments that we are made of stardust?

The Brightest Supernova in Nearly 400 Years
Lecture 56—Transcript

In the previous few lectures, I've explained how some rare types of stars literally explode at the end of their lives. We've never actually seen directly a white dwarf that then exploded as a Type Ia supernova. White dwarfs are small and faint and can't be seen unless they're very nearby. They explode only very rarely, so we've never actually seen a white dwarf that exploded. As I'll discuss in this lecture, we have seen some very massive stars that then went through the explosion process. One in particular, Supernova 1987A, was observed in great detail—both before and after the explosion. Supernova 1987A was discovered in 1987; it was the first supernova to have been discovered that year. In fact, it was the brightest supernova in nearly 400 years, so it presented a fantastic opportunity for astronomers to study the explosion of a star in great detail.

It occurred in a satellite galaxy of our Milky Way Galaxy. It's a galaxy called the Large Magellanic Cloud. Here's the Large Magellanic Cloud, shown along with the Small Magellanic Cloud. These are dwarf galaxies, small galaxies that orbit around our much bigger Milky Way Galaxy. A close-up of the Large Magellanic Cloud shows that it has a region here, a nebula called the Tarantula Nebula, which is in fact the largest region of active star formation that we know of in the local group of a few dozen galaxies close to our own. It's a massive cloud of gas in which stars are literally being born right now as we speak—and have been born in the last few million years. The arrow here points to a star in a picture taken before February 23, 1987, and that star later exploded and became so much brighter. Look at that; that incredible explosion caused the star to brighten by many millions of times, making it the equivalent of tens of millions, or even hundreds of millions, of normal suns. That's the explosion that we got to witness in February of 1987.

The Large Magellanic Cloud is about 170,000 light years away—so actually, this explosion occurred 170,000 years ago, give or take a few years. Last Tuesday, 170,000 years ago, it exploded. We can't be that precise. We don't know exactly how far away it is. But you can just think about what Earth was like back then, 170,000 years ago. That was really near the time of transition between early hominids and the very first Homo sapiens, earlier even than the Rhodesia Man's skull, shown here, corresponding to about 120,000 years ago. The before and after pictures of the star, shown here, just illustrate the brilliance of the explosion and the fantastic opportunity that astronomers were given. Indeed, it was a really big-time story. See that, "big TIME story"—pardon the pun. Anyway, "Bang, A Star Explodes, Providing New Clues to the Nature of the Universe." News like this, of an astronomical nature, doesn't reach us each day, so this was truly worthy of a giant headline like this in *Time* Magazine. Various newspapers covered the story as well, "Exploded Star Traces Reach Earth" on one, the *San Francisco Examiner*; and other newspapers in the country also ran huge stories about this celestial event.

The supernova was discovered by Ian Shelton, a student working at Las Campanas Observatory in Chile. He was using a small telescope, shown in this shed-like dome here, gathering data for Canadian astronomers. But on his nights off, when he wasn't gathering data for other people, he took photographs of the Large Magellanic Cloud, one after another, because he wanted to study variable stars in that cloud, stars that vary in brightness. Here's Ian observing the Large Magellanic Cloud with a 10-inch telescope, a rather small telescope, in this shed. He got these repeated photographs of the Large Magellanic Cloud. The one that he got on February 24 was particularly instructive. He saw in that photograph an extra little blip of light that wasn't there before. So, Ian is sitting here, comparing the February 24 photograph with previous photographs that he had observed. Indeed, around February 23, the very previous night, he had gotten a photograph of the same region of the sky. Although in this comparison here, you don't see much of a difference; in fact, if you look very, very carefully, there's an extra little blip on the right-hand side here, near the Tarantula Nebula, compared with the left-hand side there.

Ian noticed this blip on the very night that he had obtained this photographic plate. Indeed, around 2:00 or 3:00 in the morning, the wind got so strong that it blew the roof of the shed shut, and he decided to call it a night early. It was still dark outside, but he went to the darkroom and developed his photographic plate, looked at that plate, and saw this extra blip. He thought at first that it was a flaw in the emulsion, and so he went outside to just check whether he could see such a bright thing with his own naked eye, and sure enough, in the Large Magellanic Cloud, near the Tarantula Nebula, there was an extra star that simply had not been there on previous nights. Even with the naked eye, you could see this star because it was so bright compared to the surrounding stars.

It turns out that at the very same observatory, Las Campanas Observatory, there are other domes. In particular, there's a dome here where there's a one-meter telescope used by professional astronomers. There was an assistant there, Oscar Duhalde, who liked to look at the skies. He had actually gone outside and looked around that same night, February 24, and seen the star that Ian Shelton had photographed, but he didn't tell anyone about it. He went back inside the dome, and the astronomers in there needed his assistance, so he forgot to tell people that he saw what seemed like a new star up there. But when Shelton came over to that dome from his shed to tell people that he thinks he sees a new, bright thing in the Large Magellanic Cloud, Oscar Duhalde said, "Yes, yes, I saw it too, a few hours ago." Then everyone went outside there and looked up in the sky and saw this new star that was visible to the naked eye.

You will recall that I talked about Reverend Robert Evans, who has memorized the star fields around something like 1,500 galaxies, and he uses his small telescope to look for exploding stars in those galaxies. Here he is with his telescope. This was the big one that got away. He tells me that, that night, he had a premonition that there would be a bright, new supernova. He wanted to go and observe the skies with his telescope, but it was cloudy where lived. So, he called up a friend, and the friend said where he lives, 150 miles away, conditions were kind of iffy. It looks like clouds are coming and going. This would have been a two-and-a-half-hour drive each way in the middle of the night, and the Reverend Evans had things to do the next day. I don't think he had to deliver a sermon that day because it was in the middle of the week. But anyway, he had things to do, and he decided not to go to a clear spot, especially since that spot might not have been clear by the time he reached it. That was the big one that got away, basically, but he has discovered over 40 supernovae in other galaxies.

Rob McNaught, in Australia, actually had photographs of it the previous night, February 23—that is, the night before Ian Shelton, the 24th, but he didn't look at his photographs that he had taken that night. Astronomers are busy people. They take all these photographs during the night, and then they want to sleep in general during the day. We don't always examine our data right away. Too bad for Rob McNaught because he would have had it nearly a whole night, nearly 24 hours, earlier than Ian Shelton. He later had the data on film, and so the data that he has provides some important early measurements of the brightening of the supernova with time.

Then there's another story: Albert Jones, who is an astonishingly successful observer of variable stars, observing from New Zealand. He's been doing this since 1943, and in fact, he's made over 600,000 observations of variable stars for a worldwide network of variable star observers. On February 24, he reported seeing a new star in the Large Magellanic Cloud, but his report was three hours later than Ian Shelton's, so Shelton is officially credited with the discovery. Interestingly, Jones observed the same field the previous night, February 23, and didn't notice anything there. Apparently, he observed about an hour too early. That is, had he observed an hour later, the supernova would have brightened enough for him to be able to have seen it. He would have seen it, again, about a night earlier than Shelton had things gone well for him, but that's the way it goes. You have these sob stories sometimes.

Once it was discovered, everyone started looking at it with all the great telescopes because telescopes were first used for astronomical purposes in the first decade of the 17th century. There was Tycho's supernova in 1572 and Kepler's supernova in 1604, but it was only 1608 or 1609 when Galileo started using telescopes for astronomical observations. So really, we've been waiting four centuries for a bright nearby supernova to be observed with the great variety of telescopes that we have now. Here's an observatory in Chile—this one is the Cerro Tololo Inter-American Observatory, where lots of telescopes started observing the supernova at optical wavelengths. Then various telescopes in the Southern Hemisphere, such as the Parkes Radio Telescope in Australia, started observing it. Satellites were sent up above the Earth's atmosphere to observe it at x-rays and gamma rays. It was just a bonanza of observations. Everyone who had a telescope started looking at this supernova.

It turned out to be a Type II supernova, one with hydrogen. Here are my colleagues—Bob Kirshner and George Sonneborn—looking at the ultraviolet spectrum taken by a satellite known as the International Ultraviolet Explorer. Both the optical and ultraviolet spectra showed indications of a Type II—that is, there were hydrogen lines visible—but the ultraviolet showed that the object was fading very much more rapidly than it should have been. It was peculiar in that sense, and moreover, optically, it didn't brighten as much as we had expected it to. It was visible with the naked eye, but it should have been really easily visible with the naked eye had it been a normal Type II supernova. It was a Type II, but it was a weird one. This gave us a great chance—not only to test our existing theories of massive star explosions, but also to perhaps refine them and develop them further, in order to explain the peculiarities of this particular object.

Was it a massive star near the end of its life, an evolved star, a red supergiant, as in this diagram here? You will recall that I said that Type II supernovae should be red supergiants with this onion-like structure of progressively heavier elements, going all the way down to an iron core. Indeed, photographs of the Large Magellanic Cloud in the many years preceding 1987 showed many instances of this progenitor. That is, many of the photographs were of the region containing the progenitor, or the star that exploded, and there it is. We can tell what kind of star it is. Indeed, it is a massive star, a 20-solar mass star initially, near the end of its life, and it's a supergiant, so that's all good. But it turned out to be a blue supergiant, not a red supergiant. It was a blue supergiant, roughly as luminous as a red supergiant, but smaller and hotter, so it looks blue.

On a Hertzsprung-Russell, or luminosity-temperature diagram, here are the red supergiants—they're luminous and cool—and here are the blue supergiants, about as luminous, but much hotter for their surface temperatures. These kinds of stars were not expected to explode. It was thought by most astronomers that blue supergiants are on their way to becoming red supergiants and should not yet have built up an iron core. So what gives? Why did this star explode? Some astronomers even doubted that the star that had been identified as the progenitor of Supernova 1987A was correct. They said maybe there's a red supergiant hidden there somewhere, and we only think it's the blue supergiant.

Stan Woosley, at University of California at Santa Cruz, said that no, it's got to be a blue supergiant because, had it been a red supergiant, you wouldn't be able to explain some of the peculiarities of it already mentioned: the rapid decline of the ultraviolet light, and the fact that it didn't become as bright initially as people had expected it to. Here's the brightness versus time. We had expected it to become very bright, about second magnitude, or maybe third magnitude, right off the bat. But instead, it started out being roughly magnitude 4.5, and then it only gradually grew to roughly magnitude 2.5 or so. We had expected it to just go zapping up to very bright and then stay the same brightness for a couple of months, as is consistent with the light curves of other Type II supernovae that I had shown you in the previous lecture.

Woosley said this observation and other observations of its type suggests that it really was a blue supergiant, and it wasn't as bright as it should have been when it went off because, initially, it was smaller than a red supergiant. It didn't have as big a radiating area—that is, its surface wasn't as big as that of a red supergiant, so it couldn't become as bright right away. Then, after it expanded and became big, it could shine. But by that time, the expansion itself had actually used up a lot of the energy generated by the explosion. That energy was used up in expanding the star rather than in making it look bright, and that's why there's sort of a deficit of light in the first few months after the supernova's explosion. Indeed, Woosley had a computer model sitting around for a 20-solar mass star that was just sort of waiting for an observed example of the same kind. He had a computer model essentially describing this, or a similar type of process. He and his group submitted a paper for publication three days after the supernova explosion, so that's fast publication. Usually it takes months or years for you to figure things out and publish a paper, but he had already pretty much figured it out, and then this guy came along. That's fantastic.

It was an evolved massive star, but there was a wrinkle; it was a blue supergiant rather than a red supergiant. That teaches us that, in some cases, red supergiants go back and become blue supergiants, and at that time, develop an iron core. Why they do that is not understood completely yet, but we think that maybe this star did it in the Large Magellanic Cloud because the Large Magellanic Cloud is deficient in heavy elements relative to big stars in our own galaxy, which have much more heavy elements because our galaxy is able to retain the heavy elements generated by previous generations of supernovae better than a little, flimsy dwarf galaxy like the Large Magellanic Cloud—where stars blow up, and the heavy elements that they produce get ejected out of the galaxy, and so they're not around to help form new, massive stars.

When you have a star that's deficient in heavy elements, the structure of its envelope changes in such a way that it can explode as a hotter, smaller, blue supergiant rather than as a red supergiant. That's something we've learned along the way. Not all astronomers agree with that. Some say that this thing was a blue supergiant when it exploded because previously it was in a binary system, and maybe it swallowed the other star, and that changes the structure of the outer part as well, allowing it to become a blue supergiant when it explodes. We're still not exactly sure what's going on, but there are several hypotheses that are at least consistent with the observed properties of the star.

The next prediction that we wanted to test was whether a neutron star is formed during the collapse, and whether a bunch of neutrinos are emitted in the process. Indeed, neutrinos were detected by large tanks of purified water, deep underground. I've already described these tanks when I discussed the detection of neutrinos from the Sun. Solar neutrinos sometimes interact with the water and produce a signal that can be detected. In a similar way, there was a tank of water in Japan, and one in the United States, that detected the signal of a few neutrinos interacting with the water. What happens, in detail, is the following. A neutrino, or an anti-neutrino, can go zipping into the tank of water. It can actually collide with an electron or combine with a proton, producing either a high-speed electron or a high-speed positron—that is, anti-electron.

These particles can go zipping through the water faster than the local speed of light. That sounds crazy, but in fact, the speed of light in water is significantly below the speed of light in a vacuum, and there's nothing in relativity that says that particles can't exceed the speed of light in water, the local speed of light; they can. When they do that, they set up the electromagnetic equivalent of a sonic boom. A sonic boom occurs when a plane goes zipping through the sky at faster than the speed of sound. If a charged particle is going through water at faster than the speed of light, it produces the light equivalent of a sonic boom, and it's called Cerenkov radiation. You can see a cone of Cerenkov light being emitted by this high-speed electron or this high-speed positron, and that light then gets detected by a whole bunch of light detectors, which surround the boundary of this tank. Here are all these little phototubes that can detect this light.

A bunch of such detections were made with these two underground tanks of water, and here you can see the energy of the detected neutrinos inferred from the Cerenkov light—and this is in units of millions of electron volts. One million electron volts is roughly the equivalent of the energy, the rest mass energy, of two electrons. Each electron is about a half a million electron volts. Anyway, that's just some unit of energy versus time, and you can see that about 20 neutrinos were detected by Kamioka and the Irvine-Michigan-Brookhaven (IMB) experiment. Usually you'd just see zero here; there'd be just a bunch of zeroes along this plot, no detections at all. But in fact, they detected all these neutrinos. From the number of neutrinos that were detected, and knowing how rarely neutrinos actually interact with matter, you can figure out how many neutrinos were actually emitted by the supernova. You actually have to know the supernova distance as well, but we know that: 170,000 light years.

It turns out that when you do this calculation, you find out that the total energy emitted by Supernova 1987A, mostly in the form of neutrinos, in roughly one second or a few seconds, was comparable to the total amount of energy emitted by all the normal stars in the rest of the observable Universe during that one second, or during those few seconds. Here's one star, which—in a few seconds or one second—emits as much energy as all the normal, non-exploding stars in the rest of the observable Universe. That is one hunk of energy. That's a lot of energy. Supernovae are among the most powerful explosions in the Universe. The other kind of explosion, which I'll discuss later, is the so-called gamma ray bursts, and they're associated with supernovae as well.

Aside from the Big Bang itself, the creation of the Universe, these are the biggest bangs you can get. When you look at the energetics of the neutrinos, you find that over 99% of the energy of the explosion was in the form of these little, tiny subatomic particles, neutrinos. The other 1% of the explosion energy was the kinetic energy of the ejected gases, and 1/100% of energy was the visible display that we saw. Bright though it was, the optical display was just an afterthought. It was 1/10,000 of the true energy emitted by the explosion. Wow! Supernovae are bright, but that's just a sideshow for the main event, the creation of a flood of neutrinos in the case of the core-collapse supernovae, and the ejection of material, which is about 1% of the energy or so.

About one person in 1,000 experienced a neutrino interaction of the sort that the detectors in Japan and the United States had seen. Maybe you were one of the lucky few that actually detected the supernova if you were around back in 1987. You can ask, "Well, in the case of the Cerenkov cones of these IMB and Kamioka detectors, did they point back at the supernova?" In other words, if you look at the Cerenkov cones, shouldn't they point back at the direction from which the neutrino came? If that's the case, we could directly say that those neutrinos came from the supernova, although the inference is pretty clear. I mean they arrived at about the same time, and at the right time, and all that. It turns out that, in most cases, the cone of Cerenkov light does not point back toward the origin of the neutrino. That's because there are two possible reactions.

One is that a neutrino scatters off of an electron, causing the electron to go pretty much in the forward direction. That electron, then, produces the Cerenkov light, which is then detected by detectors. You could extrapolate back and find out the direction from which the neutrino came. But a far more common reaction is for an anti-neutrino to combine with a proton, forming an energetic anti-electron, or positron, and a neutron. It turns out that in this reaction, the positron can go flying off in essentially any direction. The process is called isotropic; it goes off in all directions, roughly with the same probability. It's not a forward-scattering process like the collision of a neutrino with an electron. When you go and see where the cone of light is pointing for this positron, for example, it doesn't point back toward the point of origin of the anti-neutrino; that is, it doesn't point back toward the supernova, which is too bad.

Of the 20 events that were detected, we think that all but the first were anti-neutrinos combining with protons to form positrons and neutrons, and only one—maybe the first interaction—was a neutrino scattering off of an electron. Indeed, that Cerenkov cone does point roughly back toward Supernova 1987A. You can ask yourself, "Well, what if you had a really nearby supernova in our own galaxy, where there'd be a bunch of the forward-scattering kind of neutrino interactions with electrons?" Then all of their Cerenkov cones would point roughly back toward the origin of the neutrinos. If you had enough events—that is, not just one, but thousands or something like that for a sufficiently nearby supernova—you could figure out where the supernova occurred; that is, where the explosion occurred.

Because the neutrinos arrive before the light does—not because they travel faster than light, but because they escape from the star right away, whereas the star itself doesn't brighten right away because it takes a while for the shockwave to reach the outer parts of the star and tell it to start becoming bright and all that. There's a delay of a few hours, or a day, or whatever between when the neutrinos get out and when the star starts brightening. We could have an advanced warning of a few hours or a day or so from the neutrino detectors, and they would say, "Hey, point in that part of the sky, and in the next night or so, you will see a supernova going off." That would be really cool. We hope that when the next supernova in our own galaxy occurs, there will be neutrino detectors on at that time that will be able to tell us where to look, so that we can catch it in the first few hours of the explosion. That would be really cool.

The next thing that was discovered with Supernova 1987A is that heavy elements were indeed synthesized by the explosion. The way we can tell is that gamma ray telescopes, and other telescopes observing high-energy radiation, were launched up above much of the Earth's atmosphere. From there, they could observe x-rays and gamma rays coming from the supernova. What they found were gamma rays that correspond to the specific energies at which radioactive cobalt nuclei decay into stable iron. What happened was there was explosive nuclear synthesis in the supernova, when all these neutrons were running around and all that. That formed radioactive nickel, which quickly decayed into cobalt, and then on a longer time scale, cobalt decays into iron. It does so by emitting gamma rays, having a very specific set of energies. Those are like the fingerprint of those particular nuclei. No other nuclei produce those specific energies of gamma rays, or at least that pattern of energies, and yet that's the pattern that was observed for Supernova 1987A.

Those short-lived radioactive nuclei had to have been produced by the explosion itself; otherwise, we wouldn't have seen them because they're so short-lived that they wouldn't have existed in this star that was 10 million years old at the time that it exploded. Any elements of which the star was initially made had long since decayed and become stable, whereas these guys, now, were freshly created, were unstable, and gave us this signature of gamma rays. That's what tells us that heavy elements are definitely produced by these cataclysmic explosions of massive stars. As Carl Sagan used to say, "We are made of star stuff; we are made of star dust." He didn't discover this, but he greatly popularized the notion that we owe our existence to the explosions of massive stars—one of the greatest discoveries of 20th-century physics.

More recently, we've been studying the supernova, and we've found that it has shells of gas around it that had been ejected prior to the actual explosion, in relatively gentle eruptions like those that form planetary nebulae in our own galaxy, but in this case around a massive star. Here are these shells that have nothing to do with the supernova itself, but the supernova ejecta are now expanding and colliding with these shells of previously ejected gas. Knots of dense gas in the main inner shell are now getting brighter and brighter as the outermost parts of the ejecta start colliding with this gently erupted gas. We should expect to see the supernova brighten in the next few years again as this collision proceeds.

Here, you can watch it with the Hubble Space Telescope, and at x-ray wavelengths with Chandra, and at radio wavelengths, and you can see how the main shell is brightening with time, as this supernova starts interacting with those shells. You can also see that the supernova ejecta themselves are asymmetric—so, the explosion was not spherical; it was elongated, like this. The outermost parts of the gas are now interacting with these shells, causing them to brighten. Supernova 1987A should brighten at all wavelengths again in the future, as the collision of the ejecta with the circumstellar gas proceeds. I'm sure that in the next decade or two, Supernova 1987A will have additional secrets to tell us.

The Corpses of Massive Stars

Lecture 57

In the previous lecture, we saw how supernova 1987A helped confirm our basic ideas about supernovae but showed that we also needed to refine some of our ideas. Though the progenitor star of the peculiar SN 1987A was a blue supergiant, other, more typical Type II supernovae have been found to come from red supergiant stars.

As recently as 2005, the Hubble Space Telescope observed a supernova in M51, the Whirlpool Galaxy. The progenitor of SN 2005cs was a red supergiant of about 12 solar masses, and the supernova had a more normal—or expected—spectrum than that exhibited by SN 1987A. The data for SN 2005cs strengthened the evidence that the cores of red supergiants implode and the outer parts get ejected. In most cases, an imploding core forms a *neutron star*, a very dense ball of neutrons maintained by neutron degeneracy pressure, similar to the electron degeneracy pressure experienced by white dwarfs. A neutron star 1.5 times the mass of the Sun can be only about 20 kilometers in diameter. A teaspoonful of material from such a star would weigh about 1 billion tons. A neutron star is similar to a white dwarf in that it is made of degenerate material crammed into a very small space. The pressure exerted by degenerate neutrons prevents the star from collapsing.

Neutron stars were predicted by **Fritz Zwicky** and Walter Baade in 1933. They also predicted that neutron stars were produced during cataclysmic explosions of massive stars. Neutron stars weren't actually discovered until 1967, when Jocelyn Bell detected one in the form of a *pulsar*; these objects generally can't be seen clearly, but they emit regular pulses of radio radiation. The first pulsar Bell detected had a periodicity of 1.3373011 seconds. The spacing of the pulsar's blips was regular, although the intensity, or brightness, varied considerably with time. At first, it was suggested that pulsars might be extraterrestrial communications. Shortly thereafter, however, several more regular series of pulses were found coming from other parts of the sky but with different periodicities. It was deemed unlikely that a network of intelligent species, all communicating in a similar manner, was present in our Galaxy. Moreover, there was no evidence of a periodically changing Doppler shift, indicating that the pulsars were not coming from another planet or object that was orbiting a star.

Through an interesting process of elimination, astrophysicists quickly determined that pulsars probably emanated from rapidly rotating, highly magnetized neutron stars. One clue was that most of the pulsars originated from the plane of the Milky Way Galaxy, where many of the more massive stars are concentrated. Pulsars can't arise from *oscillating* (vibrating) normal stars or white dwarfs, whose periodicity is too slow. In contrast, the expected vibration period of neutron stars is too fast. Two normal stars or white dwarfs cannot *orbit* each other so quickly, either. Two neutron stars can have such a tight orbit, but they would rapidly lose energy, and the orbital period would decrease—yet pulsar periods were observed to be very stable. The surface of such a rapidly *rotating* normal star would exceed the speed of light. Similarly, a white dwarf is disrupted if its rotation period is less than about 0.3 seconds, too slow for the rapid pulsars. Neutron stars, on the other hand, are capable of rotating about their axes at speeds that are consistent with the observed pulsar rates. We now have additional, more direct evidence that pulsars are rapidly rotating neutron stars.

Let's look at some of the characteristics of pulsars and their causes. We don't really know the details of why pulsars shine, but this characteristic is undoubtedly related to their magnetic fields. Neutron stars have magnetic fields within and surrounding them, the axes of which generally differ from the stars' axes of rotation, forming conical patterns as they rotate. The magnetic axis first points in one direction, then in another direction. This rotation can create electric fields that are strong enough to accelerate electrons to speeds close to that of light. Accelerating charged particles emit radiation along their direction of acceleration, creating (by methods still not fully understood) two oppositely directed beams of light that are visible from Earth—with their associated periodicity—as the star rotates. The effect is similar to that of a **lighthouse**: It is on all the time, but you see a flash only when the rotating lamp is pointing at you.

We think that the magnetic field is a trillion times as strong as Earth's magnetic field. (The unit of magnetism is a gauss, and the Earth's magnetic field is about 1 gauss.) It is possible that the strong magnetic field is a result of the star's collapse, forcing the star's magnetic field into such a small space that its strength increases dramatically. Why does a neutron star rotate so quickly? All stars rotate to some extent, and as they collapse, the spin rate must increase in order to conserve angular momentum.

Very young pulsars shine not only at radio wavelengths but at optical and x-ray wavelengths and at other wavelengths. As the stars get older, the high-energy forms of radiation subside; what remains are low-energy forms of radiation, such as radio waves. We have observed that one particular pulsar in the Crab Nebula, the remnant of the supernova of A.D. 1054, creates a wind, as well as jets of material, energizing the Crab Nebula and causing it to glow brightly. Over time, this neutron star has been losing energy through the production of its light beams and jets of material, slowing its pulsation rate. Quantitatively, the rate of energy gain in the Crab Nebula is equal to the rate of energy loss of the rotating neutron star, providing strong support for our basic model of pulsars.

We expect pulsars to remain turned on for only about a few million years before their rotation rates (and, perhaps, the strength of their magnetic fields) diminish so much so that light beams are no longer produced. Every pulsar is a neutron star, but not every neutron star will be visible as a pulsar. Some will have died, or some might have axes of rotation oriented in such a way that they don't allow the beams of light to cross Earth's line of sight. Typically, pulsars spin about once per second, or 10 times per second, or maybe once every 10 seconds. But some spin hundreds of times per second—these are called *millisecond pulsars*. Converting the frequencies at which these pulsars spin to audible signals will produce corresponding musical notes. Some astronomers have written musical pieces with the notes of known millisecond pulsars. We think these pulsars spin so fast because they previously accreted material from a companion star orbiting them.

In 1991, one millisecond pulsar was discovered to have at least three planets orbiting it. The key to this discovery was that, sometimes, the pulses arrived sooner than expected and, other times, they arrived later than expected. This slight deviation from perfect periodicity is caused by orbiting planets; the planets and the neutron star orbit their common center of mass, so the neutron star is sometimes slightly closer to Earth, and sometimes slightly farther away. These "planets" are quite different from those in our Solar System, though they happen to be comparable in size to the terrestrial planets. In addition, these "planets" could not have existed before the supernova that gave rise to the pulsar because the planets would have been destroyed in the supernova explosion. Therefore, it's likely that they formed from a disk of debris around the neutron star that remained after the explosion.

In the last decade, astronomers have learned some interesting things about neutron stars. Some neutron stars have magnetic fields up to 10^{15} gauss units; these are called **magnetars**, the strongest magnets in the Universe. Magnetars sometimes emit tremendous amounts of energy because, apparently, the structure of their crust changes such that it creates a kind of "starquake." Furthermore, the magnetic field changes, also releasing a tremendous amount of energy. One such magnetar was observed on December 27, 2004, in the constellation Sagittarius, creating the brightest flare ever seen from outside our Solar System. The eruption was so bright that it ionized Earth's atmosphere and activated the sensors on several satellites. Satellites transmitted the information to radio telescopes on the ground, which immediately moved to begin observing that location of the sky. These telescopes detected debris emanating from the magnetar and moving at speeds close to the speed of light. We believe this was a restructuring of the surface layers of a neutron star. After the main burst of energy, alternating flashes of light from the neutron star's north and south poles became visible as the star rotated. We believe that this type of neutron star, which gives rise to a magnetar, has an interior of liquid neutrons and other particles with a generally solid crust. The crust essentially buckles or cracks, changing the magnetic field configuration, as well. We know that magnetars survive this basic burst because some of these events have been seen to repeat after a few years. ■

"Magnetars are exciting, but don't get anywhere near one—especially if you have a pacemaker, because it'll definitely mess it up."

Name to Know

Zwicky, Fritz (1898–1974). Swiss-American astronomer; proposed that supernovae result from the collapse of the cores of massive stars, producing neutron stars and energetic particles (cosmic rays). Compiled an extensive atlas of galaxy clusters and showed that many such clusters must contain dark matter in order to be gravitationally bound.

Important Terms

lighthouse model: The explanation of a pulsar as a spinning neutron star whose beam we see as it comes around and points toward us.

magnetar: Spinning neutron star with an extraordinarily powerful magnetic field that occasionally releases a burst of gamma rays when the crust of the star undergoes a sudden restructuring (a "star quake").

Suggested Reading

Filippenko, "Stellar Explosions, Neutron Stars, and Black Holes," in *The Origin and Evolution of the Universe.*

Pasachoff, *Astronomy: From the Earth to the Universe*, 6th ed.

Pasachoff and Filippenko, *The Cosmos: Astronomy in the New Millennium*, 3rd ed.

Questions to Consider

1. Can you find any loopholes in the process-of-elimination argument used to conclude that pulsars are rotating neutron stars?

2. How did studies of the Crab Nebula pin down the explanation of pulsars?

3. Calculate the density (mass per unit volume) of a neutron star having a mass of 1.4 solar masses and a radius of 10 kilometers. Compare this with the density of a neutron or a proton (each of which has a radius of about 10^{-15} m and a mass of about 1.67×10^{-24} g).

The Corpses of Massive Stars
Lecture 57—Transcript

In the previous lecture, I showed how Supernova 1987A, the brightest supernova in 400 years, confirmed our basic ideas of how massive stars explode, by having an iron core that collapses, produces a bunch of neutrinos, and then rebounds and expels the outer layers. But that supernova also showed that we need to refine our models. In some cases, it's not just red supergiants that explode; blue supergiants can explode as well. You might ask, "Well, that's great, '87A showed the basic model, but it didn't give quite the right kind of star. Do we have evidence that some other Type II supernovae do have the expected red supergiant progenitors? Indeed, we do. In fact, recently, in 2005, there was a great example of a supernova in M51, the Whirlpool Galaxy, which isn't very far away, just a few tens of millions of light years away—a stone's throw for astronomers.

This was a bright supernova that could be observed very well with the Hubble Space Telescope; and in particular, we had pictures of that part of the galaxy, which showed the star that later exploded. Here's a photograph taken on January 21, 2005, before the explosion; and there it is on July 11, 2005, after the explosion. This region of the galaxy is right there in this little green square. Taking a close-up look at that region of the galaxy, we can see the star before it exploded and after it exploded, and we can pinpoint exactly which star it is. It turns out that this was a red supergiant, having about 12 solar masses. Indeed, in most cases, we think that it's the red supergiants that explode. This thing had a more normal light curve, normal spectra, and in other ways, this one, and other similar events that we and other groups have observed, have confirmed pretty much that the cores of red supergiants implode, leading to an explosion, as I just discussed.

In all of these cases, the imploding core should form a neutron star—a very dense ball of neutrons, held up essentially by neutron degeneracy pressure. This is similar to the electron degeneracy pressure that I had discussed for white dwarfs. Basically, it's like you have an apartment building where only two neutrons of opposite spin can occupy each level. The ones that are up here cannot jump down to lower energy levels because all of those energy levels are already occupied to the maximum possible extent. Since all these neutrons up here are moving around very, very quickly—they're very energetic—they exert an extra quantum mechanical pressure that cannot be described in the usual terms of just objects, particles moving around due to their thermal motions, due to the heat associated with them, due to their kinetic energy. This is a purely quantum effect, having to do with neutrons, like electrons, essentially not liking each other and not wanting to be in exactly the same quantum state.

When you have a structure like this, held up by neutron degeneracy pressure, it can be rather small. A neutron star having roughly 1.5 times the mass of the Sun can be only about 12 miles in diameter, so that's really small, about 20 kilometers in diameter. All the neutrons in this 1.5-solar mass worth of material are crammed into this tiny, tiny, tiny volume. In fact, the density is so high that a billion—that is, a thousand million—tons would be in one teaspoonful of this material. A billion tons of stuff in a teaspoonful; that's a far cry from white dwarfs, which had only a few tons of material per teaspoonful. If you try holding a few tons of even white dwarf material, I think you'll have a hard time. A neutron star puts white dwarfs to shame.

A neutron star has some similar properties to those of a white dwarf. Being made of degenerate material, it is smaller the more massive it is. Again, unlike bricks, where you have more of them, and they take up more volume, in the case of a neutron star, if you have more mass, the gravitational effect is stronger, and these neutrons get squeezed into an even smaller volume before their energies get high enough to keep the star from collapsing in on itself. The more massive ones have the smaller size. Neutron stars were predicted by Fritz Zwicky and Walter Baade in 1933, just a year after Chadwick's discovery of the neutron. I described in the previous lecture that Zwicky in particular was a rather interesting character. His colleague, Walter Baade, was much more friendly, and sort of got along with people better and didn't call his colleagues "spherical bastards" and stuff. Zwicky and Baade were just amazing in that they predicted the existence of these neutron stars, and even said that they are probably produced during cataclysmic explosions of massive stars. They did this in 1933, and they published a more complete paper in 1934.

You have to go forward about four decades, even more, before neutron stars were actually discovered. They were discovered by Jocelyn Bell in the form of pulsars. Pulsars were objects in the sky that couldn't be seen clearly, but from the regions in which they exist—that is, from certain regions of the sky—there appeared to be pulses of radio radiation coming that were very regular. The first one that was found by Jocelyn Bell had a period of 1.3373011 seconds. I mean every single time, 1.3373011 seconds. Here it is; here's a plot of the radio brightness versus time from this region of the sky, where there's some sort of an object producing a blip of radio radiation; then 1.3 seconds later another blip, then another blip, and so on. The spacing was very regular, although the intensity, or brightness, of the blips varied considerably with time. This was just really weird. I mean what is producing very regular blips of radio radiation from at least this one part of the sky where there was this first object discovered?

The astronomers didn't know what was going on. Jocelyn Bell said, "Here's the data," and Antony Hewish, her advisor, who built the radio telescope, didn't even really believe that it was true. He reportedly said, "Rubbish, my dear" when presented with the data. It just seemed impossible. No stars were known with such regular periods of pulsation, or oscillation, or orbital periods, or whatever—so he didn't even believe it. But then the data looked like they were pretty good, so the astronomers, for a short time, considered the possibility that these were, so to speak, little green men, communications from extraterrestrials—maybe not from Mars; this part of the sky was nowhere near Mars—but some sort of extraterrestrial intelligence that was communicating, either intentionally or unintentionally.

The trouble is, shortly thereafter, they found several more cases of these pulsars in other parts of the sky with very regular pulses, but having different periods than the first one—so maybe 0.7 seconds, or 2.1 seconds, or whatever. It seemed unlikely that there was this vast interstellar network of civilizations all communicating in the same way, using these very regularly spaced blips. Moreover, there was no evidence of any Doppler shift having to do with the planet from which the signal was originating, orbiting around a star. In other words, there were no Doppler shifts recorded of any sort, and so it looked like these things were not coming from something that's orbiting anything else. So, the little green men idea disappeared after a while—and quickly, theoretical physicists figured out that these things probably are rapidly rotating, highly magnetized neutron stars.

One of the clues was that they're found mostly in the plane of our galaxy. Here's a map of the Earth in a very special way. This is the whole Earth in this oval here; the poles are at the top and the bottom. Yet it's the whole Earth, then, projected on a finite flat sheet of paper—so, there are some distortions, but the poles are at the top; the equator runs across the center here. If you make a map of our own galaxy in the same way, you can see that it looks like this thin disk—it's a flat spiral galaxy, and we're looking through the disk here. This is essentially the Milky Way, and there's a bulge in the center. I'll talk about this a lot more later, when I discuss galaxies in detail. It was found that, basically, the pulsars come from the plane of the Milky Way. Most of them are concentrated toward the plane, and there's some up near the poles, but not many, and some at intermediate galactic latitudes. It looks like they come from the plane, and they are probably associated with massive stars because massive stars tend to be concentrated toward the plane, and lower-mass stars have a much broader distribution, up above the plane in what's called the halo, as well.

People thought they're probably massive stars because of the concentration toward the plane, and maybe there's some sort of a weird magnetic effect associated with neutron stars. The logic that they went through was essentially a very interesting process of elimination. They said, "Okay, well, there are three basic classes of models that might produce blips. There could be stars that are oscillating in size, or vibrating, getting bigger and smaller, and that might somehow lead to observed pulsations in radio brightness." If you ask yourself how does that work for a normal star like our Sun, it works; our Sun is actually undergoing small vibrations like this, but the time scale is, of order, minutes. In some cases, it's hours for other stars; in some cases, for some very dense stars, it might be just a minute or so. But the point is all those time scales are too slow compared with the one-second time scale, or the 1/10 of a second time scale, observed for pulsars—so normal stars clearly can't be what's oscillating.

What about white dwarfs? They're smaller and denser, and it turns out their natural oscillation period is about 10 seconds. That's getting close, but it still doesn't explain the one-second, or 1/10 of a second, periods for pulsars—so, white dwarfs are sort of out of the picture. Then there was the possibility of a neutron star. A neutron star is very dense, and it vibrates like this (zzzz). The trouble is that they vibrate too fast; their natural period is 1/1,000 of a second. That's too fast to be the typical 1/10 of a second to a second period of pulsars—so, oscillating stars just don't work. Then you could say, "Well, alright, maybe another class of models would be stars that are orbiting one another." How about normal stars? They can't orbit each other in one second, or 10 seconds, or 1/10 of a second, because they're too big. You can't fit them within the orbit.

What about white dwarfs? Well, they can almost do it, but they, too, are too slow. A few lectures ago, I showed you two white dwarfs that are orbiting each other in five minutes. That's really fast. That's amazing, but it's not one second or 1/10 of a second. What about neutron stars? Well, they can orbit each other very quickly. They can be very close together because they're so small. But in that case, gravitational waves are emitted, as I'll discuss more in a future lecture, and that carries away energy, which eventually causes these neutron stars to spiral in toward each other and merge. The spiraling in would cause the period of the pulsation, which is related to the orbital period, to decrease with time in a measurable way. Pulsar periods, if anything, were slowly increasing with time. A given pulsar has a period that very, almost imperceptibly, increases with time; it doesn't decrease—so, orbiting neutron stars don't really work either.

Finally, suppose a star is rotating. Maybe the rotation time about its axis is the right time scale, and there's some way of producing a pulse of light. A normal star rotating at once per second would just fly apart. If you try rotating really fast, you'll fly apart too—so, it's not dense enough. A white dwarf also rotating at once per second, or 10 times per second, would fly apart. The centrifugal forces, trying to essentially disrupt the thing, would be greater than the gravitational forces trying to hold it in. But a neutron star is so dense and small, and the gravity is so strong, that a neutron star can easily rotate up to about 1,000 times per second, but easily once per second or 10 times per second, and hence could be consistent with the time scales observed for pulsars. That's the reasoning that physicists and astronomers used to deduce that these things are probably some sort of a rotating neutron star.

This was obviously a great discovery. Pulsars showed that neutron stars exist, and they even rotate really rapidly. A lot of physics was done with these things subsequently. In fact, Antony Hewish was given the Nobel Prize for this discovery in 1974, the Nobel Prize in Physics, but Jocelyn Bell was not. Why not? She was the one who actually identified the pulsars, and Hewish even didn't believe the data initially. Hewish deserved part of the prize; he made the radio telescope and designed a lot of good techniques with which these observations could be made, but surely the person who found the pulses should have been rewarded as well. It's one of the greatest injustices ever in science, that Jocelyn Bell was robbed of the Nobel Prize in Physics. That's the way it goes. I really feel sorry for her. She took it in good stride; she doesn't feel too bad about it.

The clock mechanism is this rotation, people said. But what makes a pulsar shine? We don't really know exactly, but the basic idea is that you have a magnetic field that's in the neutron star and surrounding it. The axis of that magnetic field is not the same as the rotation axis of the neutron star—so, you get this conical sort of pattern as the thing rotates, where the magnetic axis first points in one direction, then in another direction, and so on. In some cases, you might make electric fields in this rotating magnetic field that are strong enough to accelerate electrons to speeds close to that of light along the magnetic axis. Accelerating charged particles produce radiation along their direction of acceleration—so, you might have two oppositely directed beams of electrons producing oppositely directed beams of light. Those beams of light point in different directions as the star spins around its axis of rotation.

Let me show you, then, a demo. It would be kind of like a lighthouse. The lighthouse is on all the time, but you only see it when it's pointing your way. Here I have a laser pointer. It's on all the time, as I'm pressing the button, but you don't see that light until the rotation brings it exactly across your line of sight. As it rotates around again, it can do it again, BAM, like that; you see the blast of light every time it intersects your line of sight. That's the idea with this rotating beam. It's on all the time, but you see it when it happens to pass across your line of sight. What actually produces the beam, in detail? We don't really know. That's complicated physics. Why is the magnetic field so strong? We think it's a trillion times as strong as the Earth's magnetic field. The Earth's magnetic field is about one gauss—that's a unit of magnetism—and a neutron star has a trillion gauss.

We're not sure what produces such a strong magnetic field, but one idea is that the star already had a magnetic field going through it—our Sun does, for example. As it collapses, that magnetic field gets scrunched into a smaller volume, and the magnetic field's strength increases. You can get a big increase in magnetic field strength that way. It's not quite clear yet whether you can get an increase up to a trillion gauss, but at least we're on the right track. Why is it rotating so quickly? That's a little bit more easily understood. All stars rotate to some extent, and as they collapse, the spin has to increase in order to conserve the angular momentum, the product of spin rate and mass and size. Remember the example of the ice dancer bringing her arms in, and she spins faster? In a similar way, a star that collapses will spin faster, and indeed, you can reach spin rates of one or 10 per second using this mechanism.

In the middle of the Crab nebula, the most famous supernova remnant, there's a pulsar that in fact spins and pulses, or seems to pulse, about 30 times per second. It's that star right there, and here are two photographs of it, when it was pointing at us and when it wasn't. When it's pointing at us, you can see it; and when it's not, you can't see it. This is actually a photograph taken at optical wavelengths. Very young pulsars shine not only at radio wavelengths, but at optical wavelengths, and x-rays, and things like that as well. It's only as they get older that the higher-energy forms of radiation die out, and what remains are the low-energy forms of radiation, the radio waves. In this Crab Pulsar case, detailed Hubble and Chandra views at optical and x-ray wavelengths, respectively, show a wind coming out from that region, where the rotating neutron star not only is producing these jets of material in these directions here, but also a wind of material going out along the equatorial plane. This wind and the jets are energizing the Crab Nebula, causing it to glow so brightly, as in this photograph right here. The pulsar is essentially re-energizing the nebula and causing it to glow even more brightly than a supernova remnant would have, had it not had a pulsar activating it all the time.

As this energy energizes the nebula, the pulsar had better slow down because the energy has to come from somewhere, and it comes from the rotational energy of the neutron star. As these beams of light and the equatorial winds come out, you rob the star—the neutron star—of rotational energy, and it should slow down. Indeed, astronomers observing the Crab Pulsar have noticed the marked slowdown in its pulsation period over the course of the years that we've been observing it. This is all kind of hanging together; it works out pretty well. Eventually, as the rotation period becomes too slow, the pulsar could die because, as the rotation gets too slow, the electric fields that are generated by this rotating magnetic field become insufficiently strong to accelerate charged particles—so, you don't get the beams of light.

Moreover, with time, the magnetic fields die down as well; they sort of dissipate away. We expect pulsars to remain on for only about a million, or a few million, years before both their magnetic fields and their rotation rate diminish to sufficiently low values that beams of light simply aren't produced anymore. So, every pulsar is a neutron star, but not every neutron star will be visible as a pulsar. Some will have died—after a few million years, they no longer shine as pulsars—or some might have axes of rotation that don't allow the beams of light to cross your line of sight. In that case, you won't see it either. Every pulsar is a neutron star, but not every neutron star is visible as a pulsar. Here is a neutron star in the middle of the remnant Cassiopeia A, and it's a glowing neutron star; it's a hot, young neutron star, only a few hundred years old—but, it isn't a pulsar, either because it's not pointing our way, or because it isn't on for some reason. Maybe it's not rotating fast enough, or maybe the magnetic field isn't strong enough.

Typically, pulsars spin at about one time per second, or 10 times per second, or maybe once every 10 seconds, something like that. But there are some pulsars that can spin hundreds of times per second. In fact, we call these things millisecond pulsars. They have spin periods of just a few milliseconds. If you were to take the frequencies at which they spin, like 642 times per second, and convert it to an audible signal, this would correspond to a note in about the middle of the piano keyboard. A bunch of these millisecond pulsars have been found—spinning 200 times per second, or 642 times per second, or 315 times per second. If you make audible notes out of those things, you can actually create a pulsar symphony, and some astronomers have written musical pieces with the notes of known millisecond pulsars. These guys, by the way, spin so fast because we think that they've been spun up through accretion of material from a companion star. The whole thing is rotating; it's a binary system, and the material coming into the accretion disk is also rotating. When it lands on the star, it can actually spin it up. We think these are actually very old neutron stars that have been spun up a considerable amount.

A very interesting discovery was made in 1991, where one pulsar was found to actually have planets orbiting it. That's amazing. These were actually the first extrasolar planets found, 1991, preceding even the discovery of 51 Pegasi in 1995. They were discovered by Alex Wolszczan and his group at Penn State University. They were discovered around one of these so-called millisecond pulsars, that actually has a spin period of just 162 times per second. What was noticed was that sometimes the pulses arrive a little bit too early compared to the expectation, and sometimes they arrive a little bit too late. That's shown in this graphic here, where the red dots are the expected arrival times of this very fast pulsar; 162 times per second, you get a blip. But what was observed was sometimes the blips are earlier than expected, and at other times, they're later than expected. This would be the case if the pulsar, and the planet orbiting it, both orbit their common center of mass—so that when the pulsar is on the far side of the center of mass, we see the pulse a bit too late, and when it's on the near side, the pulse appears a little bit earlier than expected.

This sort of analysis of the arrival times of the pulsars showed that this thing is probably being orbited by three planets, about 0.2 of an astronomical unit up to 0.5 of an astronomical unit from the star. The orbital periods are from 25 days to about 100 days. Indeed, their masses are comparable to the masses of the terrestrial planets in our own Solar System. That's really groovy. I mean these things not only were the first extrasolar planets found, but they are comparable to terrestrial planets in our own Solar System. There might even be a fourth planet in this system, but that's more controversial.

These are certainly not normal planets. They couldn't have existed before the supernova that gave rise to the pulsar occurred because they would have been blown away in that process. They would have been just disintegrated or launched away from the system when the star blew up. So, they probably formed from a disk of debris around the neutron star that remained after the explosion. Indeed, recently another neutron star has been detected, which presumably came from an explosion like this. Around that neutron star, there is visible a disk of debris, which could coalesce to form planets. Here's a neutron star, around which there is a debris disk that could form planets; that's amazing. If you have planets around a neutron star, you'd better also find cases of the stuff from which those planets formed. Indeed, we think we have cases like this, in at least this one situation.

There's a very interesting observation that has been made in the last decade or so—that some neutron stars have really incredible magnetic fields, up to 1,000 times stronger than even the trillion gauss magnetic field of a normal neutron star. These are called magnetars; they are the strongest magnets in the Universe. If you had one of these things pass within about half the Moon's distance from us, it would instantly erase all the data on all of your credit cards, and magnetic tapes, and all that—so, take care of your credit cards if one of these things comes nearby. They have incredible, incredible magnetic fields, 10^{15} gauss. In fact, these magnetars—an artist's impression of which is shown right here—sometimes emit tremendous amounts of energy because, apparently, the structure of the crust of the neutron star changes, and it's like a starquake. Also, the magnetic field changes in that region and releases the tremendous amount of energy—or at least part of the energy—associated with that magnetic field.

An amazing case of one of these outbursts was observed on December 27, 2004. This was the brightest flare ever seen from outside our Solar System. It occurred in the constellation Sagittarius. There was this great flare, and then it sort of pulsed a few times. Nothing like it had ever been seen before. It was really, really bright—so bright, in fact, that it affected Earth's atmosphere. Here is this rotating neutron star. Let's put it far away from us, at a safe distance. It undergoes a flare like this, sending out a pulse of gamma rays and debris; the orange stuff is the debris. The pulse of gamma rays travels at the speed of light, and it reached Earth. It activated the sensors on a bunch of satellites, and in fact, it even affected our atmosphere; it actually ionized our atmosphere. The satellites transmitted the information to radio telescopes down on the ground, which immediately moved over and started observing that location of the sky. They found a bunch of debris moving out at speeds that are a significant fraction of the speed of light. This was just an incredible outburst of energy that was not itself a supernova, but rather, apparently, the restructuring of the surface layers of a neutron star.

Here's an animation that shows what we think happened. There was this burst of gamma rays as the surface of the neutron star restructured itself. It continued to rotate, then, and we saw these flashes of light as the North and South Poles alternately came into view, and then out of view. There was a big flash, and then there were lots of these little pulsations. Overall, then, we think that this was basically a neutron star with an interior of sort of liquid neutrons and other particles—and there's this solid crust, which essentially buckled or cracked and changed its configuration. In that process, it changed the configuration of the magnetic field itself, releasing in one tremendous burst a huge amount of energy, a giant amount of energy. We know that these things survived this basic burst because some of these objects have been seen to repeat after a few years. Maybe 10 years from now, this particular one will produce an even bigger burst. But for now, this is the biggest burst this guy has produced, or any other of these magnetars has produced. Wow, are they exciting, but don't get anywhere near one—especially if you have a pacemaker, because it'll definitely mess it up.

Einstein's General Theory of Relativity
Lecture 58

Based on the idea that there is no difference between a uniform acceleration and a uniform gravitational field, Einstein's theory postulates that gravity is a manifestation of the warping of space and time produced by matter and energy; objects follow their natural trajectory through curved space-time.

In our understanding of the physical properties of neutron stars—in particular, their immense density—we need to consider Einstein's general theory of relativity. Though Newton's famous laws of motion and of universal gravitation were tremendous breakthroughs in science, the laws break down when we consider objects traveling at very high speeds or in strong gravitational fields. Einstein's special theory of relativity accounted for high speeds, as discussed in Lecture 42. Moreover, Newton never fully understood how gravity worked, and Einstein also knew that standard Newtonian gravity was inconsistent with his special theory of relativity.

The fundamental problem with Newton's theory of gravity is revealed in a thought experiment, in which Einstein tried to predict what would happen to the Earth if the Sun were to simply vanish, leaving no gravitational forces to affect Earth. Because Newtonian gravity invokes instantaneous "action at a distance," the moment the Sun disappeared, the Earth would sail along the tangent to its trajectory, no longer in orbit. With no forces acting upon Earth, it would continue moving in a straight line at a constant speed, according to Newton's first law of motion. Thus, Einstein knew that Newton's law of gravitation violated his own special theory of relativity because relativity claims that no information can travel faster than the speed of light. How can the Earth instantaneously "know" that the Sun's mass vanished?

Einstein worked on this problem for more than a decade and came up with the general theory of relativity, which deals with accelerations and gravitational fields. The theory is based on the idea that there's no fundamental difference between a uniform acceleration and a uniform gravitational field. Recall that special relativity is based on the idea that there is no difference in the laws of physics experienced in laboratories at rest and in uniform motion (constant speed and direction). This is a more restricted theory than general relativity.

Let's look at another thought experiment of Einstein's for general relativity. A person standing in a windowless elevator that suddenly accelerates upward would momentarily feel heavier. From the person's perspective, either the elevator accelerated up or a large mass was temporarily placed beneath the elevator, increasing the gravitational field and creating that momentary heavy feeling. Einstein theorized, then, that a person in an elevator moving with a constant speed would see light travel in a straight line, because that is what happens in elevators at rest. However, that person in an elevator accelerating upward would theoretically see light travel in a path that curved downward, because the light cannot "know" that the elevator is accelerating. Because accelerations and gravitational fields are equivalent, according to the general theory, light must therefore also bend in a gravitational field.

As Einstein further formulated general relativity, he found that the paths of light and particles in a gravitational field can be represented by their natural paths in curved space-time. Gravity is a manifestation of the warping of space and time; effectively, objects move along their natural paths in an intrinsically warped space. This space-time warping, or curvature, is caused by mass or energy. The warping occurs in some fourth spatial dimension, which we cannot see and to which we have no physical access. The denser and more massive the object, the more it bends the space-time around it. For example, the Sun produces a warping that causes Earth to go around it, but Earth has a little warp around it as well, so the Moon follows its natural path around Earth.

Can we actually test Einstein's theory? We know that the orbits of objects are not closed ellipses; rather, their long axis shifts, or *precesses*, with time. The rate of shift increases in stronger gravitational fields, as was first seen with the orbit of Mercury (compared with that of Venus or Earth). Many perturbations of Mercury's orbit can be explained by effects from the large planet Jupiter and other smaller influences. However, Mercury's orbit shifts by 43 arc seconds per century, which is not caused by Jupiter's gravity. Einstein explained the shift using his general theory of relativity. We also know that light from stars moving past the Sun is shifted, proving that gravity can bend light. *Quasars* are central regions in distant galaxies where we think a giant black hole is swallowing material. We can measure shifts in the quasars' positions relative to the Sun's position, indicating that their light bends through space.

> **"The global positioning system is a great practical application of relativity. If it didn't work, it wouldn't get you to the right place at the right time."**

Finally, we know that light emerging from a gravitational field is redshifted as the photons lose energy. This has been seen even in weak fields, such as Earth's, though the effect is very subtle. Einstein also predicted that time is warped by gravitational fields. We can actually measure this on Earth by using global positioning system (GPS) units. GPS satellites must have atomic clocks that are slower by 38 microseconds per day compared to clocks on Earth's surface in order for the system to work on Earth. ■

Suggested Reading

Hawking, *The Universe in a Nutshell.*

Mook and Vargish, *Inside Relativity.*

Pasachoff and Filippenko, *The Cosmos: Astronomy in the New Millennium*, 3rd ed.

Wolfson, *Simply Einstein: Relativity Demystified.*

Questions to Consider

1. If you were immersed in a gravitational field that is not uniform, how might you distinguish this from an acceleration?

2. According to the elevator thought experiment showing that light is bent by a gravitational field, does the amount of bending depend on the wavelength of light?

3. Why do the special relativistic and general relativistic corrections to GPS satellite clocks go in opposite directions?

Einstein's General Theory of Relativity

Lecture 58—Transcript

We have seen that neutron stars are formed through the gravitational collapse of the iron core of a very massive star, a star at least eight to 10 times the mass of the Sun, that either still has a massive hydrogen envelope—or, in some cases, the hydrogen envelope or even the helium layer have been stripped off through winds, or transfer to a companion star, or whatever. But, in any case, the iron core collapses and gravity wins. The thing collapses to a neutron star, releasing a lot of energy and blowing away the outer material. Being so dense, the neutron star that's formed has a very strong gravitational field. Einstein's general theory of relativity is necessary for an accurate description of a neutron star's physical properties. We'll also need Einstein's general theory of relativity for the next topic of the course, black holes, which is one of my favorite topics. Black holes are a fantastic concept. To really understand them, you need Einstein's general theory of relativity. Let me in these next two lectures discuss this magnificent theory.

Newton developed Newtonian mechanics, his famous laws of motion and the law of universal gravitation. These were a huge breakthrough in science. Newton's laws work extremely well for many, many applications: baseballs, and airplanes, and things like that. The modern world wouldn't work if Newtonian mechanics were basically wrong. However, we've already seen that Newtonian mechanics breaks down at very high speeds. In Lecture 42, when discussing the possibility of interstellar travel, we said that you have to take into account the fact that clocks slow down if they're moving relative to you, and lengths become contracted, and all sorts of other weird things happen—lack of simultaneity and all that. Newtonian mechanics doesn't work so well at high speeds, and you need this refinement, the special theory of relativity.

It turns out you need to refine Newton's gravity as well if you have a strong gravitational field. We will see examples of that in the next few lectures. The basic point is that Newton didn't know how gravity works, and Einstein found inconsistencies of Newton's theory of gravity, with Einstein's special theory of relativity. This necessitated a whole new perspective, a whole new view, on gravity. Newton said in *The Principia*: "It is inconceivable, that inanimate brute matter should…affect other matter without mutual contact…. Gravity must be caused by an agent acting constantly according to certain laws; but whether this agent be material or immaterial, I have left to the consideration of my readers." Newton didn't know. He admitted that he didn't know how gravity works, but he knew that it describes the orbit of the Moon and other things pretty well, and so use it as long as it works. If it doesn't work, and you find a situation where it's not applicable, then you've got to figure out what's wrong.

This is exactly what happened to Einstein. It became clear to Einstein that standard Newtonian gravity is inconsistent with his special theory of relativity. Standard Newtonian gravity was essentially broken. The thought experiment that Einstein had—the most famous such thought experiment, in this particular case—was one involving the Sun and the Earth, and Newton's so-called instantaneous action at a distance. To illustrate that, let's look at this diagram, where we have the Earth orbiting the Sun, and let's suppose that the Sun's mass, for some reason, were to suddenly vanish. It just vanishes from the Universe; it's no longer there. It's not that the Sun blows up or anything—not that it's going to do that—but you have to make the mass vanish completely, just a thought experiment.

According to Newtonian gravity, if the Sun's mass vanishes, POOF, then the Earth should go flying along the tangent to its trajectory, to its elliptical orbit here, at the moment that the Sun's mass vanishes because you have this instantaneous action at a distance. Newton thought that masses tell other masses to move in an instantaneous way. If there's no mass for the Sun, then the Earth has no reason to follow this curved path, and it'll continue in a straight line, going in the same direction it was going at the moment the Sun's mass disappeared. That's essentially Newton's first law, that objects continue in their state of motion unless affected by a force. But clearly, to Einstein, this was in violation of the special theory of relativity because special relativity claims that no information can travel faster than the speed of light. So, how can the Earth instantaneously have known that the Sun's mass vanished? That's the essence of Einstein's thought experiment. Special relativity and Newtonian gravity were fundamentally inconsistent. Newtonian gravity was broken.

If something is working well, like a computer, you don't need to know intrinsically how it operates, how it works. But if something is broken—if a computer is broken—you need an expert to go in and fix it, and that expert has to know how the computer works in order to fix it, or an automobile or whatever. You don't just kick it, despite the fact that that sometimes works. It's just coincidental. Usually, you have to know how the object works. Einstein, in fiddling around with Newtonian gravity, couldn't really figure out a way to make it work unless he took a completely fresh perspective. He needed to really just take it from a completely different point of view. He worked on this for about a decade before coming up with something that he really liked. That, he published in 1916. It was called his general theory of relativity because it's sort of more general than the special theory that deals with constant speeds and no gravitational field.

The general theory deals with accelerations and gravitational fields, so it's sort of more general than the special theory of relativity. It's based on the idea that there's no fundamental difference between a uniform acceleration and a uniform gravitational field. Special relativity is based on the idea that there's no difference in a laboratory at rest and one in uniform motion. That was special relativity's principle of relativity. In general relativity, Einstein said there's no real difference between an accelerated frame and a frame in a uniform gravitational field. This idea that there's no such difference between the two frames was what Einstein called his happiest thought ever. There's this profound relationship between uniform acceleration and a uniform gravitational field.

The idea, or the thought experiment, he came up with for this is, suppose you're in a windowless elevator; you're just stationary, just sitting there. Now, suppose the elevator accelerates upward. You know that you feel heavier for a moment. While it's accelerating and before it reaches a constant speed, you feel heavier. Once it reaches a constant speed, you feel the same as you always do again. But as it's accelerating you feel heavier, and when it's decelerating, you feel lighter. How do you know, in this windowless elevator, that what really happened was that the elevator accelerated? What if, instead, some massive object was suddenly placed underneath the elevator, making you feel temporarily heavier? Then it was moved out, and then you felt lighter. Maybe that's what happened. You don't know. That's what Einstein thought about, this elevator thought experiment.

He considered what you would see in an accelerated elevator, and then he deduced that you would see the same thing in a uniform gravitational field, since you don't know whether the elevator is accelerating or whether it actually has a bigger gravitational field or a smaller gravitational field under it or in it. He gained these key insights. Suppose we look at this elevator, and we first have it going along at a constant speed. Here's a constant-speed elevator with a guy sitting in there, watching a beam of light from a flashlight. Since the principle of relativity—that is, Einstein's special relativity—says that there's no difference between a frame at rest and one moving with constant velocity, clearly this fellow would see the light beam go in a straight line across the elevator, regardless of whether he's moving with a constant velocity or is at rest. We already know that those two frames of reference are identical, so the beam goes straight across.

Now, suppose the elevator is not only moving, but also accelerating upward. In that case, this fellow's eyes will be catching up with the light as he accelerates up. Once the light is in motion, once the photons have been emitted, they're just cruising along here. They don't know that the elevator—that is, the walls around the light—is being accelerated up, and so the light just keeps going in a straight line relative to absolute space, in a sense. But this guy is accelerating upward, and so relative to him, he gets closer and closer to the light beam, and it makes the light look bent. So, Einstein deduced that, in an accelerating frame, the light gets bent. But since an accelerating frame is equivalent to a frame placed in a uniform gravitational field, the deduction was that light must be bent in a gravitational field as well. What a fantastic idea, what a great thought experiment. He didn't really do this. Light moves so quickly that you actually can't measure its bending over the distance of an elevator; at least you couldn't do it a century ago. Maybe you can do it now. I don't know. But he thought about this and came up with this idea.

General relativity postulates that gravity is a manifestation of the warping of space and time—that effectively, things move along their natural paths in an intrinsically warped space. This was the geometrical idea that Einstein came up with, which was essentially equivalent to this idea of an accelerated frame. If light is bent in an accelerated frame, he said, it's bent in a gravitational field as well, and the geometrical model that he came up with was that space itself—and time, it turns out—are themselves warped or distorted by the presence of mass or energy, and other objects then move through this distorted space and time, following their natural trajectories, and in an intrinsically curved space-time continuum. The idea is that you have a rubber sheet placed across a table, and an ant goes in a straight line across that rubber sheet if it's flat—if it's not distorted or curved. That's what we call a Euclidean straight line, one that sort of follows the rules of Euclidean geometry.

But if you were to place a paperweight on this rubber sheet, in this case, Earth's gravitational field pulls on the paperweight. That's not really what Einstein was saying. Einstein was saying that the paperweight itself causes this warping of the rubber sheet. This is where the analogy breaks down a little bit. Here, it's the Earth's gravity pulling down on the paperweight, but really, in relativity, the paperweight itself causes this warping of the rubber. Regardless, you've got this warped rubber sheet, and now the ant follows a very different path from our perspective. It's still one of many natural paths that the ant could have followed. It could have gone around this warping, like that, or it could have gone inside and partway down or something. But here, it tried to go along this straight line, directly from this position to that position. In so doing, it actually had to go bend down, way down, almost to this paperweight—actually, to the paperweight, and then up again. It follows a curved path.

We illustrate this usually with what's called an embedding diagram, a diagram that shows just two of the spatial dimensions. Here's what would have been flat space, way out here near the edges of this rectangle. But then matter or energy, either one, cause a warping of this coordinate grid, and the matter or energy causing the warping is represented by this blue ball here. The point is the ball is not being pulled down by Earth's gravity; the ball itself is causing this warping. The warping is in a different dimension. Here you have a two-dimensional space, and the warping is in a third dimension. In our space, we have three spatial dimensions: X, Y, and Z, and so the warping caused by mass and energy would be in some fourth spatial dimension. This is getting kind of crazy, but this was Einstein's genius; it works.

The two-dimensional analogy is just meant to show you how this warping happens into a dimension that we can see on this sheet of paper. But in real life, the warping is into some fourth spatial dimension, which we cannot see, and we have no physical access to it. But that's the dimension—call it W, if you will—into which space warps. That's the geometrical picture that Einstein had. Then objects follow their natural path through that curved space. Let me give you an example of this. Here's a cloth sheet, stretched across this circular drum. If I hold it like this, you can see it's flat, and marbles going along it go in a straight line. Here's some marbles just going along a straight line because it's a flat space. If I were to place a golf ball inside this thing, the gravity of the Earth makes the cloth bend downward, but in Einstein's view, it's the golf ball itself, not the Earth's gravity pulling on it, that's causing this warping of the cloth.

So, you've got this warped cloth; you can see it bulging down below. Now if I take this little marble and just flick it along the surface, you can see that it goes along a curved path. In fact, it goes around this golf ball, and it eventually spirals in because it loses energy due to friction against the cloth, but basically it's going around in circles, and it's orbiting the golf ball. If the Sun were to produce a warp in three-dimensional space like this, then the trajectories of the planets could be thought of as being like this marble going along its natural path in an intrinsically curved space. It was just a marvelous idea. That was his idea, that you essentially have a warped space. The denser and more massive the object, the more it bends the space around it. So, if you have a whole bunch of these things, each of them causes a warping around it. Here are the dense, big ones, and they cause a deeper, steeper warping, and here are the little guys. They all produce a little warping. For example, the Sun produces a warping that causes the Earth to go around it, but Earth has a little warp around it as well, and so the Moon follows its natural path around the Earth. That's the idea.

Time also gets distorted or warped in Einstein's relativity. This became well known and influenced a number of people. There's this famous painting by Dali, *The Persistence of Memory*, and it's often associated with relativity among the general public because people think of relativity, warped time and all that. It's not at all clear that Dali was in any way influenced directly by relativity. To my knowledge, there's no documented evidence of that, though he may have well been aware—in fact, I think he was aware of relativistic ideas. This was a very big thing that Einstein had come up with, and he was a very famous person at this point. But it appears as though Dali was actually thinking of over-ripe Camembert cheese when he painted this famous, famous painting. He actually even said that he had dreamed about cheese, though there's all sorts of symbolism in this painting of various sorts that I'm not qualified to talk about. There are ants crawling around and devouring a gold watch, and there are various sensual-looking eyelashes and things. Who knows? People who know much more about art than I do interpret this painting, but I like to think of it as being associated with relativity. Among the general public, it is often associated with relativity because of this warping of space and time.

You might say this is kind of a weird theory. Is it right? Can we find some experimental tests for it? Even if it's basically right, that it gives some modified view of Newtonian mechanics, is it truly the correct theory? There are many versions of Newtonian gravity now that have been developed over the past century, not all of which are identical to general relativity, and they give quantitatively different predictions from those of general relativity. So, you want to have some tests. The first test, historically, that was made was the explanation of an interesting effect: the precession (or rotation) of the orbit of Mercury around the Sun. Mercury does not follow a closed ellipse around the Sun. It doesn't just go around an ellipse like that, always coming back to the same position. Rather, the ellipse rotates (or precesses). This was a well-known effect; in fact, the precession is fairly large, over 500 arc seconds per century. If you draw a line between the Sun and Mercury, that line rotates by an angle of 500 arc seconds, or about eight arc minutes, per century. That sounds like a small amount, but it was, in fact, easily measurable.

Most of that can be explained through perturbations by Jupiter and other Newtonian effects—but there were 43 seconds of arc per century (of this more than 500) that were unexplained by any known effect. Einstein knew this, and he immediately applied his general relativity to this problem of the precession of perihelion of Mercury. What he found was that the extra precession is, in fact, easily explained by general relativistic effects; and quantitatively, it was exactly what was observed, 43 extra arc seconds per century of this rotation (or precession) of the orbit of Mercury. This was considered a triumph for general relativity because he explained, quantitatively and qualitatively—most importantly, quantitatively—a known effect. But some people could say, "Well, that's great, but maybe he cooked up relativity in such a way as to explain this one phenomenon, and it doesn't really explain anything else because it's wrong." It's much better to have predictions, which can then be verified.

The first prediction that he made that was well verified was that light should be bent by a gravitational field. Indeed, I already discussed this in the lecture on solar eclipses, where I said that this bending of starlight when it is near the Sun, and due to the warping of space as predicted by general relativity, this was actually observed by Sir Arthur Eddington during a total solar eclipse in 1919. Specifically, he noted that the apparent positions of stars during an eclipse in the general vicinity of the eclipsed Sun were a little bit displaced from their true positions, or the positions that had been measured at night some other month when the Sun was nowhere near that part of the sky. This deflection that he measured was consistent with general relativity, though not to a high degree of precision or accuracy. Indeed, Eddington kind of ignored some of the data that weren't so much in agreement with the prediction, it turns out, and he selectively believed the data that seemed to agree with relativity.

Nevertheless, this was the first sort of observational confirmation of relativity, and this is the thing that made Einstein an overnight celebrity among the general public, not just physicists. Indeed, the effect is very small. Here's a little diagram. The eclipsed Sun is down here somewhere. The star was observed to be over here, and its true position, if there were no displacement, would have been here. You see this offset is very small; it's smaller than the blur circle of a star. This particular diagram has been magnified several hundred times compared to the original photographic emulsion, so it was a hard thing to measure. I'm not discrediting Eddington in any way for having made a bad measurement. He made as good a measurement as he could have, and personal bias has influenced many of us. Even scientists tend to want to believe data that agrees with some preconceived notion of the Universe, and discard data that doesn't agree. Maybe that's what was going on a little bit. We're all human, and so was Eddington.

We now measure this effect much more precisely by looking at the positions of objects called quasars. We'll look at quasars much more extensively in a few lectures, but they are the central regions of distant galaxies where we think there's a giant black hole swallowing material, and it emits radio waves and other forms of light before it actually gets swallowed by the black hole. There's all this material, all this gas, and it's emitting radio waves and other forms of radiation. You can see these things at a very great distance. Here are four of them, and they're billions of light years away. These are optical photographs. The point is we know their precise positions, and they're visible during the day with radio telescopes because the Sun isn't very bright at radio wavelengths, and you can measure their positions when the Sun is in the general vicinity of these quasars. In fact, their positions shift when the Sun is in their general direction compared with where they appear when the Sun isn't in their general direction. This has been really, really well measured with that.

The other thing that's been measured is that, if you look at starlight going past the Sun, not only is the path bent, but the path has a different length than it would have had, had it gone in what we call a Euclidean straight line. When you look at, for example, the signals from the spacecraft like Voyager when they were on the other side of the Sun, the opposite side to where we are, the signals took longer to get to us than they would have, had there not been the Sun causing this warping of space. We know exactly when the spacecraft emitted the signals, and we know exactly when they arrived. From the speed of light, you can determine the path length, and the path length agrees with an intrinsically warped space, and not with a Euclidean straight line. So, that's been measured very, very well.

Another prediction that Einstein made, the third effect—one effect was an explanation of something that was known, the precession of Mercury, and then there were two predictions, the bending of light and this second one. The second one was that light is redshifted as it comes out of a gravitational field. Here you have a mass, and if you emit blue light, if you're near the gravitational field, it's losing energy as it climbs out of that gravitational field and becomes gradually redshifted. The amount of redshifting is bigger, the bigger the gravitational field out of which the light is trying to climb. You have here a flashlight beam, where the light starts out yellow and becomes redder. You also have the light curved if it's not going away radially from the center of the star initially. The curvature effect had already been measured through the bending of starlight near the Sun, and these quasar positions, and all that, but this redshifting effect was yet another specific test.

This was proposed in 1919, but it was not measured until about four decades later by two physicists, Pound and Rebka, who used a newly discovered effect called the Mossbauer effect—with which you can measure very precisely the wavelength of a photon. In particular, you measure wavelengths of gamma rays that are emitted through radioactive decay of certain radioactive nuclei, or from a nucleus going from an excited state down to a lower energy level. A gamma ray with a very fixed energy is emitted, and the Mossbauer effect can be used to measure that energy. For example, excited iron nuclei, iron 57, that come from the radioactive decay of cobalt, emit a certain type of gamma-ray photon, having a very specific energy.

Pound and Rebka had a little emitter of this line down at the bottom of the Jefferson Physical Lab at Harvard University. Here are these iron 57 nuclei emitting this particular wavelength of gamma rays. It climbed out of the gravitational field of the Earth, going to the top of the Jefferson Physical Lab, 74 feet higher. It should have become gravitationally redshifted by a very small amount. That would mean that an iron absorber at the top end here wouldn't absorb the photon if it got redshifted because the wavelength would now be wrong; it would be the wrong wavelength because it's different from the wavelength at which it was emitted. The absorber—which only absorbs photons having a wavelength equal to the wavelength at which they are emitted—wouldn't absorb the photons because they've been redshifted.

But now if you move the source relative to the absorber, you can blueshift those redshifted photons, and the amount of blueshift that is needed to get their wavelength back to the wavelength that is needed for the absorber to absorb them is a measure of the gravitational redshift. You move this source, increase its speed until the absorber finally starts absorbing these photons again. You can do it the opposite way as well; you can have the source here and the detector there. In that case, the photons get blueshifted. The measurement was made, and the gravitational redshift is only five parts in 10^{15}—so, it's a very small effect. But the experimental result was 5.13 +/-

0.51×10^{-15} for the gravitational redshift, and theory was 4.92, which is in agreement with the experiment. Pound and Rebka measured this in 1960, and then confirmed it with 1% precision in 1963. It's a fantastic measurement of the gravitational redshift, even here on Earth. When you measure it for something like light escaping from a white dwarf, it's a much bigger effect, and you can measure that, and that has been measured. From a neutron star, it's even bigger—so, that's really cool.

There's one other thing that is related to this, and that is that time gets warped by gravitational fields. Here, I've exaggerated the effect, but if time were not warped, then near a neutron star, a clock that initially reads noon near the neutron star and noon far from the neutron star, would continue to show the same time a few hours later. But if there's a gravitational time dilation effect, then, in fact, near the neutron star, maybe two hours will have passed, whereas far from the neutron star, four hours will have passed. I've exaggerated the effect here, but the principle is correct. You can actually measure this on Earth by using the GPS system, the Global Positioning System that is used on airplanes or in automobiles to figure out where you are. You might say relativity is some bizarre thing, and it has no practical applications to everyday life, but that's not true. There is one very well known practical application, and that is that GPS devices wouldn't work if general relativity weren't correct.

The point is that you have this network of 24 satellites going with speeds of 14,000 kilometers per hour. At least four of them are visible from any point on Earth at any given time, and they are sending signals to your little unit in your car. If you know the speed of light, which you do, and if you have atomic clocks that measure time very carefully, you can figure out how far away each of the satellites is from you. If you have enough satellites, you can figure out uniquely where you are. For example, if you were told that you're 1,858 miles from San Francisco, then you know that you're somewhere on the circle having a radius of 1,858 miles. If you're also told that you're 1,161 miles from Miami, then you know that you're somewhere on this circle. You could be in Mexico, or you could be here in Chicago. If you're also told that you're 713 miles from New York, then it turns out there's only one intersection point; you have to be in Chicago.

In three-dimensional space, you need a fourth measurement, but you always see at least four of these satellites, and so you can determine uniquely where you are. This depends on relativity because, at a speed of 14,000 kilometers per hour, the clocks on the satellites are running a little bit slower than clocks on Earth; by seven microseconds per day, they're running slower. Because of general relativity, they're in a weaker gravitational field, and they're running 45 microseconds per day faster. The net effect is that the clocks in the satellites are running fast at 38 microseconds per day. That's a huge effect; it would make huge errors in our position, which would accumulate with time. What the engineers did is they took relativity into account when designing the satellites, and they made the atomic clocks up there run more slowly than on Earth, by 38 microseconds per day. Then it all works out. If they hadn't done this, GPS systems wouldn't work. Your planes would end up in the wrong places, and you wouldn't know where to drive and all that. So, there's a great practical application of relativity; it works. If it were not to work, the system wouldn't get you to the right place at the right time.

Warping of Space and Time
Lecture 59

Another effect of general relativity is the bending of light through space, also a measurable phenomenon that can help us detect the presence of brown dwarfs and black holes.

In the previous lecture, we looked at Einstein's general theory of relativity and mentioned how it applies to the global positioning system (GPS). Let's consider this in more detail to see how relativity works quantitatively and, at least in this one case, affects our everyday lives. Timing is crucial; in order for a GPS device to work, we must know exactly how far away each of the GPS satellites is. We can figure this out by measuring when a satellite's signal was emitted and how long it takes to reach Earth. The speed of light is about 1 foot per nanosecond (1 billionth of a second). But the **relativistic** effect in time difference between the satellites and our unique position on Earth is 38 nanoseconds. Such a difference, though seemingly small, would accumulate to large errors over the course of a month. If GPS designers did not take relativity into account, then after a few days, a GPS unit would begin providing quite inaccurate information. Thus, a GPS satellite's atomic clock is programmed to run at a rate that exactly compensates for both the special relativistic effect, a slowdown of 7 microseconds per day, and the general relativistic effect, an increase of 45 microseconds per day. The net effect is a compensation of 38 microseconds per day. Today, lasers, computers, and other electronic devices depend on our understanding of quantum mechanics. Thus, who knows in the future what other technological inventions will have to take into account the effects of general relativity?

Let's review Einstein's thought experiment in which the Sun's mass suddenly vanished. Then, we will consider other ways that we can test for Einstein's general theory. As we saw in Lecture 58, according to Newton's law of gravitation, at the moment the Sun disappeared, Earth would be thrown off its orbit along the tangent of its trajectory. However, Einstein claimed that Earth would not experience the Sun's disappearance until 8.3 minutes later. Thus, information about the Sun's disappearance would travel via **gravitational waves** at the speed of light, according to theory (the speed has never actually been measured). Once this information reached Earth, 8.3 minutes later, our planet would then travel along the tangent to its trajectory. An analogy would be tossing a ball into a calm swimming pool. The ball sends out concentric waves, but someone at the pool's edge wouldn't experience those waves until they reached the edge. Likewise, the removal of the Sun would create a disturbance in the warping of space, a disturbance that travels at the speed of light through the Solar System. The warp that the Sun used to produce no longer exists; space would be flat there instead, though it wouldn't happen instantaneously. Recall that the first historical test of relativity was the confirmation that Mercury's extra precession, 43 seconds of arc per century, could be quantitatively explained through general relativity. That was a great triumph, but 43 arc seconds per century is a small amount.

"We think that on large scales the theory of relatively is correct quantitatively. But it's always possible that it is false. Any new experimental test is welcome with open arms."

We now have much better evidence through our study of two neutron stars, discovered (in 1974) to be orbiting each other in just 8 hours. One of the stars is a pulsar; thus, the system was dubbed a **binary pulsar**, even though only one star is visible as a pulsar. The precession of the pulsar's elliptical trajectory is 4 degrees per year, much greater than Mercury's orbital precession of 43 arc seconds per century. Each star forms a warp in space, creating a ripple, or gravitational wave, that propagates through space at the speed of light. As these ripples travel and the stars continue to orbit around their common center of mass, energy is removed from their system. The removal of energy forces the stars to move closer to each other, due to gravity, which in turn, increases the stars' orbital speed and decreases their orbital periods.

As the stars' orbital period decreased over time, the pulsar signal also changed. The cumulative measurement of the pulsar's change over time exactly corresponded to what general relativity predicts. In 2003, an even more closely spaced binary pulsar was found, with an orbital period of just 2.5 hours, in which both of the neutron stars have visible pulsars.

Astronomers have determined that the stars' orbits shrink by 7 millimeters per day. This system, as well, strongly confirms quantitatively the predictions of general relativity.

A related effect of this warping of space occurs in a process called **gravitational lensing**. We've already encountered this in our discussion of the deflection of starlight by the Sun and other massive objects. As we've seen in a previous lecture, if we observe the light of a distant star, our Sun deflects—bends—the star's light. If we look at an intrinsically point-like light, we might actually see what is called an *Einstein ring*. Such rings have been photographed at visible wavelengths by the Hubble Space Telescope. The ring appears in perfectly symmetrical situations, such as when a black hole passes between our line of sight and a distant galaxy. In this case, the galaxy's light is actually bent such that it appears as a ring of light. If the focusing effect is caused by a foreground star, rather than by a galaxy, we call it *gravitational microlensing.*

Usually, small deviations from symmetry cause a ring to break up into smaller units, such as partial arcs or even point-like images: We can see several images of the background object that is lensed by the foreground object. Each image is a mirage; for example, quasars often become imaged into several discrete mirages when a galaxy appears in the foreground and gravitationally lenses the light from the background quasar. If the clarity of the images is not enough to show distinct mirages, we would still see an apparent brightening of the object due to the focusing of light rays toward us. The cumulative brightness of the ring, or the mirages, is increased through the gravitational focusing of light.

This brightening effect can be used to detect the presence of foreground gravitationally lensing objects, even if they aren't directly noticeable. For example, if a brown dwarf or a black hole passes along our line of sight to a background star, the focusing effect still occurs, revealing the presence of an otherwise hard-to-see object. Some brown dwarfs and wandering black holes have been detected through this process. Even a few isolated, free-floating planets have been detected by their gravitational microlensing of a background star. In 2005, a small planet (about 5 Earth masses) was discovered through this effect, which also has the potential to reveal many more exoplanets having relatively small masses.

Finally, using one more emerging test for general relativity, we are hoping to demonstrate that a rotating object drags space around it; that is, space itself rotates around a rotating object. Data gathered in 2006 from a satellite called *Gravity Probe B* are now being analyzed. The experiment involves a complex technique and the orientation of gyroscopes within the satellite. This dragging effect, which has never been measured, is believed to occur at a rotation of 0.042 arc seconds per year at the location of the satellite. By measuring the orientation of the satellite's gyroscopes, we should be able to demonstrate the effect, if general relativity is correct. ■

Important Terms

binary pulsar: A pulsar in a binary system. Often, this term is used for systems in which the pulsar's companion is another neutron star.

gravitational lens: In the gravitational lens phenomenon, a massive body changes the path of light passing near it so as to make a distorted image of the object.

gravitational waves: Waves thought to be a consequence of changing distributions of mass.

relativistic: Having a speed that is such a large fraction of the speed of light that the special theory of relativity must be applied.

Suggested Reading

Hawking, *The Universe in a Nutshell.*

Mook and Vargish, *Inside Relativity.*

Pasachoff and Filippenko, *The Cosmos: Astronomy in the New Millennium*, 3rd ed.

Thorne, *Black Holes and Time Warps: Einstein's Outrageous Legacy.*

Will, *Was Einstein Right? Putting General Relativity to the Test.*

Wolfson, *Simply Einstein: Relativity Demystified.*

Questions to Consider

1. Suppose you find several closely spaced quasars that you think are the gravitationally lensed images of a single quasar. How might you test your hypothesis?

2. Because of time dilation in the special theory of relativity, an observer on Earth sees a rapidly moving twin in a spaceship aging more slowly than he does. After returning to Earth, the traveling twin will be younger than the one who stayed on Earth. But consider this: The traveling twin thinks that he is at rest and the Earth twin is moving—in which case, the Earth twin would be younger than the spaceship twin. How do you think this famous *twin paradox* is resolved? (Hint: What does the traveling twin have to do in order to return to Earth? Does this allow the two frames of reference to be distinguished from one another? Is this similar to being placed in a gravitational field?)

3. If a specific observer sees an apparent brightening of an object due to the focusing of light rays during an episode of gravitational lensing, will this object appear fainter than expected from some other lines of sight at that same time?

Warping of Space and Time

Lecture 59—Transcript

In the previous lecture, I introduced Einstein's general theory of relativity. It suggests that objects with mass or energy warp (or distort) both space and time in their vicinity. Other objects then move through that warped space and time according to their natural paths. Clocks get messed up; distances get messed up. They don't look straight; they don't look Euclidean anymore. But apparently, this is really the way the world works. I introduced a good, everyday, practical example of relativity: the Global Positioning System. I'd like to say a few more remarks on it, just because of its importance in showing that relativity really does work quantitatively and does, at least in this one case, affect our everyday lives.

The timing is really crucial. You need to know exactly how far away each of these satellites is. The way you know this is through knowing when the signal from a satellite was emitted, and how long it took to reach you. If you know the speed of light, you know, then, how far away that satellite is, and you can do the same thing for other satellites. You can then figure out your unique position on Earth through all this timing. But the speed of light is about a foot per billionth of a second, a foot per nanosecond. A billionth of a second is a very small quantity. If the relativistic effects are as big as 38 microseconds, 38 millionths of a second, then the errors that you get in your position are, of order, miles, miles per day, and they accumulate day after day after day. Suppose you were to take a month-long vacation, and you want to drive from San Francisco to New York City, and your GPS system wasn't working the correct way; relativity wasn't taken into account correctly. You'd end up in the wrong place after a month because you'd be heading in the wrong direction.

The system designers knew about relativity, both special and general, and they programmed their atomic clocks to run at a rate that exactly compensates for both the special relativistic effect, which is only seven microseconds per day slowing down of clocks, and the general relativistic effect, which is 45 microseconds per day speeding up the clocks, for a net effect of 38 microseconds per day speedup. They took this into account, and the atomic clocks run at a different rate up there than they would have here. When relativity is taken into account, all the clocks agree, and you can figure out exactly where you are, and it's all wonderful. Relativity is important, even in a few cases of everyday life, although I must admit this is the one clear example I know of right now where general relativity is important. Special relativity is important in many other ways.

Who knows, in the future, what effects there might be where an understanding of general relativity will be important? When people thought of quantum mechanics a century ago, they weren't thinking of practical applications. They just were doing the research to understand the fundamental physics. Now lasers and the whole electronics revolution, and computers and all that, depend on our understanding of quantum mechanics. Similarly, there are effects for special relativity. Maybe general relativity, in the future, will be important in many ways, other than just the GPS positioning system.

Let's go back to Einstein's thought experiment, where the Sun's mass were to suddenly vanish for some reason. Newton would have said that the Earth goes in a straight line instantaneously, at the moment that the Sun's mass disappeared. Einstein says no, the Earth only learns about the Sun's disappearance some time later, 8 1/3 minutes later. It turns out that the information about the Sun's disappearance travels in the form of what's called gravitational waves, and they travel at the speed of light, according to theory. They've never actually been measured, but according to theory, gravitational waves travel at the speed of light, reach Earth, tell Earth that the Sun no longer exists, and then the Earth can go zooming off along the tangent to its trajectory at that moment. There's an 8.3-minute delay.

The way to think about gravitational waves is to consider an analogy like a swimming pool, where the water is at rest, just motionless, and there are no ripples in it. Now let's say I toss a ball onto the surface of the water—so there's suddenly a disturbance there. That disturbance sends a wave that propagates outward along the surface of the swimming pool. Some time later, people near the edges of the pool learn through this wave that the ball has landed on the water. Suppose their eyes were closed, and they didn't actually see the ball land, but they can learn of the ball's landing in the water through the arrival of these waves. Similarly, if the ball is just sitting there, and the water is calm, if I remove the ball, a disturbance will start passing along the surface of the water, and someone some time later on will feel it.

In a similar way, then, if you look at the Earth-Sun example, and you remove the Sun, then there's a disturbance in the warping of space that travels at the speed of light through the Solar System—because the warp that the Sun used to produce is no longer there. Now space is flat there, and you can't just remove that warp and have flat space instantaneously. That change in the warping of space travels out as a disturbance, as a ripple in the actual structure of space itself, according to Einstein's general theory of relativity. That's really cool. Well, recall that the first historical test of relativity was the confirmation that Mercury's extra precession, 43 seconds of arc per century, could be both qualitatively and quantitatively explained through general relativity. That was a great triumph, but 43 arc seconds per century is a very small amount. Even though it had been measured very well, you might still worry that that's such a minor effect. Maybe there's some other explanation, or maybe it was measured incorrectly, or whatever.

We now have a much better way of testing this same general idea, and that is through two neutron stars that are orbiting one another. They orbit each other with a period of just eight hours, and the orbits are not circular; they're elliptical. The two neutron stars are very close together because they orbit so quickly. They're just zooming around in just eight hours. One of the neutron stars happens to be a pulsar. The system, which was discovered by Russell Hulse and Joe Taylor in 1974, is called the binary pulsar (or it was initially called the binary pulsar), which is a bit misleading because only one of the neutron stars is actually visible as a pulsar. But in any case, there are two neutron stars; here's a little schematic of them. Each has about 1.4 solar masses. They're orbiting each other in elliptical orbits over the course of eight hours. Here is an artist's rendition of what they might look like. The precession of this elliptical trajectory of one neutron star relative to the other is four degrees per year—much, much larger than Mercury's orbital precession of 43 arc seconds per century. Four degrees per year is a big effect. You go around the full circle in something like a century.

The picture that we have is that each of these neutron stars forms a warp in space, like that shown here. As the neutron stars orbit one another—or more correctly, their common center of mass—this configuration—this warping of space—changes, and it creates a ripple (a gravitational wave) that propagates through space at the speed of light. It's this ripple—this distortion in the shape of space itself—in the geometry of space that travels outward and removes energy from the system. That's an important point. The wave can't travel without having some energy. Similarly, the water wave carries some energy with it. This disturbance travels outward, and it might look something like this. You have these two neutron stars orbiting each other closely, and they create this ripple that travels outward and removes energy from the system. Here's what it might look like, where these ripples are just sort of moving out.

If you have these two stars orbiting each other so closely, the gravitational waves carry out a measurable amount of energy. That means that the two stars have to approach each other more and more closely. According to Newton's laws and Kepler's laws, they have to orbit each other more and more quickly as they approach each other, and this change in the orbital period can actually be measured. The gravitational waves are removing energy from the system. That means that the two objects have to approach one another because, if their energy is being removed, then, in a sense, gravity can pull on them more strongly; not because gravity is stronger itself, but relative to the energy that the objects have, the gravity is now having more of an effect because the energy of the objects has been released through gravitational waves.

They get pulled more and more closely together, and as you saw in that animation, they approach each other. Their orbital period decreases because objects that are closer together for a given mass orbit more quickly. Mercury goes around the Sun more quickly than Venus does, and Venus goes around the Sun more quickly than Earth does. You can measure, then, this decrease in the orbital period. What's been measured is actually what relativity predicts. I'll show you the graph in a few minutes. This spiraling together of the two stars has an interesting analogy.

Recall when we discussed extrasolar planets, or exoplanets, that there were these so-called hot Jupiters, the massive planets that are largely gaseous and liquid, orbiting around the star with a period of just, say, a few days, two to four days. Here's a distribution of the number of planets discovered versus their orbital period around the star that they orbit. There's a bunch of them that are massive planets, like Jupiter or even more massive, but orbiting very close to the star. They couldn't have formed at those locations because, at those locations, the temperatures are really, really high. If you go back to our model of the formation of planets, we said that rocky, terrestrial planets can form close to a star, due to the gravitational accumulation and collisions between a bunch of small, rocky planetesimals. Gradually, these guys all converge and form planets.

To get massive planets, you needed icy objects, rocks and ices as well, and condensed gases, to accumulate into something like an Earth-like or bigger core. Then that core could gravitationally accumulate any remaining particles, and gases, and icy blobs out there, and you could build up, then, very massive planets because those cores would then attract a lot of hydrogen and helium and become massive planets. Whereas, close to the star, there was very little hydrogen and helium to attract, and moreover, it was harder to hold onto those gases because of the high temperatures and the high speeds of the gases, and so on. Massive planets should form far out, not close in. In the exoplanet case, we think they form far out, but then friction—due to sort of a rubbing of these massive planets far out—against the remaining dust and debris in that disk cause them to lose energy.

In a sense, the star, pulling on them, was able to get them to spiral in because as these planets were rubbing against other debris in the disk, they lost energy, and the star's gravitational influence became more important relative to the energy that they had, and they spiraled in. Some probably spiraled all the way into the star and got eaten. Others, as this diagram suggests, may have stopped shortly before reaching the star because, in the immediate vicinity of the star, there may not be much debris, gas and dust, with which to have any frictional effect against the planet, because the star's wind can clear out a little cavity near it, which is relatively empty of debris.

Maybe as these big Jupiters migrated in, they then stopped near the star because there was an absence of this frictional drag very close to the star. Maybe some actually did get eaten; we're not sure. In any case, this spiraling in of the big Jupiters that were formed far out and felt friction against the debris surrounding them, thus losing energy and coming closer to the star, is analogous to two stars orbiting one another and releasing gravitational waves. In a sense, this is kind of a friction. It's a gravitational friction that releases energy and causes the two stars to spiral in toward one another. I just thought I'd bring that up because it ties together two seemingly very different subjects.

As the orbits get smaller and the orbital period decreases, the pulses of the pulsar from the neutron star that you see as a pulsar start arriving at different times. If the thing were stationary, it would go beep, beep, beep, beep, beep. If it's orbiting something, sometimes it goes beep, beep, beep, that kind of a thing. But if the orbital period is changing—and in this case, specifically decreasing—then you can measure this change of beep, beep, beep, and it starts going faster—beep, beep, beep, beep. It's changing. I'm really bad about giving sound effects, so you'll pardon me, but I think you get what's going on. By measuring the pulsar, which is a great clock, they could measure the orbital period changing. Here is the effect. If you measure the time at which the two stars are closest together, and if they remain closest together in a way expected from elliptical orbits that aren't changing with time—that is, that are not getting closer and closer together—you would get a horizontal line of zero here.

But if the orbits are getting closer and closer together, then the time at which the two stars are closest together in their elliptical orbits changes. This is a measure of the cumulative effect of the change over the course of about three decades. The dots are the data, the measurements, and the curve is the prediction from general relativity. You can see that the prediction matches the data perfectly. This is a big effect, and so it's just fantastic that this was actually measured. Russell Hulse, who was a graduate student at the time that he discovered the binary pulsar, and his advisor, Joe Taylor, measured this effect over many years and showed that it exactly confirms what the prediction was from general relativity.

They received the Nobel Prize in Physics in 1993 for their discovery of the binary pulsar and for the subsequent measurement of this general relativistic effect—the single best-known test of general relativity that we have right now.

In 2003, an even more closely spaced binary pulsar was found, one with an orbital period of just two hours or so. In this case, both of the neutron stars are actually visible as pulsars. One has a period of 23 milliseconds, and the other has a period of 2.8 seconds. They orbit each other in about two and a half hours, much shorter than the original binary pulsar, where the orbital period was eight hours. You can actually measure that the orbit is shrinking by seven millimeters a day. This doesn't sound like much to you, but it's a lot. An orbit shrinking by seven millimeters per day is really quite a lot and is easily measurable. This system as well, then, has confirmed quantitatively the predictions of general relativity in a very, very strong way.

A related effect is gravitational lensing. We've already encountered this when we considered the deflection of starlight by the Sun, or by other massive objects. Here's the starlight that would have gone in a straight line, a Euclidean line, had there not been the Sun's presence there. But in the Sun's presence, it gets bent or deflected, and this is what Eddington measured. If you have an object in between us and a background object, then all the light rays from that background object traveling toward us get bent, and you get sort of a gravitational focusing of the rays toward us. Here, an observer, looking at an intrinsically point-like particle, might actually see a ring, what's called an Einstein ring—because Einstein actually predicted that this effect should happen, although he thought it would never actually be observed. He thought it was too small an effect to observe. Here, this light ray got bent toward our line of sight, so we think it's coming from over here. This one got bent toward our line of sight, and so we think it's coming from here. The ones out of the plane of the screen are also bent. In a perfectly symmetric situation, you get this ring.

Usually, the ring is not so perfect, as is shown in this extremely idealized situation. Usually, there are small deviations from symmetry that cause the ring to break up a little bit into smaller units, into little blobs. But in some cases, you do see rings like this. First, I want to show you an animation of what would be seen if, say, a black hole were traveling across our field of view. We see background galaxies back there, and the black hole's gravity distorts them, and right there, you've got an Einstein ring. Did you see that one? It was an actual Einstein ring. There was another one, where the distortion of the galaxies is so strong around the gravitational field of the black hole, that you can actually see these rings. I particularly like the one where the black hole passed right between us and this bright galaxy here. The other ones just sort of curved around. But if I stop the animation right when the black hole is between us and that background galaxy, you can see the beautiful Einstein ring.

This is just an animation, you might say. Have we really seen these things in real life? We have. Here's a case where there's a foreground object along the line of sight toward a background object, and the background object has been gravitationally lensed by the foreground object into a nicely symmetrical ring with a few little lumps and things in it. That's a beautiful ring, seen at visible wavelengths with the Hubble Space Telescope. Here's a slightly distorted ring, seen at radio wavelengths. Again, this is thought to be gravitational lensing of a background galaxy by a foreground galaxy or cluster of galaxies.

As I mentioned, deviations from symmetry usually lead to the appearance of several discrete objects. Therefore, an observer might see not a ring, but several images or mirages of the background green object that was lensed by this foreground yellow object. We will encounter this when we consider quasars later on. There have been a number of quasars where the configuration is much like what is seen in this diagram, where the background object actually gets imaged into several discrete objects or mirages; in this case, four. Here's a foreground galaxy gravitationally lensing a single background quasar, making it look like four quasars. Each of these images is a mirage of what you would have seen, had the galaxy not been in the foreground, not gravitationally lensing the light from the background quasar. That's really great.

Suppose your resolution, your clarity, isn't sufficient to show you the individual dots into which the quasar was lensed, or the ring into which it was lensed. Even if your clarity isn't great enough to see that, you will see an apparent brightening of that object because of the focusing of light rays toward you. Suppose there's a ring, or four dots, or whatever. The cumulative brightness of the ring or the four dots is greater than what would have been the case, had there not been this gravitational focusing of light. You get a brightening of the object if it's exactly along your line of sight toward a foreground-lensing object. This is a great effect because it can be used to detect the presence of foreground gravitationally lensing objects, even if you can't see them using visible light. What they do is they get along the line of sight toward some background star, focus that star's light toward us, and that star's light gets brighter because of this focusing effect.

If the focusing is due to a foreground star, rather than a massive galaxy or cluster of galaxies, it's usually called gravitational microlensing rather than just gravitational lensing. But the idea is that it travels across your line of sight, so it's a star that's moving. If it travels exactly along your line of sight to a background star, that star's light can be focused toward you and amplified, making it look like that star is brighter. This can be used to detect invisible objects like brown dwarfs. Suppose instead of a star passing between us and a background star, you have something that's hard to see, like a brown dwarf or even a black hole. There would still be this focusing effect, and you would see the background star brighten. Here's an animation that shows this. We're going to zoom into a cluster of stars. If you look at one star in particular, this one right there, and watch its apparent brightness, there, it's brightening because a dark foreground object—a black hole, or a brown dwarf, or something else that's pretty dark—passed exactly along the line of sight to that star, temporarily brightening the appearance of that star.

Indeed, some brown dwarfs have been detected this way. Some lone black holes wandering around in the galaxy have been detected this way. Even some isolated, free-floating planets have been apparently detected by their gravitational microlensing of a background star. In the case of exoplanets, you even have a situation like this, where you might have an exoplanet orbiting a star, and the star produces some gravitational microlensing, but then the planet orbiting the star also produces some gravitational microlensing. Here's what I mean. The apparent brightness of a background star rises with time and then declines as the foreground star passes across our line of sight to the background star. But then there's a sharp spike as the planet orbiting that foreground star also passes exactly along our line of sight to the background star.

Here we are; we've got this background star. First, the foreground star passes directly along our line of sight to the background star, causing this gradual brightening and fading. But later on, the planet that is right next to this foreground star passes exactly along our line of sight to the background star, and that's what causes this sharp spike. In a sense, you could call it gravitational nanolensing. Instead of microlensing by a star, it's nanolensing by a planet. People don't call it that; I just kind of thought of that right now. As I mentioned when I discussed exoplanets many lectures ago, recently, in 2005, the discovery of a five-Earth mass planet was discovered through this gravitational microlensing effect. This effect has the potential to lead to the discovery of many more exoplanets having relatively small mass. Even an Earth-mass planet passing directly along the line of sight to a background star will cause a rather noticeable sudden brightening of that star's light, due to this gravitational focusing effect. Therefore, you can find relatively low-mass planets this way.

The disadvantage is you can't just point to a particular star and say I want to find an exoplanet around that star, because the chances that that star, and whatever exoplanet is orbiting it, passes exactly along the line of sight to a background star are small. Those chances are small. But statistically, if you look at many, many stars, some of them will pass along the line of sight to background stars and cause this effect. You can statistically find many exoplanets this way, as the technique gets refined.

Finally, there's another test that I want to tell you about general relativity—that is that a rotating object drags space around it, drags space-time. In other words, space itself rotates around a rotating object. There's a satellite that completed its data-gathering in 2006, after just a couple of years. It's called Gravity Probe B, and the data are now being analyzed. The satellite knew its orientation at all times by having a telescope that pointed at a particular star called IM Pegasi. Within the satellite, there were some gyroscopes, which are these spinning balls. You can tell the orientation of the gyroscope very precisely using a special technique, using superconducting magnets and all that. If the gyroscopes are changing their orientation, you can measure it. There are two effects; one is called the geodetic effect, which is 6.6 arc seconds per year. That's well known already.

The other, which has never been tested, is this frame-dragging, which is a rotation of 0.042 arc seconds per year. It's a very small effect, and you can see here just 4/100 of an arc second per year is the rate at which the orientation of this gyroscope should be changing as a result of this frame-dragging, of the fact that the gyroscope, the spinning top, finds itself in a space around the rotating Earth, which itself is dragged by the rotating Earth. That would be really cool to measure. They're measuring the orientation of these gyroscopes by having the gyroscopes be superconducting—they have a superconducting layer of niobium on the gyroscope, and it generates a magnetic field. You can measure the direction of that magnetic field with a superconducting loop and a SQUID—a Superconducting Quantum Interference Device.

This orientation can be measured. The effect should be measurable given the precision with which the physicists made this spacecraft and its contents. I'm really hoping that they will confirm this effect of general relativity soon. If they do, it'll be great. If they don't, I actually believe in relativity so much that I'll actually possibly even think that maybe there was something wrong with the experiment. I know that's not the way you're supposed to think, but I hope I've shown you that there have been so many tests already of relativity, that we really do think that on large scales, it's true; it's correct quantitatively. But it's always possible, and we should keep our minds open to the possibility, that it might be false. Therefore, any experimental new test is welcome with open arms.

Black Holes—Abandon Hope, Ye Who Enter

Lecture 60

Our discussion of general relativity is motivated in part by the existence of neutron stars, very dense stars that form when a massive star collapses. But there exists a phenomenon that is even stranger than a neutron star—a black hole.

A black hole is a region of space where material is compressed to such a high density and the local gravitational field is so strong that nothing—not even light—can escape. If we could shine a light on a black hole, nothing would be reflected; also, no light is emitted from within a black hole. To understand how black holes form, we must recall that a neutron star, held up by neutron degeneracy pressure, can have only a certain maximum mass before it collapses. We don't know the exact limiting mass for a neutron star because we don't yet fully understand the structure of matter at nuclear densities. The limiting figure could be 2 to 3 solar masses. If we consider rotation, a neutron star could be stable at up to about 5 solar masses. Beyond 5 solar masses, barring the existence of some yet unknown form of matter, a typical rotating neutron star would collapse to form a black hole.

The most massive star whose mass has been reliably measured has about 60 solar masses. Some stars may be as large as 100 solar masses; beyond this point, the radiation pressure of a star itself would tear it apart and prevent its formation. Yet the most massive stars have strong winds, so their outer layers essentially evaporate away. These stars (and others) can also lose mass through transfer of matter to companion stars if they are in binary systems. Therefore, it is possible that the most massive stars do not give rise to black holes because much of their mass is easily lost and they end up with relatively small cores that become neutron stars instead. It is also possible that a massive star's core could collapse to form a black hole. Some astronomers believe that stars having initial mass between 20 and 40 solar masses are the most likely to form black holes. Below 20 and above 40 solar masses, a neutron star is more likely to form at the end of a star's life. There is some evidence to support this hypothesis.

Let's look at why an object compressed at a high density would appear black, as well as some characteristics of a black hole. Newton's law of gravity states that $F = GM_1M_2/d^2$, in which F is force, G is Newton's constant of universal gravitation, M_1 and M_2 are the masses of two objects, and d is the distance between them (more precisely, the distance between their centers of mass). From this, we can derive an object's escape velocity—that is, the speed at (or above) which a projectile would have to travel in order to completely escape from the object. If the radius of the object is compressed but its mass remains the same, then the escape velocity increases. That is, the projectile would have to travel even faster to fully escape from the object.

NASA/Goddard Space Flight Center

An artist's rendition of a black hole.

Such an argument was proposed in the late 18th century independently by John Mitchell and Pierre-Simon de Laplace to suggest that there may be objects in the Universe that are so dense that not even light can escape—black holes. This is an example of a Newtonian plausibility argument. The Newtonian argument provides a formula for determining the radius to which an object would have to be compressed in order for its escape velocity to reach the speed of light: $R_S = 2GM/c^2$. The derivation is not rigorously correct, but fortuitously, it agrees with the result from the general theory of relativity. The relativistic calculation was done by the German physicist Karl Schwarzschild in 1916, and this radius is now known as the **Schwarzschild radius** in his honor.

The Schwarzschild radius of the object is directly proportional to its mass. The more massive the object, the larger the minimum radius to which it would have to be compressed in order to become a black hole. For example, the Schwarzschild radius for a 10-solar-mass star is 30 kilometers; thus, if the star were compressed to a radius of 30 kilometers or less, it would form a black hole. What would the Earth's radius have to be in order to form a black hole? Of course, this isn't possible, but for the sake of illustration, Earth would have to be compressed to a radius of about 1 centimeter. A 60-kilogram person would have to be compressed to a radius of 10^{-23} centimeters, 10 orders of magnitude smaller than a proton, in order to become a black hole!

The **event horizon** of a non-rotating black hole is the imaginary spherical surface with a radius equal to the Schwarzschild radius. It is called an event horizon because we cannot see events that occur beyond it, and nothing can escape from within it. Once matter is inside a black hole, gravity still acts on it, and thus, that matter continues to collapse. Theoretically, the matter would reach a point of infinite density, called a *singularity*. However, quantum mechanical effects will surely modify this, which we will discuss later when we talk about string theory. Despite the correct equation for the Schwarzschild radius given by the Newtonian argument, the only way we can truly understand black holes is through general relativity. The Newtonian formula, $F = GM_1M_2/d^2$, obviously breaks down where black holes are concerned because light doesn't have mass. Thus, light is not trapped by the "gravitational force" but, rather, by the extreme curvature of space-time around a dense object.

"It's possible that there are types of stars that are smaller and denser than a classical neutron star, yet not truly a black hole, not smaller than the event horizon."

Recall our example of a ball distorting a rubber sheet, making the sheet (or space) bulge. As the ball (or celestial object) increases in density, the bulge increases in its depth, making it more difficult for light to escape that depth. Indeed, light coming from a strong gravitational field is bent and redshifted. For example, light shining tangent to the surface of a collapsed star having a radius of 1.5 Schwarzschild radii, can be bent so much that it actually goes into orbit around the star, creating a **photon sphere**. If the star were to contract even more, most of the light would be bent so much that it would be absorbed back into the star. Only a small amount of light could escape in a narrow beam, called the *exit cone*, which has a certain opening angle. Once the star was sufficiently contracted (to a radius equal to the Schwarzschild radius), the exit cone's opening angle would shrink to zero and no light could escape.

Far from a black hole, the properties of space are basically normal; the idea that black holes are giant cosmic vacuum cleaners, sucking up everything around them, is a misconception. Once a black hole has reached equilibrium—after the star has fully collapsed—its properties are simple. According to the famous *no hair theorem*, a black hole is described completely by its mass, electric charge, and angular momentum (total spin). The nature of the objects thrown into the black hole is irrelevant.

Some physicists have suggested the possible existence of material on a stellar scale even denser than that of neutron stars, yet that is not an actual black hole. There is some tentative observational evidence for this, but nothing conclusive. ■

Important Terms

event horizon: The boundary of a black hole from within which nothing can escape.

photosphere: The visible surface of the Sun (or another star) from which light escapes into space.

Schwarzschild radius: The radius to which a given mass must be compressed to form a nonrotating black hole. Also, the radius of the event horizon of a nonrotating black hole.

singularity: A mathematical point of zero volume associated with infinite values for physical parameters, such as density.

Suggested Reading

Begelman and Rees, *Gravity's Fatal Attraction: Black Holes in the Universe*.

Ferguson, *Prisons of Light—Black Holes*.

Kaufmann, *Black Holes and Warped Spacetime.*

Pasachoff and Filippenko, *The Cosmos: Astronomy in the New Millennium*, 3rd ed.

Thorne, *Black Holes and Time Warps: Einstein's Outrageous Legacy.*

Will, *Was Einstein Right? Putting General Relativity to the Test.*

Questions to Consider

1. How would the gravitational force at the surface of a star change if the star contracted to 1/5 of its previous diameter without losing any of its mass?

2. Why doesn't the pressure from electrons or neutrons prevent a sufficiently massive star from becoming a black hole?

3. If someone close to (but not inside) a black hole were shining a blue flashlight beam outward, how would the color that you see be affected if you are far from the black hole?

4. The average density of an object is its mass per unit volume. If the volume of a non-rotating black hole is proportional to the cube of its Schwarzschild radius, show that its average density is inversely proportional to the square of its mass. (Of course, all of the mass in a black hole is actually concentrated at the singularity—either the classical or the quantum variety.)

Black Holes—Abandon Hope, Ye Who Enter

Lecture 60—Transcript

My discussion of general relativity had been motivated in part by the existence of neutron stars, very dense stars that form as a result of the collapse of the core of a massive star. I said that near a neutron star, relativistic effects are important because it's so dense, and the gravitational field is so strong. As weird as neutron stars are, it turns out there's something even weirder, even more dense, with even stronger gravity, and that's the black hole. A black hole is a region of space where material is compressed to such a high density, and the local gravitational field is so strong, that nothing—not even light—can escape; hence the name: black hole. If you shine a light on it, nothing gets reflected. No light gets emitted from the black hole. Nothing gets transmitted through the black hole; it's just black.

Here is my award-winning photograph of a black hole. Normally, this goes for $10,000, but for you, I'll sell it at a special price. Just contact me, and you get a special price. Or you can just make one yourself, if you want. Keep the lens cap on your camera if you have a traditional camera, while you're taking the exposure; it'll be black. If you're using a computer software program like I did, just choose a black slide background, and you'll get this prize-winning photograph. Don't contact me; I won't sell it to you.

Anyway, black holes appear very frequently, of course, in pop culture, and especially in science fiction books and movies. They're all over the place, and people use them to travel from one universe to another. I'll discuss that in a few lectures. It doesn't really happen, but I'll discuss why mathematically one might think that it happens. I've seen many cartoons featuring black holes. One of my favorites came on a greeting card. It said something like this card is late. There's a guy standing in a room, and a black hole suddenly appeared for no reason in his room—and his TV set, and his dog, and the couch were all sort of being sucked in by this black hole. That was his excuse for sending the card late. There are all these cartoons and stuff, and I'm sure you've seen them.

Why would a black hole form? I mentioned that a neutron star—held up by neutron degeneracy pressure—can have only a maximum mass. It's kind of like a white dwarf's maximum mass, which is the Chandrasekhar limit. In the case of the white dwarf, it's the mass beyond which electron degeneracy pressure simply can't hold it up anymore. Neutron degeneracy pressure suffers from the same problem. You can't just keep on piling up more and more material and expect these neutrons to keep on supporting this against collapse. The neutrons themselves don't really exert any pressure. It's their configuration in these lowest energy levels of the apartment diagram that I've shown several times now that causes the majority of them to have significant amounts of energy, and hence pressure, keeping the star from collapsing. But if you pile up more and more material, there's only a certain limit to which these neutrons can exert this weird quantum mechanical degeneracy pressure, and the whole thing becomes unstable to collapse.

Exactly what the limiting mass is for a neutron star is not really known. We think it's something like two to three solar masses. Rather general arguments suggest that it's around three solar masses, but we don't know more specifically than that because we don't really know the structure of matter at nuclear densities very well. The best we can measure in laboratories are atomic nuclei, but they're small things, sitting there all alone. What if you had a whole bunch of neutrons, like in a neutron star, 1.4 solar masses of them sitting in a sphere like this, the diameter of a city. That may behave very differently from single atomic nuclei. We just don't know how they behave in great detail.

People use these general arguments, and they say, "Well, they probably can't exist beyond about three solar masses." But if you include rotation, you can actually get a neutron star to be stable, perhaps up to something like five solar masses. Beyond five, nearly all physicists agree that, unless you have some really weird form of matter unlike that of a typical neutron star, the thing will have to collapse to form a black hole, to form a region where the gravity is so strong that nothing, not even light, can escape. You can call this gravity's ultimate victory. Nothing has enough pressure or energy to prevent eventual gravitational collapse if the thing is massive enough.

What kinds of stars ultimately give rise to neutron stars at the end of their lives, and what other kinds of stars give rise to black holes? We're not sure. Clearly, it's the massive stars that give rise to both. I mentioned that we have stars of 10, 20, 30, 40 solar masses initially. It turns out that the most massive star ever found definitively—with a definitively measured mass—is about 60 solar masses. There are some other stars that we think have masses up to about 100 solar masses, and so that's about the limit that we think we've observed. Theoretically, they should not be any more massive than about 100 solar masses because, beyond that mass, the radiation pressure of a star itself tends to tear it apart and prevent its formation. As the material is gravitationally accumulating, if there's too much material, the radiation being emitted by that material prevents more material from being accumulated. That's sort of the limiting mass of a star initially, about 100 solar masses.

People, for a long time, thought that maybe the stars that have 10 to 30 solar masses initially form neutron stars at the end, having a mass of 1.4 solar masses, and the rest gets blown away. Maybe the ones between 30 and 100 solar masses initially are the ones that finally collapse to become black holes. That was the standard line for a long time. But we now know that the most massive stars have strong winds. They generate this pressure, which tends to evaporate away the outer layers. Here's an actual picture of a star whose winds are blowing away the outer layers of the star, in a roughly spherically symmetric way in this case—or in a more bipolar way in the case of Eta Carina, a very massive, bright star in the Southern Hemisphere.

Massive stars lose mass over the course of their lives through winds, and they can also lose mass through accretion, or transfer of matter, onto a companion object. I've illustrated this several times already. There's a number of ways in which stars can lose mass. It is now thought that mass loss through winds occurs best for the most massive stars because their radiation pressure is the strongest. Maybe the most massive stars are the ones that are most able to get rid of the majority of their envelope and end up with a relatively small core that could conceivably become a neutron star rather than a black hole. Maybe it's the less massive, sort of pretty massive, stars that aren't able to get rid of as much material, and end up with a more massive core than the more massive stars. Then that core might collapse to form a black hole.

Some people think that maybe there's a window, between 20 and 40 solar masses initially, where a black hole forms at the end. Below 20 solar masses initially, a neutron star forms at the end; and above 40 solar masses initially, maybe a neutron star forms at the end. That's sort of speculative, but it might be this 20 to 40 solar-mass window where the star ultimately collapses to a black hole. There is some recent, although still controversial, evidence for this. There's a cluster that's been studied called Westerlund 1. In this cluster, the Chandra X-ray Observatory has discovered a neutron star; there it is, right there. Yet in this cluster, there are also some stars that are 40 solar masses. If all the stars formed at the same time, and if there's already a neutron star there, that neutron star came from a star more massive than 40 solar masses. The more massive stars have a shorter evolutionary time scale; they burn up their fuel more quickly because they're more luminous.

If there are 40-solar mass stars observed in this cluster, that means that whatever produced the neutron star was initially more than 40 solar masses, and it produced a neutron star—not a black hole. That's suggestive of this idea that the most massive stars get rid of a lot of their mass and form neutron stars. Maybe not all the stars formed at the same time. How do we know that exactly all of them formed at the same time? It's still controversial, but it's interesting evidence. There's also evidence from Supernova 1987A that stars at about 20 solar masses might form a black hole. The reason that we think that is that Supernova '87A had a core that collapsed and, at least temporarily, formed a neutron star.

We really do know that because this burst of neutrinos was emitted from the hot neutron star that was formed, at least for a while. Even at a distance of 170,000 light years from us, there were so many of these neutrinos that 50 billion of them passed through every square centimeter of your bodies in 1987, and every square centimeter of these large underground tanks of water, through which the neutrinos passed, had 50 billion of them. Yet only a few were detected in each tank—only 10 were detected in each tank. From these numbers, you can calculate how many total there were, knowing the interaction cross-section of a neutrino and all that. You figure out that there were something like 10^{58} neutrinos. Some unbelievable number of neutrinos was emitted by the object inside Supernova 1987A when the core finally collapsed. That just couldn't have been a black hole. You have to have a neutron star. Only a neutron star emits this gargantuan number of neutrinos.

Definitely, a neutron star formed in Supernova 1987A; yet now, when we look at the remains of Supernova 1987A, we see neither a pulsar in the middle, nor any other evidence for a neutron star accreting (or gathering) material from its surroundings. There's just nothing much going on in the center. Maybe the neutron star is there, but it's not pulsing yet. It's not a pulsar that's on. That's conceivable, but you'd still expect some accretion of material by the neutron star from its surroundings, and we just don't see that happening. Maybe some sort of a wind or something from it prevents the accretion. We're just not sure yet. But I, and many other people, think that there's a 50/50 chance that the neutron star was temporarily formed in Supernova 1987A, but subsequently collapsed to form a black hole.

The progenitor star of Supernova 1987A was a 20-solar mass star; so if, at 20 solar masses you get a black hole, and at 15 maybe you get a neutron star, then again, this suggests that there may be this window between 20 and 40 solar masses, which at least for single solitary stars, leads to black holes at the end, rather than a neutron star. If the star is a companion to another star, then there can be transfer of matter. Life gets more complicated when you have a companion sometimes, and that's the case for stars. Life becomes better as well, if you have a companion, and maybe that's the case for stars as well. I'm not a star, so I don't know. In any case, the calculations are more complicated when you have a companion star interacting with it. In the case of '87A, maybe there's a black hole there. We just don't know. We need to do some more research.

So why should an object that's compressed to a high density appear black? Why should no light be emitted from it? You can consider a Newtonian plausibility argument, looking at Newton's law of gravity, F, the force, is equal to G times the mass of one object, times the mass of the other object, divided by the square of the distance between their centers of mass. Suppose M_1 and M_2 are constant, so suppose M_1 is the Earth and M_2 is a ball, and there's a distance between them. There is a certain force acting on the ball. If I were to throw the ball up, the force brings it back down. Now I could, in principle, throw the ball faster and faster and faster, and it would go higher and higher up. If I threw it really fast, had I eaten my Wheaties this morning, I could maybe throw it at a speed at which it would never come back down to the Earth. That's called the escape velocity; the minimum speed that is needed to get an object to escape from—that is, never come back down to—the Earth or whatever object you're tossing it from.

We consider this escape velocity argument, and then we go back to the diagram, and we say suppose we were to squish the Earth down to half of its present radius, keeping its mass the same. Then M_1 and M_2 remain constant in that equation, but the distance went to 1/2 of the previous distance; $(1/2)^2$ is 1/4, and the reciprocal of that is four, so the force is four times greater. It turns out I have to throw the ball faster by a certain amount to get it to escape. The escape velocity increases. If I then go and squish the Earth down even more, keeping its mass the same, the force increases still more, and the escape velocity increases as well. The escape velocity actually increases in proportion to one over the square root of the radius of the massive object—so even though the force is quadrupled, if I made the object half as large, the escape velocity only goes up by something like one over the square root of the distance, so it goes up by the square root of two.

That gets mathematical, but I don't really care about the math here. I just want to illustrate that, with a more and more compressed object, the escape velocity increases, and you might conceive of an object that's been compressed so much that the escape velocity reaches the speed of light. Such an argument, indeed, was used in the late 18th century by John Michell and Pierre Simon de Laplace, to suggest that there may be objects in the Universe that are so dense that nothing can get out, not even light. That would be a black hole. They used sort of a Newtonian plausibility argument.

The Newtonian plausibility argument gives, fortuitously, the right formula for the radius to which you would need to compress an object in order for its escape velocity to reach the speed of light. You have to compress an object to some radius or smaller for its escape velocity to be at the speed of light or bigger. That radius is called the Schwarzschild radius. It's given by 2 times G, times the mass of the object, divided by the square of the speed of light. As I said, the Newtonian argument just gives you the right answer, but to do it right, you have to use general relativity. This was first done by the German physicist Karl Schwarzschild in 1916, shortly after Einstein published the general theory of relativity. This Schwarzschild radius sort of marks the boundary of a black hole. You can't see anything that's beyond that boundary, that's within that radius. Curiously, "Schwarzschild," in German, means "black shield." You can't see beyond this black shield, and I actually doubt that's why Karl Schwarzschild was doing these calculations, but it's funny that it turns out that his name is extremely relevant to what this object ends up being.

The Schwarzschild radius is the radius beyond which, if you compress the star, nothing—not even light—could get out. The formula $R_s = 2GM/c^2$ shows that the radius of this object is proportional to its mass. The more massive the object, the bigger is the minimum radius to which you need to compress it in order for it to become a black hole. You can calculate how big that would be, say, for a 10-solar mass star. It turns out that the Schwarzschild radius is about 30 kilometers. If you took a 10-solar mass star and compressed it down to 30 kilometers or less, as gravity might do, then it would form a black hole. Some black holes are very massive. We'll discuss later black holes in the centers of galaxies. They might have a mass up to 1 billion solar masses. In that case, you'd have a 3 billion-kilometer radius for the Schwarzschild radius. That's comparable to the distance between the Sun and Uranus. The Schwarzschild radii of these super-massive black holes in galaxies are comparable to the size of the Solar System.

You can also go down to small masses and simply ask yourself what would the Earth's radius have to be in order for the Earth to be a black hole? The Earth isn't going to be a black hole, but suppose you had some giant cosmic vise, and you squeezed the Earth. How much would you have to squeeze it to turn it into a black hole? It turns out to a radius of about one centimeter. The Earth's mass is pretty big; it's 6×10^{24} kilograms. You have all this mass, and you have to squeeze it down to one centimeter for it to become a black hole. That's unlikely; in fact, there's no physical process we know of that would do that, but it's fun to play with these numbers. A human is, say, 60 kilograms. To what radius would you have to squeeze a human to make a black hole? It turns out to be a radius of 10^{-23} centimeters. That's 10 orders of magnitude smaller than a proton. Now a proton is yea big, and I exaggerate a lot, so you'd have to squeeze me down, 60 kilograms of me, to 10^{-23} centimeters to get me to be a black hole. You'd need nearly a trillion kilograms to have a black hole the size of a proton, so black holes are really small.

The spherical surface having a radius equal to the Schwarzschild radius is called an event horizon. It's a horizon because you can't see beyond it, and it's an event because, if you were beyond this horizon, you would never get out. It's bye-bye to you; that's a really serious event in your life if you go beyond the event horizon. Once you're inside, you're always inside. You can't get out because nothing—not even light—can get out. You could say, "Abandon hope, all ye who enter here," quoting Dante's *Inferno*. It's kind of like if you go to Oakland Raiders games, you'll notice that there's a zone called the black hole zone. You're sort of drawn into it, and then you can't get out because none of the fans will let you out once you're in the black hole zone. Here they all are, "The Black Hole" and "Atomic Moss." They're kind of funny.

This matter, once it's inside the black hole, can't get out, and gravity is still acting on it, so it keeps on collapsing. Technically, it goes down to a point of infinite density, a point called a singularity. Quantum mechanical effects will surely modify this, and I'll discuss the basics of string theory later on, and I'll tell you a little bit about what physicists are thinking nowadays. It's not that we know for sure what's going on inside a black hole, but it probably doesn't become a point of infinite density. It might become some little package of energy of very small size and not quite mathematically infinite density. But at least according to classical general relativity, that's what happens to stuff inside a black hole.

It's interesting that William Jefferson Clinton, the 42nd president of the United States, was interested in questions like what's inside a black hole. He said in an address on the 21st of January, 2000, "There are so many more questions yet to be answered. And so I wonder, are we alone in the Universe?" We've already discussed that. "What causes gamma-ray bursts?" I'll discuss that later. "What makes up the missing mass of the Universe?" That'll be another topic of my course. And then, "What's in those black holes, anyway?" Maybe the biggest question of all, "How in the wide world can you add $3 billion in market capitalization simply by adding .com to the end of a name?" I'm not going to discuss that in this course. We don't know quite what will happen to material in a black hole, but we can sort of answer Clinton's question. It'll turn into a singularity.

Despite the correct equation given by the Newtonian argument, the only really correct way to treat a black hole and its surroundings is to use general relativity because the gravity is so strong. Really, in relativity, what happens is that light is trapped not by the gravitational force, GM_1M_2/d^2, because light doesn't have any mass, M, so that formula really breaks down. Instead, light is trapped by the extreme curvature of space-time around a dense object. So, we go back to our favorite diagram of a ball distorting a rubber sheet, making it sort of bulge out into a dimension different from the original dimensions of the sheet. If you make the ball denser and denser, that bulge becomes deeper and steeper, and it's harder and harder for light to get out. Indeed, light coming from a strong gravitational field gets gravitationally bent and redshifted. Here's a flashlight emitting a beam of light from the surface of a star that's collapsing. That beam of light gets gravitationally redshifted—I discussed that in the previous lecture—and it also gets bent.

If you collapse the star some more, you can get to a situation where if you point the light straight up, it doesn't get bent because it's going straight up, but it does get gravitationally redshifted. If it goes at an angle, it gets bent. But if you shine it tangent to the surface of the star—that is, in this direction here—it gets bent so much that it actually goes into orbit around the star, and it'll hit you in the back of the head here. This is called the photon sphere. If you're at this point, which happens to be one and a half times the Schwarzschild radius, and you shine a flashlight off of a mirror, or your bald spot, or whatever, it reflects off, goes around, all the way in a circle, and you could see your bald spot in front of you by shining the flashlight off of your bald spot there. Not that that's a very practical application of a mirror, but this is sort of a gravitational mirror. Light can orbit at this photon sphere.

Now let's contract the star even further. It turns out that most of the light starts getting completely absorbed by the star because the gravity bends it so much that it just falls into the star. The only light that's able to get out is the narrow beam of light that comes out through what's called the exit cone, pointing basically upwards, but having a certain opening angle. That's the only light that would actually get out. Once you contract the star enough so that the exit cone's radius squishes down to zero, then no light can escape whatsoever, and you have a black hole. That happens when the star collapses to the Schwarzschild radius or below. At that point, you have an event horizon, surrounded by a sphere of photons that go around in circles at about one and a half times the radius of the event horizon. This would be the structure of a collapsed star.

As the warping gets stronger and stronger, as shown in this diagram, you get these steeper and steeper sides to the well. When the sides are vertical, then you get a black hole, and that's where the event horizon is, as shown in this little picture here. The light gets trapped, either by going around sort of in circles like this (around, and around, and around), or you can think of it another way. You can say that maybe the light is going along this direction, the radial direction, trying to get out. But in so doing, it loses all of its energy due to gravitational redshifting. The gravitational redshift is infinity, and so by the time the light gets out, there's no energy left. In a sense, the light didn't get out. The light gets trapped by a black hole because of this extreme curving of space-time.

Here's a picture of a black hole in the Large Magellanic Cloud, where you can see sort of this representation of the warped space-time, where the space in particular, shown here, becomes so steeply warped that light can't get out because it would be like trying to climb up the sides of a well that's vertical, and there's oil in the well. You just can't do it. Maybe you can do it if you attach some footholds or something, but it's such a steep well that, by the time you get to the top, you've lost all of your energy. If you're light, you cease to exist.

Far from a black hole, the properties of space are basically normal. If the Sun were to turn into a black hole—it won't, but if you had some cosmic vise that did this to the Sun—first of all, it would have a Schwarzschild radius of only three kilometers, so that would be the size of its event horizon. Far from a black hole, you wouldn't notice that the gravity is any different from what it was when the Sun was just the Sun, a gaseous ball. The formula $F = GM_1M_2/d^2$ already tells you that. The distance hasn't changed between you and the center of mass, and the mass of you and the Sun hasn't changed. You've just turned the Sun into a black hole, but you don't care at this big distance. It's not true that black holes are some giant cosmic vacuum cleaners, sucking up everything around them. They're often portrayed that way in the comic strips, but that's not actually what happens. Far from a black hole, the nature of space and time is basically normal, and you don't notice the effects unless you're actually quite close to the black hole, or have atomic clocks, like the ones used in the GPS satellites.

There's another interesting aspect of black holes, and that is that once a black hole from a collapsing star has reached equilibrium—that is, during the collapse, all sorts of weird stuff might be going on. But once, in a sense, you've settled down and formed a black hole that's just sitting there and not experiencing the collapse of a star anymore, that black hole is very easy to describe. It only has mass, charge, and spin (or angular momentum). It has no other details. Physicists say that this is the no-hair theorem—a black hole has no hair—in the sense that there are no details that you need to describe the properties of a black hole. It is completely described by its mass, its charge, and its angular momentum. So, no black holes with hair—very simple objects.

You might say, "Okay, but these black holes are kind of weird. Are there any alternatives to them? Do we really know that they form?" First of all, we think that neutron star material is about as dense as it can get, but some physicists have recently suggested that there could be even denser material, kind of like that of a neutron star, which leads to an object smaller than a normal neutron star, but nevertheless bigger than the event horizon of a black hole. Some physicists have suggested such material, and there's been some observational evidence for it. For example, an object called 3C58 is a neutron star-like thing, but the temperature derived for this neutron star, from Chandra X-ray Observatory data, is lower than you would have anticipated for a classical neutron star.

You can get a lower temperature if the object consists of some sort of weird material, not just a bunch of neutrons. Maybe it consists of things called kaons, and pions, and other weird kinds of particles. They could achieve a denser configuration than a neutron star and avoid being a black hole. This is still controversial, but maybe there are pion and kaon stars. Those are weird particles, and you can take The Teaching Company *Particle Physics* course if you want to learn more about them. More controversial still is an object that was called a quark star some time ago. This is an object that, again, looks like a neutron star, but the data suggest that it's considerably smaller than a normal neutron star, and the astronomers suggested that maybe it consists not of neutrons, but of the constituent particles of neutrons, called quarks.

Here's a bunch of neutrons. Each neutron consists of three quarks, the constituent particles, and they're kept a bit far apart from each other, in a sense because each neutron is a little sphere, a little ball of these quarks. That gives rise to a neutron star having a radius of, say, 10 kilometers. If the quarks, instead of being confined to neutrons, were free, then they could get closer together, and you would have a denser object with a smaller radius. There are much more mundane explanations for the data that explain the data pretty well, so I actually don't think that the quark star hypothesis has been substantiated with experimental data. Nevertheless, it's possible that there are types of stars that are smaller and denser than a classical neutron star, yet not truly a black hole, not smaller than the event horizon. Nevertheless, if you have enough mass, even those objects would become unstable, and at least theoretically should collapse to form a black hole if they have enough mass. In the next lecture, I will show you that, indeed, we have abundant evidence for the existence of black holes in the Universe.

The Quest for Black Holes

Lecture 61

A major prediction of general relativity is the physical possibility of a black hole, a region of space where there's so much material in such a small volume that the space-time curvature is sufficiently strong to trap everything, even light. But if we can't see black holes, how do we know they exist?

We don't see black holes directly, but we can detect them through their gravitational influence on other objects. Recall a binary system, in which two stars orbit their common center of mass. If one star is more massive than the other, that star is closer to the common center of mass than the less-massive star. We can detect the stars' orbits and deduce their masses. If the larger object isn't visible, we can still detect its mass by measuring the wobble in the smaller star's spectrum. Recall that this method is used to detect the presence of extrasolar planets orbiting suns. If we have a wobbling star and the cause of that wobble (the other object) is not visible, we might conclude that the other object is a black hole if its mass exceeds a certain amount. For example, if the object shows no absorption lines in the spectrum, is not visible, and is 10 solar masses (a star this size would be visible), we might conclude that it is a black hole. Other observations would verify whether the object really is a black hole.

The best place to look for black holes is in a spectroscopic binary system. Owing to the number of such systems, how can we narrow down our search to those that might yield black holes? Observations at x-ray wavelengths can provide such a clue. If a black hole or a neutron star is orbiting around another star, it can steal material from the other star. This material would glow as it settled into an accretion disk; the accretion disk's gases come so close to the compact object—either a neutron star or black hole—that they heat up in the strong gravitational field. One such object found with x-ray telescopes is Cygnus X-1, the brightest x-ray source in the constellation Cygnus the Swan. This object appears to be a star orbiting a black hole, creating an accretion disk that glows at x-ray wavelengths. The black hole is at least 7 solar masses, but it could be as large as 16 solar masses. If it were a star, it would be easily visible, but it's not. The problem that arises in this case is that the visible star is giant, possibly as large as 33 solar masses. Using Newton's version of Kepler's third law, 33 solar masses makes the black hole's mass small and uncertain by comparison. In addition, the mass of the giant star itself is uncertain, making the companion's mass even more uncertain. Thus, the evidence for a black hole is less conclusive than one would like.

The best case for determining whether or not an invisible object is a black hole is to find a low-mass star, say a K or an M main-sequence star, orbiting the candidate black hole. In this case, the mass of the orbiting star would be irrelevant. If a low-mass star is orbiting a black hole, then a measurement of the orbital speed (V) and period (P) yields the mass of the black hole. Mass = $V^3P/2\pi G$, and we can measure V and P from spectra. Thus, we can deduce M, the minimum possible mass of the invisible object. This is only a minimum possible mass because we don't know the inclination of the system—that is, whether or not we're observing it edge on or face on. If this minimum mass exceeds 5 solar masses (the maximum for a rapidly rotating neutron star) or even 3 solar masses (the maximum for a non-rotating neutron star), there is a good chance the object is a black hole.

Let's take a closer look at low-mass stars orbiting black holes and see what information our observations can reveal about black holes. Sometimes, the accretion disks that form around the compact object (the black hole or neutron star) develop "blobs" that rapidly fall inward and release tremendous energy. At x-ray wavelengths, this energy appears as a flare, a giant burst similar to a nova. The flaring accretion disk also emits ultraviolet radiation and optical radiation in its cooler regions. The whole disk is hot, but not all parts are equally hot; thus, different parts emit different forms of radiation. X-ray satellites find these x-ray novae and alert optical astronomers. We then wait for the radiation to fade, and eventually, we might see the faint normal star in the system. We can then take spectra of the star. A sophisticated analysis of the spectra over time would show the wobbling of absorption lines, indicating that the star is orbiting around something that is gravitationally tugging on it. From this, we can deduce the minimum mass of the invisible object.

How do we get an idea of the system's inclination—whether we are viewing it edge on, face on, or somewhere in between? We look at the brightness of the visible star over time, its light curve. If the brightness varies over time (as a result of how we observe the tidally distorted star), then we are viewing the system inclined to our line of sight. If the light curve is flat—that is, no change in brightness—then we are viewing the star system face on. In the past decade, about 20 such binary systems—wherein a star orbits a probable black hole—have been found and accurately measured.

Some black holes spin (rotate), and gas particles in an accretion disk from the orbiting star can get closer to a spinning black hole (closer than 3 Schwarzschild radii) than they could to a non-rotating black hole. We know this by looking at the shape of the x-ray spectrum. A rotating black hole with a spinning accretion disk can also show high-speed jets of particles emitted along the rotation axis of the black hole and the disk. These jets can reach speeds of more than 90% of the speed of light. For a black hole with an orbiting star, we notice that the accretion disk isn't bright in the central region during times of *quiescence*—that is, prior to an outburst, when it is faint. If the star were orbiting a neutron star instead of a black hole, the accreting material would glow in the center, where it hits the surface of the neutron star and releases its energy of motion. If the center is actually a black hole, it is likely that the accretion material is being swallowed by the black hole; hence, there is no hard surface to hit, and the material doesn't glow as much. Though still somewhat tentative, this observation suggests that we really have seen evidence for material going beyond the event horizon and being swallowed by a black hole. ■

Suggested Reading

Begelman and Rees, *Gravity's Fatal Attraction: Black Holes in the Universe*.

Ferguson, *Prisons of Light—Black Holes*.

Kaufmann, *Black Holes and Warped Spacetime*.

———, *The Cosmic Frontiers of General Relativity*.

Pasachoff and Filippenko, *The Cosmos: Astronomy in the New Millennium*, 3rd ed.

Thorne, *Black Holes and Time Warps: Einstein's Outrageous Legacy*.

Questions to Consider

1. Under what circumstances does the presence of an x-ray source associated with a spectroscopic binary suggest to astronomers the presence of a black hole?

2. If a visible star orbits an object at least five times the mass of the Sun in a period of only 8 hours, can you think of anything the object could be besides a black hole?

3. Based on what you've learned previously in this course, is it possible to detect a black hole that is not part of a binary system or is not surrounded by stars and gas in a galaxy? If so, how?

4. Given that objects can fall into a black hole, what do we mean when we say that the interior of a black hole is cut off from the Universe?

The Quest for Black Holes
Lecture 61—Transcript

In the past few lectures, I've been discussing Einstein's general theory of relativity, which tells us more accurately than does Newtonian gravity what happens in the vicinity of massive or energetic objects. It's a weird theory. There are a lot of counterintuitive aspects to it. I only had a chance to describe some of its fundamentals. If you want to learn more, there's a Teaching Company course on general relativity, and indeed the quantum revolution, which you can view at your pleasure and learn many more details of all these weird theories of modern 20th-century physics. It's fantastic stuff, and what, to me, is amazing is that not only have these various theories been substantially confirmed from experimental measurements, but also that they do have an effect on our everyday lives, especially quantum mechanics and special relativity, but to some degree, even general relativity.

A major prediction of general relativity is the physical possibility of a black hole, a region of space where there's so much material in such a small volume that the space-time curvature is sufficiently strong to trap everything, even light. Light can't get out; the thing appears black, and then it's a black hole. That's a theoretical possibility, but does nature actually choose to produce such weird objects? That is, are black holes science fact or fiction? Just because the equations say that they're physically possible doesn't necessarily mean that nature chooses to produce such objects. We'd like to find out whether there really are black holes. Light can't escape from a black hole, so you can't really see one.

You might say, "Well, just look up into the sky for dark blobs, and those are black holes," but you can't do that. You can't just say there's a dark spot in the sky; that's a black hole. That could just be a space between stars, where there aren't any stars or galaxies, and that's not a black hole. There are cartoons that illustrate this. Some astronomers are looking at a photograph, and there's sort of a circular region that looks dark. They say, "Wow, it's black, and it looks like a hole, I'd say it's a Black Hole." It's not that easy. You have to have much better evidence before the scientific establishment accepts your proposition that you found a black hole.

We don't see black holes directly, but we can detect them through their gravitational influence on other objects. Here, for example, to remind you, is a binary star system where the two stars orbit their common center of mass. If one of them is more massive than the other—in this case, by a factor of 3.6—then it's closer to their common center of mass than the less-massive object is. Suppose you had a black hole as the more massive object, and a regular star orbiting it. You could detect the orbit of the small star around the center of mass, and deduce from the amount of this orbit—that is, from its period and the speed—the mass of the object that's pulling on it. This is very reminiscent of what we've already talked about when discussing exoplanets. The wobble in the star's spectrum allows us to deduce the presence of an extrasolar planet orbiting that star.

You look for what's called spectroscopic binaries, stars where the spectrum shows a shift in the absorption lines. If you have a spectroscopic binary, and you measure the orbital speed and period of one of the stars—let's say only one star is visible—and you deduce that the other star has some minimum mass that's above three or five solar masses, yet is not visible, but should be visible if it were any kind of a normal star, you might deduce that, "Gosh, it must be a black hole," by the process of elimination. That's the idea. But let's go back and review some of these concepts.

Here is a binary star that you're looking at with a telescope, and you might only see it as one star, even through the telescope. But a series of spectra shows the absorption lines moving back and forth in a periodic way. One set of absorption lines is associated with one star; the other set of absorption lines is associated with the other star. If you increase the orbital speed, you can see that these absorption lines wobble back and forth more than in the case of the smaller speeds. Suppose you have a system like this, but you only see one set of absorption lines. Let's say the one that's going back, forth, back, forth, back, forth, like that. You don't see the other set of absorption lines, so you're only seeing one of the stars. You would deduce that the other one is too faint to see. You can't see it, so you can't see its absorption lines.

From the orbital period and the amount of the shift, you can deduce what its mass must be in order to produce the observed motion of the stars whose absorption lines are visible. If you measure the star whose absorption lines are visible—let's say it's a G-type main-sequence star like the Sun—then you know its mass; it's one solar mass. Maybe it's a B-type star, having 10 solar masses or something like that. In other words, if you measure the mass of a visible star, you have some information on the total mass of the system, contributed by that one star. The measurement of the wobble back and forth gives you another set of information about the total mass of the system. Recall Kepler's third law, written in Newton's form. It said that the sum of the masses, M_1 plus M_2, times the square of the orbital period, is equal to $4\pi^2/G$ times the semi-major $axis^3$.

If you make measurements of the one set of lines in the spectrum that you see—that's called a single-lined spectroscopic binary because you only see one set of lines—and use Kepler's law, you get some information about the sum of the masses. But if you also know something about one of the stars, the visible stars—because it's a G star, or a B star, or whatever—you know how much that one visible star contributes to the sum of the masses, and so you can get some idea of the mass of the invisible object. That's the general principle. I'm just trying to give you briefly what astronomers do. It's somewhat more complicated than that, but this is the essence of what we do. The properties of a visible star can tell you something about how much it's contributing to the total mass of the system.

Maybe you'll find that the visible star is such a pipsqueak that it contributes so little to the total mass of the system that, to a good first approximation, the entire mass is the mass of the black hole. Arguments like this can give you some idea of the mass of the invisible object; that's what I'm trying to tell you. If that mass exceeds a certain limiting value, yet the object is not visible—suppose it's 10 solar masses. If it were a 10-solar mass normal star, it would be easily visible, but it's not. That kind of argument, the process of elimination, can lead you to deduce that it's probably a black hole. That's a good black hole candidate. Then you want to do other observations that verify whether it really is a black hole or not.

You want to look for spectroscopic binaries that might yield the presence of a black hole, but where do you look? There are lots and lots of stars in the sky. You can't just take spectra night after night of all those stars; it would be prohibitively time consuming. We need a clue as to which star would be the preferable one to look at, that yields the biggest chances of finding a black hole. Observations at x-ray wavelengths can provide such a clue. If a black hole, or a neutron star, is orbiting around another star, it can steal material from that other star. As in this animation, that material can glow because it settles into an accretion disk, and the accretion disk's gas gets so close to the compact object—either a neutron star or black hole—that it heats up a tremendous amount in the strong gravitational field around that neutron star or black hole.

All this material is spiraling in, hitting all the other material, and there's a lot of friction. It heats up a lot, and it can glow at x-ray wavelengths, not just at optical or ultraviolet wavelengths, as would have been the case for an accretion disk around a lower gravitational field star, like a white dwarf or a normal star. They have accretion disks that glow at visible or ultraviolet wavelengths. But the strong gravity of a neutron star or a black hole creates accretion disks that glow at x-ray wavelengths. In fact, this heart of the accretion disk is the black hole, which itself is black, but the disk around it is glowing brightly at x-ray wavelengths, and at other wavelengths. The picture that I showed reminds me, every time I see it, of Joseph Conrad's masterpiece *Heart of Darkness*, written in 1902. The heart of the system is the black hole; it's this dark object. But surrounding it is gas that's glowing a lot and can be easily seen.

In particular, at x-ray wavelengths, there's this distinctive signature of a strong gravitational field. You look at the sky with x-ray satellites like the Chandra X-ray Observatory, pictured here. There have been many other x-ray satellites. Indeed, Riccardo Giacconi is an astronomer who, in the 1960s and 1970s, developed x-ray satellites, and he was sort of the pioneer in x-ray astronomy. The development of these x-ray detectors and satellites led to the discovery of all sorts of exotic objects. He actually was awarded half of the 2002 Nobel Prize in Physics for his development of x-ray astronomy and his studies in x-ray astronomy.

One of the most interesting early objects to be found with x-ray telescopes was an object called Cygnus X-1, the brightest x-ray source in the constellation Cygnus the Swan. It appears to be a star orbiting a black hole, creating an accretion disk that glows at x-ray wavelengths. That black hole has a mass of at least seven solar masses, but more likely something like 16 solar masses. That's way above the neutron star limiting mass, be it three or five solar masses; white dwarfs can only reach 1.4 solar masses. A normal star having a mass of seven to 16 solar masses would be glowing brightly, and yet we don't see it. This observational evidence for a black hole orbiting around the visible star in the Cygnus X-1 system was hailed as the first observational evidence for black holes in our galaxy. The data were gathered and published in 1971 by Tom Bolton.

The conclusions for this particular case depend a lot on what you assume for the mass of the visible star. The problem is that the visible star is a giant massive star, having something like 33 solar masses. If you put 33 solar masses for, say, M_2 in Kepler's third law, and you put in the orbital period and the distance between them and all that, you get a value for M_1, the mass of the putative black hole, but that value of M_1 is uncertain because the value of M_2 being so large, 33 solar masses, is itself also uncertain. We're not sure that that star has a mass of 33 solar masses. Maybe because of its interaction with a companion, its structure has changed, and maybe it's only a 15-solar mass star. Maybe it's a 50-solar mass star, in which case the companion wouldn't need to be very massive and might not be a black hole. The problem with Cygnus X-1 is that it doesn't provide a definitive case. Most of us think it really is a black hole, but you can weasel your way out of it and say it's something else. The visible star has such a high mass, and such a large uncertainty associated with that mass, that it kind of messes up your conclusions about the invisible object and makes them less certain.

What we'd really like to have is a very low-mass star, say a K or an M main-sequence star, orbiting a black hole. In that case, the mass of the star is irrelevant. Let's go back to Kepler's third law, as written by Newton; M_1 plus M_2, times the orbital period^2, is $4\pi^2/G$ times the semi-major axis^3. If M_2, the mass of the visible star, is far, far less than the mass of the object that it is orbiting, the putative black hole—that is, if M_1 is much, much greater than M_2—then you can essentially ignore M_2 in this equation. It doesn't really matter whether it's half a solar mass, a third of a solar mass, or one solar mass. It doesn't make much difference. Yes, it makes some difference in detail, but it doesn't make much difference. Let's just ignore it all together. In that case, the equation simplifies. It becomes $M_1P^2 = 4\pi^2/Gr^3$. For a circular orbit, the circumference, 2π times the radius of the orbit, is equal to the time it takes to complete that orbit, the period, multiplied by the velocity, V, so distance = rate x time. If you simplify this and solve for r, you get r is $VP/2\pi$. Plug that into your other equation, and you get that $M_1 = V^3P/2\pi G$.

If you don't care about the math, don't worry about it. I'm just doing this for those who want to see more of the details. The point is the following. If a low-mass star is orbiting a black hole candidate, then a measurement of the orbital speed, V, and the period, P, gives you the mass of the thing around which the star is orbiting. That mass is just $V^3P/2\pi G$. You can measure V and P from spectra. Plug it into some equation, this equation, and deduce M_1, the mass of the invisible object. What you really get is a minimum mass. You don't know the inclination of the system. Is it edge-on to your line of sight; is it face-on, or nearly face-on?

This ambiguity is the same one that affects the determination of masses of exoplanets. If you don't know the inclination of the system, you don't know what the correction factor is. But what you do know is that the measured gravitational influence gives you a minimum possible mass for the unseen object. If that minimum possible mass exceeds, say, five solar masses—the maximum for a rotating neutron star—you've got a good bet that this is a black hole. Even if it exceeds three solar masses—the maximum for a non-rotating neutron star—you've got a pretty good bet that this is a black hole, although it might be a neutron star.

How do you find good candidates for low-mass stars orbiting black holes? Again, you look in x-rays. Sometimes these accretion disks that form around the compact object develop blobs, and there's a rather rapid accretion event where a blob moves into the accretion disk, traverses a big distance, and releases a large amount of energy in a short time. You get a flare at x-ray wavelengths, and the x-ray telescopes notice those bursts; they're called x-ray novae, like a nova, but at x-ray wavelengths. Of course, the accretion disk gets hot all around, and so it emits not just x-rays in the hottest parts, but ultraviolet radiation in the cooler parts, and optical radiation in the still cooler parts. The whole disk is hot, but not all parts are equally hot, so different parts emit different forms of radiation. The whole thing brightens at all wavelengths, not just x-ray wavelengths, but you notice it at first at x-ray wavelengths because there's this accretion disk, and a blob formed.

X-ray satellites find these x-ray novae and alert optical astronomers, such as myself, to their existence. We then wait for all the radiation to fade. It gradually fades with time after the accretion event, and eventually, we might see the faint normal star that's in that system. My group did this in 1995, the first of many such objects we studied. It was an object called GS2000+25—that's just its phone number or something—also known as QZ Vulpeculae. It was discovered by a Japanese satellite called the Ginga satellite, and it became bright at all wavelengths. Then it faded with time, and eventually a picture showed this faint star at the nominal position of the x-ray source. I went to the Keck Observatory with my students and measured the spectrum of this faint star, and you can see the following. There's hydrogen emission—that comes from whatever gas still remains in the accretion disk—but most of the light is dominated by the light from the faint star.

You can see that it has absorption lines in it. These are similar to the absorption lines in normal stars, an example of which is shown below the spectrum of GS2000+25. There are these absorption lines, and if you take a series of several spectra, or many spectra over the course of the night, these particular spectra look noisy, but nevertheless, the information is there. A sophisticated analysis of the spectra shows that the lines are wobbling back and forth in wavelength over the course of the night. Indeed, here is the radial velocity curve that we derived for this object over the course of one night. Initially, the visible star was moving away from us at 520 kilometers per second. Four hours later, it was moving toward us at 520 kilometers per second. It then started zooming away from us, reaching a maximum speed of 520 kilometers per second, and so on. This is a sinusoidal pattern, indicative of a circular orbit of that faint star around something that's tugging on it, something that gravitationally pulls on it and allows it to be in this orbit.

When you put in the speed, 520 kilometers per second, and the orbital period, roughly 8.3 hours, into Kepler's third law, you find out that the minimum mass of the invisible object is five solar masses. Any normal five-solar mass star would be easily visible. A white dwarf can't be that massive; even neutron stars probably can't be that massive unless they're a very special kind of neutron star. Therefore, this was a good case for a black hole. Indeed, there was a headline in the *San Francisco Chronicle*, "UC Astronomers Observe Black Hole in the Milky Way." The subtitle was, "It may be 14 times as dense as the sun." They didn't mean 14 times as dense; they meant 14 times as massive. We came up with the number 14 after taking into account the probable inclination of the system. The minimum of five solar masses turns out to be more like 14. Subsequent better analysis of those data, and additional data that we gained, suggests that the black hole has a mass of eight to nine solar masses, so we revised our number after this article came out.

This velocity curve gives the minimum mass, but if we want to get some idea of the inclination, we can look at the brightness versus time, or the light curve, of the visible star. If you look at brightness versus time, in this particular case, we saw that the visible star got somewhat brighter and then fainter, and then brighter and fainter. This can be understood to be the result of the star presenting a different shape, as seen by us, during various parts of its orbit. Here's the black hole that it's orbiting. When it's in this configuration, the star looks basically circular. But when it has moved over to this part of its orbit, the star looks more like a pear, or a teardrop. Remember, the black hole is stealing material away from the star, causing it to have this deformed shape. Since it looks bigger in this position than in that position, it looks brighter here than there. That causes this variation in observed brightness.

Notice, if the star had been orbiting perpendicular to our line of sight, its shape would have been constant, independent of the position in its orbit—so, the light curve would have been flat. Clearly, between these two extremes, either edge-on view or a face-on view, there are many possibilities. The amplitude (or height) of the light curve variation, of the brightness variation, tells us something about the inclination. You can also learn about the relative masses of the two stars, the visible star and the invisible star, by noting that it's not just the visible star that orbits the invisible star; they both orbit their common center of mass. It's just that the more massive thing doesn't move very much, but it moves a little bit.

This motion can be detected because the accretion disk produces an emission line, as you can see in these spectra. In a series of spectra taken over the course of the night, the emission line moved back and forth, almost imperceptibly. You have to do a rigorous analysis of the data to show that it's moving around, but the data do show that the thing is moving around. That gives you the relative masses of the putative black hole and the star. Again, if you know something about the mass of the visible star, and you know the relative masses, then you can get information about the mass of the invisible object. When you do all this kind of stuff, the point is that you can come to a pretty good estimate of the mass of the invisible object. In this particular case, we found that the inclination is 65 degrees, and the invisible object has something like eight to nine solar masses. We did this work in 1995 and in the subsequent few years.

Since then, there have been about 20 such systems found and measured with very well measured masses for the invisible object. Here's a diagram showing schematically a bunch of these systems. Here's Cygnus X-1, the first one. You can see the bloated star, distorted into sort of a teardrop shape by the black hole that it's orbiting. Then there are other ones here as well, where there's the star, and the black hole and its accretion disk. For comparison, you can see the size of Mercury's orbit around our Sun. Mercury orbits our Sun at about 4/10 of an astronomical unit. You can see that a bunch of these x-ray binary systems that are thought to have black holes in them are considerably smaller than 4/10 of an astronomical unit—so that's really kind of cool.

It turns out that we can even tell that some of these black holes are probably spinning because material, gas in the accretion disk, can get closer to a spinning black hole than to a non-spinning black hole. It turns out if you're a gas particle near a non-spinning black hole, you can get down to three Schwarzschild radii, and you can have a stable orbit there. But if you get closer in than three Schwarzschild radii, you rapidly get sucked into the black hole. It just eats you up; it's like this cosmic vacuum cleaner that I said the Sun would not be if it were a black hole. If things are close enough to the black hole, closer than three Schwarzschild radii, they get sucked in.

For a spinning black hole, it turns out that material can spend a substantial amount of time closer in than three Schwarzschild radii, but not yet sucked into the black hole. You can spend more time close to a spinning black hole than to a non-spinning black hole. Here's a representation of that. Here's a non-rotating black hole, and there's all this stuff orbiting it. The last stable orbit is at three Schwarzschild radii. There's some amount of gravitational redshifting of the light seen from particles of gas at that last stable orbit. That gravitational redshifting can be measured from the spectrum, in a way that I won't go into because it's technically complicated. If you have a spinning black hole, material can get closer in. This is actually related to the frame-dragging effect that I discussed in the previous lecture. It can get closer in, and the gravitational redshift suffered by photons escaping from these inner orbits is greater than the gravitational redshift from gas farther away. That can be discerned from the x-ray spectrum of the star.

Here, you have the x-ray flux, or brightness at x-ray wavelengths, versus energy in some units, thousands of electron volts. The point is that the shape of this x-ray spectrum differs from that around a non-spinning black hole, and thus you can discern that this particular black hole is probably spinning. Now let's consider the formation of relativistic jets in the vicinity of a black hole. Here's an animation of a spinning black hole with an accretion disk. What's illustrated is that, if you have a spinning black hole with a spinning accretion disk around it, high-speed jets of particles can go zooming along the rotation axis of the black hole and of the disk. You get these relativistic jets of particles moving at a substantial fraction of the speed of light.

Many newspaper articles say that the jet is somehow emerging from within the black hole and zooming out at speeds close to the speed of light. That's not true. Nothing is coming out of the black hole, at least not classically. I'll talk about black hole evaporation in a future lecture, but that's not what's going on here. Here, the jet is formed outside of the black hole, outside of the event horizon—in the vicinity of the black hole, to be sure, where the gravity is strong and things can get accelerated to high speeds, but not from within the event horizon. Just be sure that that's clear when you read these articles that sometimes say that a jet has been seen emerging from a black hole—what they mean is from the vicinity of a black hole.

Here's an excellent example of one of these objects. It's an object called SS433 that's been known for a long time. For many years, astronomers measured the speed of this jet of particles coming from some compact region in the middle of an accretion disk. In this case, the speed is a quarter of the speed of light. That's how fast these things are coming out. Some other jets show speeds of 90% of the speed of light; it's amazing. For a long time, we didn't know whether the source of this jet was gravitational energy near a neutron star or a black hole. Many people thought it was a neutron star, but recent evidence suggests that there's a 16-solar mass black hole in the middle of this SS433 object, and that's really cool. Mind you, we don't yet understand how the particles get accelerated to incredible speeds in a jet, but we do seem to see the presence of jets near compact black holes.

Finally, the evidence that's very interesting and recent is that, if you look at an object that is thought to be a black hole in a binary system, you find that the accretion disk isn't very bright in the very central region. What we think is happening is that material in the accretion disk is being swallowed beyond the event horizon in these cases, but is hitting the hard surface of the neutron star in the cases where there's a neutron star rather than a black hole.

Basically, what's going on is, if material is being accreted onto a neutron star, it hits a hard surface and glows because it gives up its gravitational energy and starts glowing. If, instead, it gets swallowed by a black hole, it never hits the hard surface and doesn't give up its energy of motion and produce light, so it appears dimmer. In quiescence—that is, outside of an x-ray outburst, when there are no clumps in the accretion disk—the objects that have black holes in the center are dimmer than those that are thought to have neutron stars in the center, suggesting that we really have seen evidence for material going beyond the event horizon and being swallowed by a black hole.

Imagining the Journey to a Black Hole

Lecture 62

"What you, yourself, would see and experience if you were to go into a black hole, and what an outside observer would see as you are going toward the black hole and into it ... the two perspectives yield very, very different results."

What's a black hole really like, and how could we simulate the experience of going toward and into one? Recall that a non-rotating black hole has a spherical event horizon with a radius known as the Schwarzschild radius ($2GM/c^2$). Within the event horizon, the black hole has a tight grip on everything, including light. An outside observer can't see any events that occur at or inside the event horizon. If an object is outside this radius, on the other hand, it can, in principle, escape from the vicinity of the black hole. A massive black hole has a stronger gravitational field over a larger area than a small black hole. The event horizon has a radius that is directly proportional to the mass; thus, a massive black hole has a bigger event horizon than a low-mass black hole. Intuitively, larger black holes seem more dangerous to approach than smaller ones. However, we could theoretically get closer to a supermassive black hole than a smaller one before being torn apart.

What would happen to us if we got close to a black hole? If we fell feet first, we would be stretched along the length of our bodies and squeezed along the width by the hole's tidal forces. The stretching (informally known as *spaghettification*) occurs because the gravitational pull on our feet (closest to the black hole) significantly exceeds that on our heads. The tidal effects on a human body (say, 2 meters in height) would be smaller for a supermassive black hole than for a stellar-mass black hole because the length of a human is smaller relative to the hole's Schwarzschild radius. Outside a stellar-mass black hole, the tidal spaghettification is so extreme that it would tear a human apart.

What would we see if we could safely approach a black hole? We would see a highly distorted sky because of the strong curvature of space-time around us. Starlight traveling in one direction is bent toward us and enters our eyes such that the star appears to be in a completely different direction. The degree to which this would happen depends on the exact path of the light. Light that passes closer to the black hole is bent more than light that passes farther away. Indeed, we might see multiple images of the same stars. Assuming that we were traveling at a certain speed toward a black hole, the stars would begin to appear to move away from the black hole. They would also get brighter, and eventually, we would see secondary images of the stars.

If we circled the black hole from a distance of 10 Schwarzschild radii, we would see a distorted view of stars whirling around the hole, reminiscent of the Einstein rings we talked about in a previous lecture. From the photon sphere at 1.5 Schwarzschild radii, the sky would be behind us, and we would see nothing but black as we peered inside the hole. If we looked up from the photon sphere itself, while stationary, we would see the stars, but they wouldn't be circling. If we circled while at the photon sphere, we would see a wild view of millions of stars whizzing past and all around, with dramatic distortions of space. Continuing farther down, all we would see as we looked up would be a point of light representing the whole sky of stars squashed into one tiny spot. As we got closer and closer to the event horizon, that circular patch would get smaller and smaller. Beyond the event horizon, no one really knows what we would see. But we might see much of the future of the Universe squeezed into a tiny amount of time, before we hit the singularity.

"You can get closer to a supermassive big black hole without being torn apart than you can to a small black hole."

What would we experience while falling into the black hole? The closer we came to the event horizon, the more time would slow down as seen by an outside observer. In a strong gravitational field, clocks run more slowly than in a weak gravitational field. The closer we get to the event horizon, the greater is this apparent slowing down of time, called *time dilation*. At the event horizon, time would stop as seen from an outside perspective. Nothing appears to actually reach the event horizon, because it would take infinite time to do so. In fact, anything that has ever fallen into a black hole, from the outside perspective, is actually hovering infinitesimally above this imaginary surface we call the event horizon. Indeed, there's a paradigm for studying black holes, called the *membrane paradigm*, in which nothing ever falls in. But from the perspective of the person falling in, he or she really would cross the event horizon and crash into the singularity in a finite amount of time. From the outside perspective, a collapsing star would reach some minimum size—essentially the size of the event horizon—at which point, it would not get any smaller.

If we could get close to a black hole, then escape, our clocks would have registered a much shorter time period than those clocks outside the black hole. In this way, we could theoretically travel to the future without having aged much. From the outside perspective, the light emitted from a flashlight outside the event horizon would experience **gravitational redshifting**. As the photons escape the gravitational field, they lose energy and exhibit progressively longer wavelengths. The emitted light would also fade because photons appear to be emitted at slower and slower rates as gravity increases (time dilation). Similarly, a star collapsing to form a black hole fades because of gravitational redshifting, time dilation, and another effect in which less and less of the star's light is visible as the exit cone shrinks. Finally, someone falling into a black hole would quickly be torn apart by gravity, but not before this traveler noticed the surrounding stars as being blue—the opposite effect of redshifting caused by light coming into a black hole rather than trying to escape from it.

The previous effects all apply to non-rotating black holes, and most of them apply to rotating black holes as well, with some modifications. Let's now look at rotating black holes, whose structures are more complex. For a given mass, the event horizon of a rotating black hole is smaller than that of a non-rotating black hole, but it remains spherical. A region called the *ergosphere* surrounds the event horizon, where space is inexorably dragged around the black hole. It is impossible to remain stationary relative to the outside world if one is in the ergosphere. As the rotation of a black hole increases, its event horizon shrinks until the black hole reaches a maximum rotation speed; at this point, the event horizon has a radius equal to 0.5 of the normal Schwarzschild radius.

In fact, we will see in the next lecture that two event horizons form, one inside the other. At the maximum rotation rate, the two horizons meet and vanish, giving rise to a *naked singularity*, or a singularity that is not cloaked by an event horizon. Because there is no indication of the event horizon, we wouldn't know we were approaching a naked singularity until we hit it. Some theorists think that naked singularities cannot occur because they believe that a black hole can never reach its maximum rotation rate. This has never been completely theoretically proven. ■

Important Term

gravitational redshift: A redshift of light caused by the presence of mass.

Suggested Reading

Begelman and Rees, *Gravity's Fatal Attraction: Black Holes in the Universe.*

Ferguson, *Prisons of Light—Black Holes.*

Kaufmann, *Black Holes and Warped Spacetime.*

Nemiroff, *Virtual Trips to Black Holes and Neutron Stars*, antwrp.gsfc.nasa.gov/ htmltest/rjn_bht.html.

Pasachoff and Filippenko, *The Cosmos: Astronomy in the New Millennium*, 3rd ed.

Pickover, *Black Holes: A Traveler's Guide.*

Thorne, *Black Holes and Time Warps: Einstein's Outrageous Legacy.*

Questions to Consider

1. As measured by distant observers, nothing ever enters a black hole because time slows down near the event horizon. Where does the material go?

2. If you were an astronaut in space, could you escape (a) from within the photon sphere of a non-rotating black hole, (b) from within the ergosphere of a rotating black hole, or (c) from within the event horizon of any black hole? Explain each of these cases.

3. Explain why the tidal stretching (spaghettification) is smaller near a supermassive black hole than near a stellar-mass black hole.

Imagining the Journey to a Black Hole

Lecture 62—Transcript

We've seen that both the theoretical and observational evidence for stellar-mass black holes appears to be quite convincing. We've got these binary systems where one star is tugging on another star. The star that's doing the tugging is dark, and yet very massive, and so it seems like it must be a black hole. In a future lecture, I'll show you that there's really compelling observational evidence for the existence of supermassive black holes in the centers of galaxies. That evidence is even stronger than the evidence for the existence of stellar-mass black holes. Still, it's hard to get an intuitive feeling for what it would be like to go into a black hole and actually watch what you're seeing. What's a black hole really like? How do we experience one? How do we get a gut feeling for one?

You can do that with modern computer programs that trace the light rays as they go through bent or curved space-time, the highly distorted space-time around and in a black hole. I'll show you in this lecture some of what you see when you run these computer simulations. I'll also explore what you, yourself, would see and experience if you were to go into a black hole, and what an outside observer would see as you are going toward the black hole and into it. We will find that the two perspectives yield very, very different results. The outside observer sees completely different phenomena than those that you see. It's simplest to consider, at least initially, a non-rotating black hole. A lot of what I say can be applied to rotating black holes as well, with some modifications, but it's just simpler to consider a spherical, non-rotating black hole. I'll have more to say about rotating black holes later in this lecture, and in another lecture.

Let me remind you that, for a non-rotating black hole, there is a radius known as the Schwarzschild radius, within which the black hole has a tight grip on everything, including light. Not even light can get out. If an object is outside the Schwarzschild radius, it can, in principle, escape from the vicinity of the black hole. But if it's within the Schwarzschild radius, given by $2GM/c^2$, where M is the mass of the black hole, then you can't get out. The spherical surface, or the spherical region, at that radius is called the event horizon because an outside observer can't see any events that occur within that distance. It's a horizon; you can't see events beyond it. As I joked last time, it would be a real event in your life if you were to go past the Schwarzschild radius, past the event horizon, because you wouldn't be able to get out. That wouldn't be so great.

A massive black hole has a stronger gravitational field over a larger area than a small black hole. The event horizon has a radius that is directly proportional to the mass, so a big black hole has a bigger event horizon than a little black hole. You might think that the big black holes are more dangerous. That's true in the sense that they occupy a bigger volume. On the other hand, it turns out you can get closer to a supermassive big black hole without being torn apart than you can to a small black hole. Near a small black hole, the tidal forces, or the forces trying to stretch you apart, are much greater than those near a big black hole. In that sense, a big black hole is the safer bet because you can get closer to it and observe it in more detail without being torn apart.

Let me review tidal forces, particularly in this context. Suppose we have a black hole down here. Here you are, rather large compared to this black hole. You're coming down, and the force of the black hole exerted on your feet is denoted by this large, downward-pointing arrow. The force of the black hole on the top of your torso, for example, is denoted by this smaller arrow. Because your torso and head are farther from the black hole than your feet are, the force on your head and torso is smaller than the force on your feet. Relative to the center of your body, the feet are being stretched away. You get stretched like this, and indeed, the ends of your arms get pulled toward the black hole like this, along radial directions, and so, too, do other parts of your body get pulled toward the center of the black hole along radial directions.

Not only do you get stretched, but you get squeezed as well. You're being squeezed down, in like this, and stretched along the length of your body. As you get closer and closer to the black hole, this tidal stretching and squeezing become greater and greater, and you eventually become elongated into a big, long structure. We call this process, informally, "spaghettification" because you sort of become a strand of human spaghetti. In fact, your body can't withstand such forces. It actually snaps apart—so you'd snap in two. Then each of those halves would snap in two, and each of those halves would snap in two. You'd just go snap, snap, snap, snap, snap, and you'd be torn apart; the spaghetti wouldn't last very long. If you want to go and explore the environment of a black hole, go find a supermassive one like those that exist at the centers of many galaxies, as I'll discuss in a few lectures. Don't go toward a stellar-mass black hole because those are the ones that will tear you apart.

Suppose you do find a supermassive one, toward which you can safely go, at least for a while, without being torn apart. Let's ask ourselves what you would see if you were to look at the sky. You would see a very highly distorted sky because of the strong curvature of space-time around you. Starlight that's going in one direction might normally have missed your eyes, but it gets bent toward you and enters your eyes, and so the star appears to be in some completely different direction. This is happening all around you, and the degree to which it's happening depends on the exact path of the light. The light that passes closer to the black hole is bent a lot more than the light that passes farther away, so you get this highly distorted image. Indeed, you can get multiple images of the same stars.

Let's take a look at some simulations of what you would see. These are based on the general theory of relativity. They do assume a certain speed for your spacecraft, and what you would see does depend, in some detail, on exactly what you're doing, what your speed is, and all that. Let's not go into all those details right now; let's just assume we're flying toward the black hole and get a general idea of what you would see. First is sort of the first frame of the first movie, where I want to set my bearings, in a sense. You're looking at the sky toward a black hole—the black hole is over here—and you see the familiar winter sky. Of course, I've assumed that the black hole is in our galaxy, in our vicinity. But I just told you to find a supermassive one in some other galaxy, so forget about that complication.

Suppose it's in your galaxy, and it's along the line of sight toward the winter constellations, and it's a safe one to get close to. Here's what you would see. There's Orion, with Betelgeuse, Rigel, the Belt. Here's Sirius in the Big Dog, Canis Major; Aldebaran in Taurus, Castor and Pollux in Gemini, and Procyon in the Little Dog, Canis Minor. Right there, you can see fainter images of

Betelgeuse, Procyon, and Sirius. These are sort of the counter images caused by the distortion of space around the black hole. Though the main images are very bright and distinct, these counter images—at least initially, when you're far from the black hole—are faint. But you will see as you get closer to the black hole that those images get brighter. Indeed, all of the stars will start getting secondary images. Let's now watch the movie.

There's the winter sky, and now we're getting closer to the black hole, and you can see how, first of all, the stars sort of move away from the black hole because of this distortion of their trajectories. All of the stars, if you look closely here, start having secondary images opposite, or along a line through, the center of the black hole and to the opposite side. You get sort of a very distorted view of the Universe as you're going toward the black hole; that's kind of fun. Let's suppose you get down to 10 Schwarzschild radii from the black hole, and you start looking around. Here's what you would see. You're now at 10 Schwarzschild radii, and suppose you want to circle the black hole. You start circling the black hole, and you get this very distorted view of all these stars whizzing around. Let's do that again. You're circling around the black hole at 10 Schwarzschild radii, and that's what you would see.

This is reminiscent of what you would see if you had a black hole passing across your line of sight, toward a bunch of distance galaxies. Remember, I showed that a few lectures ago. Here's this black hole passing across the line of sight toward distant galaxies. You get these Einstein rings forming, and you get this apparent rotation of the galaxies around the black hole. That rotation is reflected in this animation as well, where you see the star sort of rotating around the black hole as you circle the black hole. It's this weird relativistic effect. Now let's suppose you get down to the photon sphere which is 1.5 Schwarzschild radii from the center of the black hole. Suppose you start at 10 Schwarzschild radii, and you travel down toward the photon sphere at 1.5 Schwarzschild radii. You've got, first, this view of the black hole, as we saw in the last frame of the previous movie. Now you're going down toward the black hole. Finally, when you get to the photon sphere, the whole thing appears just black because you're looking into the heart of the black hole now. The rest of the sky is now behind you, and toward you is the black hole. Obviously, it looks dark. That's what you would see as you approach the photon sphere.

Now you're at the photon sphere itself, and you're looking down into the black hole, and you see just this blackness. But as you lift your gaze higher up and start looking out toward the sky behind you, you would see the sky rise like this, and the stars would appear. But they're not moving around now, left and right the way they were in the previous movie, because I'm not circling around the black hole now. I was just sort of at the photon sphere and letting my gaze drift upwards. Now let's actually circle the black hole while we're at the photon sphere and see what happens. You get all these stars whizzing around, with multiple images in some cases, and you can see the dramatic curvature and distortion of space because the images of the stars go flying off all over the place, and curving off toward the side, and doing all sorts of weird things. You'd get a wild view; look at that. It would just be a wild view as you're looking around at stars traveling at the radius of the photon sphere. To really understand this, you have to study these simulations in much more detail and watch the trajectories of individual stars. We just don't have time to do that right now, but I wanted to give you some idea of what you might see.

Now suppose you continue going down farther, after crossing the photon sphere. Recall that the exit cone of any light that you yourself try to shine out—using a flashlight, say—becomes narrower and narrower. If you angled your light at too big an angle, it would just go into the black hole; and if you angled it nearly straight up, it would escape. In a similar way, starlight is coming in toward your eye through this exit cone, but now it would be, I guess, called an entrance cone. Starlight from over here, say, would bend around, go into the exit cone, or the entrance cone, and hit your eyes. Starlight from over here would do the same thing, and from over here it would do the same thing. The whole sky would enter this entrance cone, and you would then see the whole sky as a small circular patch above you. Everywhere around, there would be darkness.

The whole sky would be scrunched into a little circular patch above you. As you got closer and closer to the event horizon, that circular patch would get smaller and smaller—nevertheless encompassing the entire sky—because all photons from all stars can reach you because of this gravitational bending, but they're all squished into appearing to come from this narrow exit cone. That's really a weird effect. Finally, when you reach the event horizon, the exit cone squishes down to zero, and so the whole sky is essentially in a point, but you can't distinguish the stars at that point because they're all squished together. Then when you go beyond the event horizon, no one really knows what you would see at that point.

It would be a very highly distorted view, for sure, because photons entering the black hole from different directions would bend around a lot and reach your eyes—so you'd see some complicated mess. Unless you happened to be in the center of the black hole and looking out along radial directions; then I think it would be not such a mess, although I'm not sure exactly, and I haven't found anyone who's actually done the calculations. I think more research is needed in this area. That gives you some idea of the very distorted view you would see if you were to try to travel toward a black hole.

Now suppose you want to stay comfortably outside the black hole, but you want to throw your mortal enemy toward the black hole because this creature is doing terrible things to the world, and you just want to get rid of them. Suppose you're, for example, Yoda, and you want to throw Darth Vader into the black hole. You do this with great gusto. You say, "Vader, die, you wicked fiend. We're getting rid of you and the dark force forever." And Vader, on his way in, is saying, "But I'm Luke's father. Don't do this." It would high drama indeed. But let's go back to relativity and physics and examine what we would actually see, and what Darth Vader would experience while falling into the black hole.

First of all, Yoda would see Darth Vader's clocks slowing down because Darth Vader is going into a strong gravitational field. In a strong gravitational field, clocks run more slowly than in a weak gravitational field, out here where Yoda is. Suppose you start out, and it's noon in both places, on both clocks; you've synchronized them. Then far from the black hole, where Yoda is, four hours might pass; whereas closer to the black hole, where Darth Vader is, only two hours may have passed. The closer Vader gets to the event horizon, the greater is this apparent slowing down of time, time dilation. That is, Yoda, looking at Darth Vader's clock, thinks that Vader's clock is ticking more and more slowly relative to his, Yoda's, clock. This is a very real effect, this time dilation. Indeed, we've even detected it here in the Earth's very weak gravity because the GPS satellites and the whole positioning system wouldn't work if this weren't the case.

In fact, it turns out that, as Vader gets extremely close to the event horizon, the clocks slow down to a halt. In fact, they come to a stop when Vader reaches the event horizon. From Yoda's perspective, Vader never crosses the event horizon because it would take an infinite amount of time to do so. Yoda sees Vader getting progressively closer to the event horizon, with Vader's clocks slowing down more and more, but it would take an infinite amount of time to even reach the event horizon. So, Yoda doesn't have the pleasure of seeing Vader actually cross the event horizon. Indeed, from an outside observer's perspective, nothing ever crosses the event horizon of a black hole because it would take infinite time, from our outside perspective, to do so.

In fact, anything that has ever fallen into a black hole, from our perspective, is actually hovering infinitesimally above this imaginary surface called the event horizon. It's like a membrane. Indeed, there's a whole paradigm for studying black holes called the membrane paradigm, where nothing ever falls in; it just hovers, from our perspective, an infinitesimal distance above the event horizon. Anyway, that's kind of a cool effect. From Vader's perspective, he really does fall in, and it takes just a finite amount of time. But from the outside world's perspective, nothing falls in.

If a star were collapsing, from the outside perspective, it would get smaller and smaller and smaller, and then it would reach some minimum size—essentially the size of the event horizon—at which point it would become frozen. It would no longer get any smaller. Indeed, Soviet astrophysicists, initially studying black holes, gave them the name "frozen star" because they appear to come to a halt as they're collapsing. But that name didn't stick; the Princeton physicist John Archibald Wheeler came up with the term "black hole." That was just something that people could really appreciate more than frozen star, I guess, and so that's what they're called now. But in a sense, frozen star would have been a good term as well.

From Vader's perspective, he falls in. Clocks are ticking normally for him; he doesn't know that anything is going on. In your local frame of reference, everything is always occurring in a way that's independent of your frame of reference because that's actually what relativity is all about. It's only from other frames of reference that things look different. Vader crosses the event horizon and does die in a finite amount of time. He hits the singularity, which classically is a point of infinite density. But if you include quantum mechanics, it probably has some small, fuzzy, fluctuating size. I'll talk more about quantum mechanics and string theory later on. We really don't know what the singularity looks like, but in any case, it's probably very dense, but not infinitely dense.

Suppose Vader were to get close to the black hole, but then turn on some rockets and escape without having crossed the event horizon. Then, because Vader's clocks ran more slowly during the time when he was in the vicinity of the black hole, once he emerges and comes back out to fight Yoda, way out here where space-time is normal, his clocks will have registered much less time than Yoda's, so Vader will have aged much less than Yoda aged during this time. This is a way of jumping into the future. Less time passes on your clocks than on those of the outside observers. That doesn't mean you read more books, or see more movies, or fight more of the various creatures they fought in those movies and books. You don't increase your actual life span, but you jump into the future relative to other people, or other creatures, having aged much less than they did. That's kind of an interesting effect.

The other thing that happens is that Vader's light signals—suppose he's got a flashlight or a light saber emitting light. Yoda would see those light signals getting progressively more and more redshifted as Vader gets closer and closer to the black hole. That's because the light signals experience this gravitational redshift that I talked about before. You've got, in a sense, the light trying to come out of a very steep well. In so doing, it loses its energy; it progressively gets redder and redder and redder. It's not an instantaneous effect. You don't go from blue light to radio waves; rather, you go from blue to green, to orange, to red, to infrared, to radio. As the photon is trying to get out of this gravitational field, it becomes progressively longer and longer wavelength, and hence smaller and smaller energy—so you get this gravitational redshift effect.

You also get a fading effect; not just because the photons drift from the optical window to the infrared and radio, and hence fade, but also because the photons are being emitted by Vader's flashlight atoms. In the little filament of the light bulb, there are all these little electrons and other things glowing, and those are like clocks. All clocks are slowing down, and so the photons, the light, are being emitted at a progressively slower and slower rate. Therefore, there's a fading due to that effect as well. For yet another reason, Yoda wouldn't have the pleasure of seeing Vader actually cross the event horizon—because by the time Vader gets close to the event horizon, he will have faded beyond visibility. He just sort of blinked out, basically.

We've seen this effect, we think, in the vicinity of a rotating black hole, the object called Cygnus X-1 that I mentioned when I discussed observational evidence for black holes. There's this blob that was observed by the Hubble Space Telescope—not in such detail, but the light from it apparently was seen. The blob faded as it got closer to the black hole, and became redder. It became redshifted, and it faded. We think this effect, then, has been observed, although the interpretation of the data on which that animation is based is still a little bit controversial. If you were to look at a star that's collapsing down to form a black hole—not the Sun because it's not massive enough, but a more massive star might do this—you would see it fade as well because of the gravitational redshifting effect, and the time dilation effect, and another effect.

That effect is that less and less of the light is getting out from the star because only that part that's going out within the exit cone can possibly reach your eyes. The light from this edge of the star wouldn't reach your eyes up here because that light is going in this direction, and it would be captured by the black hole. Less and less light is coming out of the exit cone. For three reasons, a star would fade as it is collapsing down to form a black hole. That's the way it goes. You never see the matter cross the event horizon—and moreover, it fades from view. The clocks slow down and it fades from view, so you never actually see the formation itself of the black hole. You only see the events leading up to it.

What about Vader? Well, as I said, clocks, to him, are running at the normal rate. He crosses the event horizon in a finite amount of time, and he hits the singularity and gets crushed in a finite amount of time. In between coming close to the black hole and being squished in the singularity, he would have been stretched by this spaghettification effect. A lot of bad things would happen, and that's the way it goes. On his way in, he would see a lot of blueshifted light from stars out in space because their light is coming into a strong gravitational field and is getting gravitationally blueshifted rather than redshifted. He sees lots of stars in a highly distorted view, as I showed in those earlier simulations. The stars appear bluer than normal because light coming toward a strong gravitational field gets blueshifted in the same way that light coming out of a gravitational field gets redshifted. There's nothing Vader can do to stop his eventual destruction. He will get eaten by the singularity. That's the case if you go toward a non-rotating black hole.

If you go toward a rotating black hole, it is at least mathematically possible that you might traverse it and end up somewhere else. I'll discuss that in the next lecture. Let me introduce, then, rotating black holes. The structure of a rotating black hole is considerably more complex than that of a non-rotating black hole. For a given mass, the event horizon is smaller, but always spherical, as in a non-rotating black hole. Here's a non-rotating black hole, and then for a given mass, if you give the thing some rotation, the event horizon actually shrinks, and you get a region around it called the ergosphere, which I'll talk about more in a few minutes. If you rotate the black hole more and more and more, the event horizon shrinks more and more, and you reach a maximum rotation speed, at which the event horizon has a radius equal to one-half of the normal Schwarzschild radius.

In fact, what happens, if you look at the mathematics—and I'll show you this in the next lecture in more detail—is that two event horizons form. When you have no rotation, there's just one event horizon. If you give the black hole a little bit of spin, then a second event horizon forms close to the center, and the outer event horizon moves in a little bit. If you then rotate the black hole even more, the outer event horizon moves in, and the inner event horizon moves out. Finally, at the maximum rotation rate, the two horizons meet and actually vanish. When the two horizons meet and vanish, you would have what's called a naked singularity. A naked singularity is one that's not clothed. It's not cloaked by an event horizon, so you would be able to see it just sitting there. Since its gravity is so strong, there'd be nothing really to see; no photons would escape.

The danger of a naked singularity is that you wouldn't know that you've hit one until you're there. You'd start getting stretched, of course, tidally, but the point is there'd be no event horizon that would signal the presence of something really dangerous. Naked singularities are not clothed by an event horizon. A lot of theorists think that naked singularities cannot occur in nature. People say that nature abhors a naked singularity. They actually think that a black hole can never reach this maximum rotation rate at which the event horizon disappears, or vanishes, but that's never been completely theoretically proven. Indeed, some physicists have found contrived ways in which you might be able to form a naked singularity.

Let's go back to this picture that we had of a rotating black hole. I had mentioned that the event horizon gets smaller, but then there's this region, indicated in gray, around the event horizon called the ergosphere. Erg is like work or energy, and sphere is a sphere, but it actually looks kind of squashed, elliptical, here. That's a region where the dragging of space around the black hole is so great that no rocket could be devised, not even one traveling close to the speed of light, in such a way as to keep you stationary. In other words, space is dragged around a rotating object—the Gravity Probe B is attempting to measure this around the Earth—but around a rotating black hole, the dragging of space is so extreme that no rocket ship would allow you to remain stationary. You'd have to be dragged around.

The interesting thing about this is that, if you look at the top view of a rotating black hole rather than the side view, as was the case in the previous diagram, in the top view, you've got this rotating black hole, a circular ergosphere region around it, and you can actually derive energy, or work, from that ergosphere. For example, you could have a civilization with both a garbage problem and an energy problem. They have too much garbage, and they have too little energy. They could solve both problems by sending garbage trucks into the ergosphere, dumping the garbage into the black hole at the appropriate moment, and that would then fling the garbage truck out with greater energy than the energy it had coming in.

That garbage truck could then hit a windmill or a garbage truck mill, turning it, generating a current, which would then light up all the light bulbs in this city's civilization. This is a practical application of a rotating black hole, where you're tapping or stealing some of the rotational energy of the black hole in such a way that you actually decrease the energy of the black hole. It's not rotating as much anymore, in a sense, but you've used that energy, then, to help your own people, or aliens, solve their energy crisis. Admittedly, it would take a pretty advanced civilization to achieve this, but this at least provides nice fodder for sci-fi books and movies.

Wormholes—Gateways to Other Universes?
Lecture 63

"There are some mathematical studies that suggest that it might, in principle, be possible to traverse a black hole—in particular a rotating black hole, or possibly a charged black hole—and end up either in a very distant part of our Universe through a shortcut, or possibly even in another universe altogether."

We've seen what can happen, theoretically, if a person were to fall into a black hole. Let's now take a look at the structure of black holes. First, recall how space curves from our diagram of a rubber sheet distorted by a paperweight in the middle. The two-dimensional sheet was initially flat, but the presence of mass or energy causes it to warp, curving the space around a third mathematical dimension (but retaining the two-dimensional character of the sheet). In flat space, traveling from point A to point B while always keeping an arrow pointing in the same direction, parallel to itself, leads to a final arrow orientation that is independent of the path we take. In curved space, on the other hand, the final arrow orientation depends on the adopted path. It is fairly obvious that the surface of a sphere is a curved space, not a flat space, but what about the surface of a cylinder? A cylinder is actually flat space, which is perhaps counterintuitive. A cylinder is not distorted in the same way as a sphere; it still retains the mathematical properties of flat space. There is always some distortion when we try to project a sphere onto a flat two-dimensional map. Thus, as we look at flat diagrams of curved space or warped space-time, keep in mind that the maps do not reliably portray all aspects of the curvature. This is evident in Mercator projections of Earth onto a flat map, for example, but these maps still have important uses.

Warped-space diagrams, or maps, show the geometrical structure of space—a two-dimensional slice of space—outside of a massive body that is causing this distortion. If we make the massive body denser, the local curvature of space and time becomes more severe (more warped), resulting in what appears to be a stronger gravitational field. When the object becomes sufficiently dense, space around it is curved so much that it becomes vertical, marking the presence of a black hole. But space is so curved that, as seen from the outside, it appears as if the black hole joins up with another part of space through a passage formally named the *Einstein-Rosen bridge*, or more often called a **wormhole**. Such maps make it appear as if we could travel the long way around from point A to point B or simply go through the wormhole and make the journey far shorter. They also suggest that different universes are connected by wormholes, providing a way to enter other universes. Though these maps nicely illustrate the geometry of space outside a black hole, they don't actually tell us what's going on inside. To see this, we need another kind of map, and it, too, will have certain distortions.

As shown using the *Kruskal-Szekeres diagram* of a non-rotating black hole, we could not traverse a wormhole because to do so, we would have to travel faster than the speed of light. The horizontal axis of the diagram denotes one dimension of space (for example, the x, or left-right, direction). The vertical axis of the diagram denotes time. This is an example of a *space-time diagram.* The diagram is set up in such a way that light travels along 45° lines. We would have to travel along lines less than 45° to the vertical because our speed must not exceed that of light. The diagram also shows that after passing through the event horizon, there is no way to escape from a black hole—inevitably, we reach the singularity at some point in the future. The diagram also proposes the possibility of *white holes*—theoretical regions of space from which matter could emerge into our Universe, although such matter would not be able to return to its origin. (In contrast, matter could not escape from a black hole.) We don't believe white holes exist because we've never seen any evidence for them. The singularity in the Kruskal-Szekeres diagram is basically a horizontal structure rather than a particular point. This implies that, as seen by us within the black hole, the singularity covers much of space instead of being a specific point, and we would hit it at some time in the future. In essence, we can't avoid the singularity because it's everywhere around us. The time at which we hit the singularity would depend on

"It looks like you could go through the ring, or, in a sense, between these two event horizons, and end up in another universe."

exactly what trajectory we took into the black hole. That time is always finite and, indeed, is quite short.

The structure of rotating black holes is very different. The Kruskal-Szekeres diagram of a rotating black hole shows an infinitely repeating pattern of additional universes. Thus, in principle, we could cross a series of event horizons as we moved forward in time and emerge into another universe. The structure of such rotating black holes allows for travel across universes without having to move faster than the speed of light. The singularity of a rotating black hole is vertical, running parallel with the direction of time; therefore, we could avoid hitting it, depending on our trajectory. Recall that we said a cylinder represents flat space. If we take a piece of paper (flat space) and roll it into a cylinder, we've changed its overall topology but not its local geometry. Now imagine the flat Kruskal-Szekeres diagram of rotating black holes rolled into a cylinder. We could take all kinds of interesting journeys through space into and out of other universes. Mathematically, then, it is possible to take such a journey and arrive back at Earth at a time before our initial departure!

What if we decided to alter history in our travels to prevent certain events from happening? Physicists call this a *violation of causality*, a disturbing prospect. We can't go back in time, physicists think, and change the history of the Universe in such a way as to prevent our existence—for example, by preventing our parents from ever having met and, thus, precluding our own birth. How could we do so if we weren't born? It is possible that the geometry implied by the diagram is valid only for an idealized black hole into which no material is falling or has previously fallen. In other words, as soon as an object actually tries to traverse the wormhole, it closes. If a wormhole results from a black hole that was formed by the collapse of a rotating star, there is enough matter in the central regions to crunch up the wormhole and squeeze it shut. Thus, for the case of idealized black holes and wormholes, the Universe would need to be born with these objects already present. But even if we traveled into an idealized black hole, we would gain so much energy that it would cause the wormhole to squeeze shut. Thus, a wormhole in an idealized black hole may present a passage to another universe (or to a distant part of our Universe) as long as we don't actually attempt to travel through it. If we do so, our own mass—and the energy associated with our motion—would curve space enough to close the wormhole.

Some physicists have wondered what would happen if we could travel in a spaceship made of antigravitating material that would prevent the wormhole from closing. In this case, we still have the problem of the violation of causality. Perhaps we can take solace in the fact that no such antigravitating substance has been found, though we will see later that *dark energy*, perhaps a property of space, has properties reminiscent of antigravity. ■

Important Term

wormhole: A hypothetical connection between two universes or different parts of our Universe. Also: *Einstein-Rosen bridge*.

Suggested Reading

Begelman and Rees, *Gravity's Fatal Attraction: Black Holes in the Universe.*

Hawking, *The Universe in a Nutshell.*

Kaufmann, *Black Holes and Warped Spacetime.*

———, *The Cosmic Frontiers of General Relativity.*

Pasachoff and Filippenko, *The Cosmos: Astronomy in the New Millennium*, 3rd ed.

Pickover, *Black Holes: A Traveler's Guide.*

Thorne, *Black Holes and Time Warps: Einstein's Outrageous Legacy.*

Questions to Consider

1. Given that in a non-rotating black hole, the singularity takes up much space at a particular time, rather than a point in space over a long period of time, does it make sense to think that space and time have, to a certain extent, reversed roles inside a black hole?

2. What sorts of problems could be produced by the violation of causality—that is, if you could travel through a wormhole and return before your departure?

Wormholes—Gateways to Other Universes?

Lecture 63—Transcript

I've discussed how someone falling into a black hole would be ripped apart by tidal forces, the "spaghettification" process, and finally crushed in the singularity, the very dense region in the very center of the black hole; maybe infinitely dense, maybe not. We need some sort of a quantum gravity theory to tell us, really, the nature of the singularity. In any case, it wouldn't be pleasant. But there are some mathematical studies that suggest that it might, in principle, be possible to traverse a black hole—in particular a rotating black hole, or possibly a charged black hole—and end up either in a very distant part of our Universe through a shortcut, or possibly even in another universe all together; a disconnected region of space and time, like a bubble, that's different from our own space and time, our own bubble. There might be these passages called wormholes between universes, and you could traverse those wormholes by jumping into a rotating black hole. This idea has appeared quite often in science fiction movies and books.

Let's just explore it here. It has some basis in mathematics, but as we will see, unfortunately, it turns out probably not to be possible to make these journeys. Let's explore the structure of black holes. First, I need to carefully define what I mean by curved space. We've already had some diagrams that try to illustrate it. A flat piece of rubber, if distorted by a paperweight in the middle, would appear curved like this. Here I've left out the paperweight because I want to emphasize that it's the mass or the energy itself that causes this distortion. It is not the gravity of the Earth pulling on that paperweight that causes the sheet to distort. I've left out the paperweight in this diagram, but you can see that the thing curves down to a third dimension. The sheet was initially two-dimensional, and then the presence of mass or energy causes it to warp, and curves the space around it.

Let's see if we can come up with a better mathematical definition of what curved space really means. Let's first consider flat space. In flat space, it turns out that if you move a little arrow along different paths from point A to point B, the final orientation of the arrow is independent of the path that you chose. Let me illustrate. Here's a flat sheet of paper. We start at point A with an arrow pointing this way, and we want to end up at point B. If we choose this path here, always keeping the arrow parallel to itself—so you take tiny, little steps (little, infinitesimal steps) along the way, making sure that the arrow is always parallel to what its direction was at the previous location—you end up with the arrow pointing in this direction when it reaches point B.

If instead you take some other path and do the same thing, always keeping the arrow parallel to itself, you find that the final orientation of the arrow is independent of the path that you chose. All paths lead to a final orientation that is the same. That is the property of flat space. If we now, instead, go to a curved space, we will find that if we take two different paths, always keeping the arrow parallel to itself along small, successive steps along that path, then we will end up with an orientation that is dependent on the precise path taken. When you find that to be the case, you know that you're experiencing a curved space. Let's take a look at this on a sphere. Most of us have some intuitive feeling that a sphere is indeed curved; it's not flat. There are all sorts of distortions if you try to flatten out a sphere onto a sheet of paper.

I'm going to start down here with the pen, and I'm going to draw an arrow pointing in this direction. Now I'm going to go along the equator of this sphere, always keeping the arrow along the same direction. Here it is, pointing like this right now. I've kept it parallel to itself along this journey. Now let me go along a line of longitude up to the pole, always keeping the arrow parallel to itself. In the end, it points at me. Let me rotate it over. You can see that it's pointing at me if I take this path to the final position. If instead I start down here at the equator, the arrow is pointing this way, but now I displace the arrow along a path here, always keeping it pointing the same way, and then come up a line of longitude up to the pole, again always keeping the arrow pointing the same way. You see that along this path, in the end, the arrow basically points at you, not me. The final orientation of the arrow was dependent on the path taken, even though the initial orientation of the arrow was the same in both cases. This tells you what you already intuitively know—a sphere is a curved space, not a flat space.

Now that we've looked at a sphere, let's take a look at a cylinder and ask ourselves whether a cylinder is flat or curved. I can take a sheet of paper and perform this experiment, and find that the sheet of paper is, as intuition tells you, flat space. If I now bend this sheet of paper into a cylinder like this, and if I were to try to draw little arrows in different paths along that cylinder—you can try this at home—I would find that the final orientation of the arrow is independent of the path taken. In other words, a cylinder is flat space, which is perhaps counterintuitive because it looks like it's rolled up. That's true; I've rolled up the flat space, but I haven't crinkled it in any way. I haven't distorted it. I haven't mathematically warped it. I've simply wrapped it around itself, but I've retained the mathematical properties of a flat space. Keep that in mind because we're going to come back to it later.

Now let's look at two-dimensional maps of curved space. There's always some distortion, and that's what I want to emphasize here. Whenever you look at a two-dimensional map of a three-dimensional curved space, there's going to be some distortions. When you look at this top diagram here, and you ask people on the street what that is, they tell you that's the Earth. Wait a minute. How can that be the Earth? It's rectangular. It's got four corners. It's got edges. I mean that's not the Earth that I know. It's got this Antarctic continent that, first of all, is broken up into two pieces and, second of all, looks even bigger than it should, as does Greenland, which looks huge here. It's a pretty big thing, but it's not as big as all of North America. So, there's all these distortions when you try to take an intrinsically curved, or spherical, Earth and map it onto a two-dimensional flat sheet of paper.

We're used to such distortions. We say, "Okay, there are aspects of this map—this so-called Mercator projection—that are useful, and there are aspects that are not." You have to keep in mind which things are distorted and which are not. It turns out the Mercator projection is very good for navigation because north and east directions are preserved at each location. If you're just moving from one place to another, instantaneously north, east, west, and south do point the right way on a map like this. But clearly, the areas and shapes of continents are grossly distorted on a map like this, and no one would use such a map to compare the relative areas of Greenland and Egypt or something like that. You have to keep in mind the distortions. All of the maps that I'll show you in this lecture have one distortion or another, but you have to keep that in mind. They also have certain useful properties, and that's why we use these maps. I will try to point out where the maps are giving you the wrong impression and where they are useful.

Now let's take a look at this warped space that we've seen in a number of lectures now and ask ourselves, what is this really showing us? This is showing us the geometrical structure of space—a two-dimensional slice of space, not the full three dimensions—outside of a massive body that is causing this distortion. If we make the massive body denser and denser and denser, the local gravitational field—that is, the local curvature of space and time, though time isn't indicated here—all get more warped, more curved. As seen from the outside, then, it kind of looks like, when you have such high density that you have a black hole, and the black hole actually joins up with another part of space through a bridge here, formally called the Einstein-Rosen bridge, often called a wormhole. Here's the structure of a black hole, as seen from the outside, and it looks like it joins up with another similar structure somewhere else in the Universe. Indeed, that region of the Universe could be a long way away if you went along the normal path, but it's just a short distance if you go through the Einstein-Rosen bridge.

You can even consider the possibility that a black hole joins you up with another universe somewhere else. These two sheets of paper are not joined at the edges here. They are two separate sheets of paper that look like they each have a black hole joining up somewhere between them, in some third dimension, which is sort of inaccessible to the creatures that live in flat space. But this wormhole forms if you have a sufficiently compressed object. These are the kinds of diagrams that led some science fiction writers to seriously consider the possibility of using black holes to travel to other parts of our Universe, very distant parts of our Universe, or perhaps, better yet, even to other universes. This appears in the movie and the book *Contact*, for example.

The problem with those maps that I just showed you is that they nicely illustrate the geometry of space outside the black hole, but they don't actually tell you what's going on inside. To look at what's going on inside, you need another kind of map—and it, too, will have certain distortions. But other aspects of the maps I'm about to show you will better illustrate the interior of a black hole. These maps are based on what we call space-time diagrams. If we have two axes—time going in the vertical direction, distance or space going in the horizontal direction—we can define forward being in this direction, positive X, for example; backward is negative X. Time moving forward is in the upper direction—that's the future, and the past is down here.

The diagram is set up in such a way that light travels along 45-degree lines. If you have one year from point O to C up here, then one light year would be the same distance along the forward direction, or the X-axis. Light travels one light year in one year, so it travels along this diagonal line, this 45-degree line. It travels 10 light years in 10 years. Again, it travels along this 45-degree line. It travels one light second in one second; so again, a 45-degree line. Any material object can't travel as fast as the speed of light, so it covers less distance in a given amount of time than light would have. It travels along a path that is steeper than the path of light, the 45-degree line. If you're sitting in one place, and you don't want to move anywhere at all, then time simply goes forward for you, and you go up vertically like this.

If you're lying on your bed, at a given instant your body would be a little line segment like this. It would be at several different places in X at a given time—call it zero. Events that occurred in your past can send a signal to you if they're anywhere within this 45-degree, half-angle cone. A signal could come from event A to you because that signal, traveling through regular postal mail or whatever, could reach you. A bicycle could get to you if the bicycle is traveling fast enough. But a signal from point B out here, in this region that we call "elsewhere," could not get to you because, even traveling at the speed of light, the signal would only go along a 45-degree line and would, at best, get to point C. To get from point B to O point, where you are, the line would have to be shallower than 45 degrees, meaning that the signal would have to travel faster than light, and that's not permitted. These areas here are called "elsewhere." Signals from events in those areas cannot reach you—whereas signals from the past can reach you, and you can send signals at the speed of light—or in a car or a rocket ship, or just staying in one place—into the future, or you can move forward into the future.

With this diagram in mind, let's draw a special kind of diagram of a black hole, called a Kruskal-Szekeres diagram, after Martin Kruskal and Peter Szekeres, who first developed these diagrams. This will show you the interior of a black hole. It looks kind of weird, but once again, space is to the right or left, time is up and down, and this diagram has the property that light travels along 45-degree lines. First of all, note that at the left here, there is our Universe. On the right, you see the other universe—the one that I talked about in those diagrams where it kind of looked like there is an Einstein-Rosen bridge joining up two universes. There they are. There's ours, and there's the other one. By the way, the whole Universe, even if it's infinite, is compressed into a finite area square in this diagram. That's a distortion if I ever saw one. But in any case, the map will be useful, as you will see.

Let's take a look at other parts of this diagram. Here's the interior of the event horizon. This is the black hole, and here's the singularity. It's drawn fancifully with these teeth, like it's a shark or something. The singularity is going to eat you if you get there. Here's the event horizon. There's also a singularity back here, corresponding to another black hole, but as we'll see in a few minutes, that's a weird black hole—it's called a white hole. Anyway, let's look at possible trips that we could take from this location here. We could take trip A, from our part of the Universe to some other part of the Universe. Let's say we go out to dinner at a local restaurant. We go basically forwards in time and a little bit away in space, and so we end up at our restaurant. That would be a perfectly valid and good trip to take.

We could take trip B, beyond the event horizon of the black hole, but that would be stupid because, once you cross the event horizon, there's no way you can get out again. To do so, you would have to follow a trajectory that is more shallow than a 45-degree line. The event horizon is already a 45-degree line, and only light, at best, can travel along that line. You could go, at best, along that line or anywhere in the forward direction, or along this line here if you were light. But basically, you can only go forward along your forward future cone, and that cone always intersects the singularity, so you'd definitely get squished. There's no way to escape because, to escape, you would have to travel faster than light.

You might say, "What if I want to traverse the wormhole and end up in this other universe? Can I do that?" You can't even do that because, once you enter the event horizon, or pass the event horizon, again, you can only go always within a cone with a 90-degree opening angle. You can basically only go forwards and a little bit left and right. You can't get to this other universe because, to do so, you would have to follow a trajectory that's shallower than

45 degrees, and that would mean that you're traveling faster than the speed of light. You can't get to this other universe. For a non-rotating black hole, which this diagram is a representation of, you can't actually traverse that wormhole because, to do so, you'd have to go faster than the speed of light, and that's not permissible.

Let me get back to this other singularity here. If you had particles in this region here, this blue region here, they could move basically forward in time, or anywhere within a cone of opening angle 90 degrees. You can see that they could enter our Universe, or they could enter the other universe, but they could never go back into this region here. This is actually the time reversal of a black hole. It's a region from which stuff emerges, but it can never go back in, just like a black hole in our Universe is a region into which stuff can go, but it can never come back out. Theoretically, mathematically, there might be these regions called white holes, where we would see things coming out from no real apparent source. There's no factory there or something; it's just stuff appearing. We've never seen that in the Universe, so we think that this mathematical white hole doesn't actually exist. It's sort of just a mathematical oddity, but the white holes apparently don't exist.

You'll notice an interesting aspect of the singularity in this Kruskal-Szekeres diagram. The singularity is basically this horizontal structure. It's not a particular point in space, as seen by you within the black hole; rather, it covers much of space, and you hit it at some time in your future. That time depends on exactly what trajectory (what path) you decide to take. But there's no way to avoid it because, in a sense, it's everywhere around you, and you will hit it, regardless of what you do, at some particular time in your future. That time can be shorter or longer, but always finite, not infinite. The singularity is no longer a point in space; it's an event in your future, in time, and there's no way to avoid it. That's the problem with the non-rotating black hole, and the Kruskal-Szekeres diagram illustrates that very well.

Next we will see what we have with a rotating black hole, and you'll see that the structure is very different. In particular, mathematically at least, the singularity can be avoided. In a rotating black hole, or a charged black hole, you get a very different structure. We don't expect charged black holes to actually exist in nature because, if they were charged, they'd quickly steal the opposite charges from their surroundings and neutralize themselves. Electromagnetic forces are very strong. We don't expect to see charged black holes, but rotating black holes we do expect to see because they form from the collapse of rotating stars. If you look at the Kruskal-Szekeres diagram of a rotating black hole, first it looks really hairy and complicated, but let's just look at it slowly and try to understand it.

Once again, there's our Universe over here, and the other universe that was familiar in the Kruskal-Szekeres diagram of a non-rotating black hole. Then there are these other universes as well. Here's another universe; there's another universe. Indeed, there's an infinitely repeating pattern of this sort, where you have big squares like this, two sides of which are additional universes. It looks like, in a sense, the rotating black hole joins you up with an infinite progression of universes. There are other interesting aspects here as well. This line here is the outer event horizon of the rotating black hole. That line there is the inner event horizon. You will recall that I said in the previous lecture that a rotating black hole has two event horizons. Here you can see them. There's one of them; there's the other.

You can go into this region here, and then go close toward the singularity, and even hit it. You could, in principle, cross the outer event horizon, then cross the inner event horizon, then cross another inner event horizon forwards in time, and then back out through the outer event horizon and into another universe. What an amazing trip; going forwards in time, never at an angle greater than 45 degrees, you could make a journey, trip C here, that in fact avoids the singularity and allows you to end up in another universe—either this one, or, if you changed course somewhere along the way, you could end up in that one. Then you could traverse additional wormholes and end up in some other universes. What an amazing trip that would be.

The stupid trip would be trip B. You'd cross the outer event horizon, then the inner event horizon, and then you sort of head toward the singularity. You can easily avoid the singularity in this case, at least mathematically, because now the singularity is basically a vertical structure. It's something that exists at a particular point in space, and moves forward in time; in distinction to the singularity in the non-rotating black hole, which was basically everywhere in space beyond the event horizon, and you hit it at some particular time. Here, the singularity is basically in the time direction, and you can avoid it by choosing your spatial location wisely. Then you can take a really boring trip within your own universe, going to a restaurant here. Then, of course, you can't get to that universe because, as in the case of the non-rotating black hole, it would require a speed greater than the speed of light.

This kind of diagram suggests that you can actually travel through a rotating black hole to another universe. What a trip that would be. Another way to think about it is that there's kind of a ring singularity in the middle of this black hole. When you have a rotating black hole, the mathematics, which I won't show you today, kind of indicates that the singularity is a ring, and you could actually go through that ring, avoiding the dense parts. Now you'd have to worry about being tidally disrupted through this spaghettification process. You'd need to worry about that. But forgetting that minor complication, it looks like you could go through the ring, or, in a sense, between these two event horizons, and end up in another universe.

Recall what I said earlier when I said that you could take a flat sheet of paper and roll it into a cylinder, and you haven't changed the local geometry. This is still flat space. You've changed what's called the topology. You've taken something that goes out in all these directions, is flat, and doesn't join in on itself, and now you've joined it in on itself so that you could actually sort of go around in circles and stuff. You've changed the topology, but you haven't changed the local geometry. Now suppose I take this Kruskal-Szekeres diagram of the rotating black hole—I've got it on a sheet of paper, and I wrap the paper around, making a cylinder. Here's the diagram that I would get. This looks complicated, but it's not so bad.

You've got the usual Kruskal-Szekeres diagram wrapped around itself to form a cylinder. Consider what sort of a journey you might take. You could start out here from Earth, go through the wormhole of a bunch of rotating black holes, all the way around, and come back to your own universe. In this case, the journey ends up at a place sort of forward in time from where you started here on Earth. But you might imagine starting somewhere here, let's say, going into the rotating black hole, going around all these universes here, and then ending up here, at a position in time that precedes the time at which you left for your journey. This diagram suggests that you could go through these wormholes and come back to your own universe, or go to other universes, at a time before you left. This is bad news. What if you decided to go off and prevent your parents from ever meeting before you were born? So, your parents never meet. They go off, marry someone else, and have other kids or whatever. How could you, then, have been born to make this journey in the first place and prevent your parents from having ever met?

This is what physicists call a violation of causality. There's always a cause and effect relationship. Whatever effect you have, there's a cause that preceded it, cause and effect. You can't go back in time, physicists think, and change the history of the Universe in such a way as to have prevented your existence and your going back and doing all these things that changed the history of the Universe. This appears in science fiction books and movies all the time, *Back to the Future* and all this kind of stuff. But this is a real problem for physicists because all of physics falls apart if you violate causality. This looks like a real problem. It's often said that you go back in time and kill your parents, but I don't want to be that violent. Just prevent your parents from ever meeting before you were born. How could you have done that if you weren't born? That's the essence of the problem.

Okay, so how do we deal with this? One way in which physicists deal with this is by saying that there's something about the violation of causality that's so fundamental that that fundamental aspect of physics prevents any real journey through black holes of this sort, even though it appears to be mathematically possible. The other way to deal with this is to say these diagrams that I drew are actually for idealized black holes. By that, I mean that these are black holes that existed with the birth of the Universe. The Universe was born with these black holes and wormholes already present during the time of birth. These are not wormholes or black holes that formed as a result of the collapse of a real star.

The problem is that, if the wormhole results from the formation of a black hole that resulted from the collapse of a rotating star, then it turns out there's enough matter in the central regions to crunch up the wormhole and squeeze it shut. Even if there's not that material—even if you've got an idealized black hole to begin with—if you decide to make this journey into it, you will, in the process, gain so much energy and contribute it to the black hole, that that will squeeze the wormhole shut. It's sort of like a catch-22. The idealized wormhole and black hole may present a passage to another universe, or to your own universe, as long as you don't actually choose to make the journey. If you choose to make the journey, your own gravity, and the energy associated with your motion, squeezes the wormhole shut. That may be the way to get out of this conundrum. You can't violate causality because the wormhole squeezes shut.

Then some physicists have countered and said, "Well, what if there's some sort of locally anti-gravitating material out of which you could make a spacesuit that would prevent the wormhole from squeezing shut?" If that's the case, then maybe you could make this journey, and we're back to this problem of the violation of causality. Perhaps we can take solace in the fact that no such anti-gravitating substance has been found that you can locally harness and create an anti-gravity field around you. I will talk later about dark energy that has anti-gravitating properties and fills our whole Universe. But it does not appear as though we know of any kind of material that you could locally harness and collect, and create a spacesuit out of, and prevent the wormhole from squeezing shut. It looks like, though mathematically you might be able to make these journeys through idealized black holes, in practice it just doesn't work. Indeed, the science fiction stories are just that—science fiction, not fact.

Quantum Physics and Black-Hole Evaporation

Lecture 64

"Though each of them works well in its own realm of applicability, when you try to put them together, and you try to describe the properties of very small particles, or even of space itself, over very small spatial scales, the results from quantum physics and general relativity are completely at odds with one another."

For many decades, physicists thought that black holes were truly black. But now, in attempts to unify general relativity and quantum mechanics, we have found that perhaps black holes may exhibit characteristics never before imagined. Even though the two theories (general relativity and quantum mechanics) work well in their own applicable realms, when we try to describe the properties of small particles or even of space itself over tiny spatial scales, quantum physics and general relativity are at odds. We don't yet have a consistent unified theory for general relativity and quantum physics, but it is thought that a generic property is the peculiar evaporation of black holes.

Why might we expect a black hole to evaporate? We will answer this question in general terms, initially ignoring quantum effects in favor of classical physics. Recall that the mass of a non-rotating black hole cannot decrease; it can only stay the same or increase by accumulating material falling into it. As its mass increases, so does its Schwarzschild radius, increasing its surface area. Because the mass cannot decrease, the surface area cannot decrease. It can only remain the same or increase. For a rotating black hole, again, the surface area of the event horizon can only increase or remain constant. However, the mass can actually decrease under certain circumstances because its rotational energy is tapped when matter enters it in a certain way through the ergosphere.

The second law of black-hole dynamics states, "In any natural process, the surface area of the event horizon of a black hole always increases or, at best, remains constant; it never decreases." This resembles the **second law of thermodynamics**—the study of the relationship among heat, work, and other forms of energy—which plays a significant role in governing the physical Universe. This law states, "In any natural process, the entropy of a closed system always increases or, at best, remains constant; it never decreases." A *closed system* is one from which nothing escapes; *entropy* is a measure of the amount of disorder in the system.

Black holes behave in ways that can be described similarly by the other laws of thermodynamics, not just the second law. The correspondence between the laws led the physicist Jacob Bekenstein to propose that a black hole's surface area is proportional to its entropy. The entropy of a black hole is related to the no hair theorem (Lecture 60), in which black holes in equilibrium can simply be described by their mass, spin (more precisely, angular momentum), and charge (if any). Because matter devoured by a black hole is no longer identifiable, we don't know what the black hole consists of. This loss of information corresponds to a gain in entropy. Therefore, black holes must have a tremendous amount of entropy.

"The Schwarzschild radius is proportional to the mass of a black hole, so the surface area of a non-rotating black hole is proportional to the square of the mass of the black hole."

The analogy with the laws of thermodynamics suggests that a black hole is physically a thermal body. Indeed, it closely resembles what we called a black body (in Lecture 43). Recall that a black body doesn't transmit or reflect any radiation; it only absorbs it, being a perfect absorber. It also emits radiation through thermal motions of its constituent particles in a manner dependent only on its temperature. Similarly, a black hole doesn't transmit or reflect any radiation; it only absorbs photons. If a black hole can be thought of as a thermal body, then it must have a temperature. If there's a temperature, then it must radiate, or shine. The idea that a black hole shines is paradoxical if we ignore quantum effects. In classical physics, nothing can escape from within a black hole. Yet the whole shining process of a thermal body is really a quantum-mechanical property. Classically, nothing can escape a black hole. For this reason, Bekenstein didn't pursue his idea any further. But **Stephen Hawking** realized that the conclusion that nothing can escape a black hole might be false if we consider quantum-mechanical processes.

Hawking's theory that black holes can evaporate via quantum-mechanical processes, regardless of the details of any unifying theories, is related to the creation of particles and antiparticles. The basic idea for black-hole evaporation is also related to Heisenberg's uncertainty principle (Lecture 20), according to which there is uncertainty in any measurement. We cannot know with arbitrary precision the energy of a system or the time at which we made that measurement; the product of their uncertainties cannot be equal to zero, or less than the quantity $h/2\pi$, definitely not 0. Similarly, there is a relationship between the product of the uncertainties in the position of a particle and its momentum, its mass times its velocity. Thus, we can know the position of any real object but not its momentum with certainty. Or we can know its energy (momentum) but not its position with certainty.

It is possible for pairs of **virtual particles** to form spontaneously out of nothing. They now have some energy and they exist for a nonzero amount of time, but the product of the uncertainties in energy and time is *less than* $h/2\pi$. Quantum mechanics allows for (and even demands) such violations of classical laws in which something (the pair of virtual particles) appears out of nothing. This is called a **quantum fluctuation**. One way of thinking about this process (though not the only way) is that the positive energy created by the pair of particles creates a negative-energy hole, so that the particles and the hole cancel each other out; the net energy is still zero. However, one could also say that there is no negative-energy hole. Quantum mechanics is consistent with a small violation of the classical law of conservation of energy. Thus, we begin with nothing, have something for a short time, and then the particles annihilate each other, leaving a net energy of zero.

These quantum fluctuations are virtual in that we can't directly measure them because they don't last long enough. But they occur everywhere, and they do affect the Universe, as we will see. Near the event horizon of a black hole, a particle (or antiparticle) can sometimes escape, while the other particle enters the black hole with *negative energy* (from our outside perspective). The escaping particle takes positive energy with it. This decreases the mass of the black hole. One can also think of this as a *quantum tunneling* effect, with particles emerging outside from inside the black hole. As particles escape, they accelerate past each other, emitting photons, and annihilate each other, creating even more photons. The result is a thermal distribution of energies called **Hawking radiation**—the evaporation of black holes.

In summary, outside the event horizon, particles and photons are emitted, the black hole loses energy, and the temperature is proportional to the inverse of the black hole's mass. As the black hole evaporates, its mass decreases, its temperature rises, and the rate of evaporation consequently also increases. As the mass nears zero, the evaporation rate approaches infinity, and the black hole explodes. The rate at which stellar-mass and supermassive black holes evaporate is utterly negligible because they accrete material from their surroundings much more quickly. A high rate of evaporation occurs only for miniature black holes, maybe a billionth of the mass of the Earth or less. Hawking has suggested that such tiny black holes formed shortly after the birth of the Universe. If that's the case, then all of the ones born with less than 10^{15} grams have already evaporated. Those with initial masses of 10^{15} grams, having an event horizon roughly the size of a proton, are now evaporating. Because the rate of evaporation approaches infinity as the mass approaches zero, we end up with explosions. Most of the released photons from the explosions should be emitted as gamma rays. Have we ever detected gamma-ray bursts that might be indicative of the evaporation of miniature black holes? We'll see in the next lecture. ■

Name to Know

Hawking, Stephen (1942–). English physicist, best known for his remarkable theoretical work while physically incapacitated by Lou Gehrig's disease (ALS). His prediction that black holes can evaporate through quantum tunneling is an important step in attempts to unify quantum physics and gravity (general relativity). He is Lucasian Professor of Mathematics at Cambridge University, as was Newton.

Important Terms

Hawking radiation: According to Stephen Hawking, the thermal radiation emitted by black holes because of quantum effects.

quantum fluctuations: The spontaneous (but short-lived) quantum creation of particles out of nothing.

second law of thermodynamics: In any closed system, entropy (the amount of disorder) never decreases; it always increases or remains constant.

virtual particle: A particle that flits into existence out of nothing and, shortly thereafter, disappears again.

Suggested Reading

Ferguson, *Prisons of Light—Black Holes.*

Hawking, *A Briefer History of Time.*

Pasachoff and Filippenko, *The Cosmos: Astronomy in the New Millennium*, 3rd ed.

Pickover, *Black Holes: A Traveler's Guide.*

Shu, *The Physical Universe: An Introduction to Astronomy.*

Thorne, *Black Holes and Time Warps: Einstein's Outrageous Legacy.*

Questions to Consider

1. Are you surprised to learn that black holes aren't completely black after all?

2. If an evaporating black hole behaves like a black body (specifically, the Stefan-Boltzmann law), and if the temperature of a black hole is proportional to the inverse of its mass while the Schwarzschild radius is proportional to its mass, to what power of the mass is the black hole's luminosity proportional?

3. Given that an outside observer has no knowledge of the composition of a black hole (that is, what kinds of materials or objects were thrown in it), can you argue that a black hole has enormous entropy (disorder)?

Quantum Physics and Black-Hole Evaporation

Lecture 64—Transcript

Given the extreme curvature within a black hole, for many decades, physicists thought that black holes are truly black. Nothing can escape from within the confines of this highly curved space-time. But now, in attempts to unify the two great pillars of modern physics—general relativity on the one hand, the physics of the very large; and quantum mechanics on the other hand, or quantum physics, the physics of very, very small things—it has been found that perhaps black holes actually do shine, gradually radiating away their mass and ending up as gargantuan explosions. That's a pretty amazing thought, black holes shining. Why would that happen? This is a generic feature of attempts to unify quantum physics and general relativity. You might ask, "Why should they be unified if they work so well?"

It turns out that, though each of them works well in its own realm of applicability, when you try to put them together, and you try to describe the properties of very small particles, or even of space itself, over very small spatial scales, the results from quantum physics and general relativity are completely at odds with one another. They are in violent disagreement. The two theories are utterly incompatible and give nonsensical answers when you try to apply both quantum mechanics and general relativity to high-energy particles, or very small regions of space, or things like that. We don't yet have a fully self-consistent unified theory of general relativity and quantum physics together. Attempts are being made all over the place at such a theory, and the leading candidate is string theory. I'll have a little bit to say about string theory later on in the course.

What I want to emphasize here is that, regardless of the specifics of the unification theories, of which there are several, it is thought that a generic property will be this interesting evaporation of black holes. That is, it seems to be so fundamental that it'll end up being a characteristic, or a prediction, of any specific unified theory, whatever it ends up being. The basic idea is that this evaporation occurs due to the presence of quantum fluctuations in and near the black hole's event horizon. It works best for tiny, little black holes with a lot of curvature, where particles can actually escape from the black hole in a matter actually analogous to quantum tunneling. It's kind of like if I hit myself against the wall enough times, I'll end up on the other side some day, unscathed. The odds for that are very low, so I'm not about to do it. But if you have enough particles trying to do this, and if the wall, the barrier, isn't insurmountable, then they can actually travel through it and get out to the other side.

Why might we expect a black hole to evaporate? I'm going to give you the general argument, and it gets pretty complex. This is really hairy stuff, kind of like traveling through a wormhole, which I discussed in the previous lecture. I just want you to get the gist of the idea. Don't worry about the details or the mathematics. It's really quite complex stuff, but I thought it was just too good to pass up in a course like this. In the next lecture, I'll get back to the observations and things that go bang, that you can photograph and see and stuff, but here we're getting a little bit of theoretical physics, and it's fun. My account will follow the historically correct order so you can get some idea of how these notions developed with time. Initially, we will ignore quantum effects entirely. We'll say there's just classical physics. There's no quantum mechanics, no quantum fluctuations, nothing weird like that.

We've said that, classically, the mass of at least a non-rotating black hole cannot decrease with time. It can only stay the same or increase with time. I said that nothing can get out of a black hole. Matter and energy (photons) can be added to a black hole, increasing its mass. That increases its Schwarzschild radius, and it increases the surface area of the black hole as well, because the surface area of the event horizon, being a spherical structure, is just 4π times the Schwarzschild radius^2. If you're adding material in, and the Schwarzschild radius is growing, the surface area will grow as well, in proportion to the square of the Schwarzschild radius. The Schwarzschild radius is proportional to the mass of a black hole, so the surface area of a non-rotating black hole is proportional to the square of the mass of the black hole.

Since M cannot decrease, the surface area cannot decrease either. It can, at best, remain the same or, if stuff falls in, it increases. Now let's consider the analogous situation for a rotating black hole. For a rotating black hole, it turns out that the surface area, again, can only increase or stay the same; it can never decrease. The mass can actually decrease under certain circumstances because, as I said, you're tapping the rotational energy when you send garbage in a special way toward a rotating black hole. You can fiddle around with the spin, and the mass, and the other properties, but the thing that cannot decrease is the surface area of the event horizon of a rotating black hole. That can only remain the same or increase.

This idea that the surface area of either a rotating or a non-rotating black hole can only increase or remain the same got stated as the second law of black hole dynamics by Stephen Hawking. He said, "In any natural process, the surface area of the event horizon of a black hole always increases, or at best, remains constant; it never decreases." That's the second law of black-hole dynamics. Worded in this way, it closely resembles the second law of thermodynamics. Thermodynamics is the study of the relationship between heat, and work, and other forms of energy—what happens to the temperature of a gas if you pump on it, like in a bicycle tire. All that is thermodynamics, and it's a very important field in physics. It has four laws: the zeroth, first, second, and third. The second law is by far the most important. It plays a huge role in governing the physical Universe. The second law states the following: "In any natural process, the entropy of a closed system always increases or, at best, remains constant; it never decreases." That sounds a lot like the second law of black-hole dynamics, where it's the surface area of the event horizon that always increases or, at best, remains constant.

What are entropy and a closed system? Entropy is just a measure of the amount of disorder in a system—things jumbling around. Are there many different ways of organizing something, or are there only a couple of different ways? How much disorder is there? You start out with a clean desk, maybe at the beginning of the week—I never have a clean desk—but by the end of the week, it's all jumbled up; that's entropy. Entropy of the Universe basically increases unless you take pains to decrease it. If you do that, the entropy somewhere else has to increase. A closed system; what's that? A closed system is one from which nothing escapes; all are present and accounted for, so to speak. You haven't lost touch with any components of the system.

Stated in this way, the second law of thermodynamics and the second law of black-hole dynamics sound quite similar. It turns out that there are ways of stating the zeroth, first, and third laws of thermodynamics in a way that's analogous to the properties of black holes. That is, black holes behave in ways that can be described by sentences that resemble very much the zeroth, first, and third laws of thermodynamics, not just the second law. This resemblance, this correspondence, between the laws of black-hole dynamics and thermodynamics, led in 1972 to a proposal by the physicist Jacob Bekenstein that the surface area of a black hole is basically proportional to its entropy. You have the event horizon, and it has some surface area, which can never decrease in size. He said let the surface area be proportional to its entropy. There's some disorder in a black hole that can be measured in some way, but a nice way of just quantifying how much disorder there is, is by simply saying here's the surface area of the event horizon. The bigger the surface area, the greater is the entropy of the black hole.

You might say, "Why is there any entropy at all in a black hole? What's all this disorder? We don't see anything inside it, so it's not like my cluttered desk." The entropy of a black hole is related to the no-hair theorem that I mentioned in Lecture 60. Remember, I said that a black hole has no hair. A black hole in equilibrium, which is not still collapsing, but has pretty much reached an equilibrium configuration, can be described by three, and only three, properties: its mass, its angular momentum or spin, and its charge. In fact, we don't even expect most black holes to be charged—or even any, for that matter—because they rapidly steal the opposite charges away from their surroundings and neutralize themselves, even if they happen to have been born with some charge. A black hole is a very simple structure. It has no details; it has no hair. It only has a mass and a spin, and maybe a charge.

That's kind of weird because there was a lot of information that went into making the black hole. You've got this black hole, and you throw books into it; they're full of information. You throw sculptures in it. You throw yourself into it; that wouldn't be so smart. But anyway, you throw things into a black hole that clearly have information. They have properties, and you know those properties. Yet after the black hole has devoured them, you can't tell anything about what the black hole consists of. You don't know what was thrown in there. You've lost all that information. A loss of information corresponds to a gain in entropy. Therefore, indeed, black holes have a tremendous amount of entropy because of this loss of information.

Next step—the analogy with the laws of thermodynamics suggests that a black hole is physically a thermal body. It's like a body jiggling around, and emitting radiation, and all that. Indeed, it resembles very closely what we called a black body earlier in this course. Remember, a black body is one that doesn't transmit any radiation, and it doesn't reflect any radiation; it only absorbs it, and it heats up. A black hole doesn't transmit any radiation—a black hole isn't transparent—and it doesn't reflect any photons; it only absorbs them. It really looks, in that sense, like a black body. It absorbs stuff; it doesn't transmit it or reflect it. If a black hole can be thought of as a thermal body of this sort, then it must have a temperature associated with it—because any thermal body has a temperature associated with it. If there's a temperature associated with the black hole, then, as in the case of any black body at a non-zero temperature, it must radiate; it must shine, right? Any thermal object with a non-zero temperature has things jiggling around, and they shine; they emit radiation. That's what a black body does.

The idea, then, would be that a black hole, being like a thermal body (like a black body) shines. That's weird. That's paradoxical right away because remember, I said that we're ignoring quantum effects. In classical physics, nothing can get out of a black hole. Yet the whole shining process of a thermal body is really a quantum-mechanical property. So, you've got this conundrum where you started treating the black hole as a classical body, and suddenly you conclude that it shines because it's got a temperature. But shining is an inherently quantum-mechanical process, so you're forbidden from using that. Plus, nothing can get out of a black hole anyway. We were taught that since we were on our mother's knee, or at least since you heard me a few lectures ago.

Therefore, Bekenstein basically didn't pursue the idea any further. He felt that this is where the analogy between black holes and normal black bodies must fail. He didn't pursue it. But Stephen Hawking realized that the conclusion that nothing can come out of a black hole might not hold if you consider quantum-mechanical processes. He said that maybe stuff can come out. For example, if you have automatic creation and annihilation of little particles, and antiparticles, and things like that, maybe there's a way for them to get out in some cases. In fact, then, in 1975, Hawking published an article where he proposed that black holes can evaporate through a quantum-mechanical process, regardless of the details of whatever theory finally emerges unifying quantum mechanics and general relativity in a fully self-consistent way. He said that this should happen regardless of the details because it's a fundamental process having to do with creation of little particles and antiparticles.

He describes some of his work in a very popular book published in 1988, *A Brief History of Time*. It's an interesting book. It's heavy going at times; it's not conceptually easy. Indeed, he's written some books since then—*The Universe in a Nutshell*, and *A Briefer History of Time*—where the concepts are explained a little bit more clearly. Plus, more time has gone by, and new things have been discovered. The original book, *A Brief History of Time*, has one equation in it, $E = mc^2$. Apparently, his publishers told him that sales would drop by one-half for every equation that he includes in the book. He apparently wasn't willing to have sales drop too much, but $E = mc^2$ is so important that you can't have a book on physics without including $E = mc^2$.

As I said, this book is interesting, but a tough read. It's been purchased by a lot of people. It was a number one bestseller, but I kind of like to joke that it's the most purchased, but least read through to completion, book in existence. People like to have it on their coffee tables and stuff, regardless of whether they really understand it. Visitors come by, and they say, "Ooh, you read Stephen Hawking's books. You must be really brilliant." I'm not denigrating anyone here if you don't understand his book. It's hard stuff. It's difficult quantum mechanics and general relativity, which he's trying to explain to a layperson. He does an admirable job, but I think a better job in his later books, which were written also with the help of people who have a lot of experience explaining things to the general public.

Here's the basic idea for the evaporation. It goes back to the Heisenberg Uncertainty Principle, which I had already introduced in Lecture 20, when discussing the wave particle duality of light. You will recall that Heisenberg was this great physicist who thought of things while other physicists were also considering quantum aspects of the world. Heisenberg in particular said that there's a certain uncertainty in any measurement. If you measure, for example, the energy of a physical process over a certain amount of time, then the product of the uncertainty in your measurement of the energy and the uncertainty in the time at which you made that measurement, which is essentially the length of time the measurement took, has to be bigger than some number. It's a small number. Here you can see $\Delta E \, \Delta t$ is bigger than, or approximately equal to h, Planck's constant, after Max Planck, a very small number—divided by 2π. He said that you cannot know with perfect precision the energy of a system or the time at which you made that measurement, such that, say, the product is zero or less than this quantity, $h/2\pi$.

Similarly, there's a relationship regarding the product of the uncertainties in the position of a particle, let's say, and its momentum, its mass times its velocity. You can't know both quantities with perfect precision. $\Delta x \, \Delta p$—p is momentum, and Δ means uncertainty—that product, $\Delta x \, \Delta p$, is at least as big as Planck's constant over 2π. I have this T-shirt from Purely Academic T-shirts that says, "Heisenberg says $\Delta x \, \Delta p \geq \hbar$"—$\hbar$ is $h/2\pi$—"The Equation knows best." Indeed, this is one of the fundamental equations of all of quantum physics. Heisenberg contributed to the ideas that many of his contemporaries were giving out. But his uncertainty principle, I think, is one of the most essential aspects of quantum theory. You really can't have quantum theory without something like the uncertainty principle.

You can know the position of something really well, in which case you don't know its momentum very well at all—its mass times its velocity, or just its velocity let's say—or you can know the energy really well, a small uncertainty in the energy, but in that case, you don't know the time very well. Or vice versa—you can know the time really well and not the energy, or you can know the speed or momentum really well, but not the position. These are complementary quantities. There's this great joke about Heisenberg. He was driving a car, and he was breaking the speed limit by a lot. He was stopped by a police officer, and the police officer said, "Do you know how fast you were going?" And Heisenberg said, "No, sir, but I know where I am." He knew where he was, but he didn't know at all how fast he was going, so he appealed to the uncertainty principle. I think he was still issued a ticket, but I don't really know. No, it's a joke.

Let's get back to the energy and time version. $\Delta E \, \Delta t$ is greater than $h/2\pi$ whenever we make an observation of a real system. However, there can be cases where virtual particles form spontaneously out of nothing, with an energy—E or ΔE; ΔE is just their energy. They start out with nothing, and now they have some energy, so that's a change in energy, ΔE, and they exist for a certain amount of time. Boom, Δt! If that time is short enough, the product, $\Delta E\ \Delta t$, can be less than Planck's constant/2π. $\Delta E\ \Delta t$ is less than Planck's constant/2π. This is a temporary quantum violation of the classical law of conservation of energy because something appeared out of nothing. But that's okay; quantum mechanics allows violations of classical laws. That's what quantum mechanics is all about. Another way to think of it is that the positive energy created by this little pair of particles that gets created leaves behind sort of a negative energy hole, so that they cancel out, and the energy is still zero. There are different ways of thinking about it, and I'll talk about this more later when I talk about the dark energy of the Universe.

In any case, this creation of virtual particles is going on everywhere in this room, everywhere inside of atoms, everywhere all the time. They are called quantum fluctuations. You can look at them here. You've got, for example, a neutrino gets created out of nothing, and so does an anti-neutrino. A little while later, they annihilate. A proton gets created out of nothing, and so does an anti-proton. A little while later, they annihilate. You start with zero; you end up with something for a short time, and then finally, they annihilate, and the final state is zero once again. Quark and anti-quark; there was zero to begin with, and then for a while there were both of them—and then a little bit later, there's zero in the end, or electron and positron.

These are all matter/antimatter pairs, particles/antiparticles. What do you call matter and what do you call antimatter? It doesn't really matter, if you get my drift. We could have called antimatter matter, and then the opposite thing would have been antimatter. They're just the opposites. When they hit each other, they annihilate. We've already encountered positrons, which are anti-electrons, when we considered nuclear reactions in the Sun and other stars. These quantum fluctuations are virtual in that you can't directly measure them because they don't last long enough. But they do affect the Universe; they affect the energy levels of the hydrogen atom, and there are other things called the Casimir effect. I'll discuss this all more when I discuss this dark energy in the Universe that appears to be making the Universe expand at a faster and faster rate. I'll get to this in more detail in the cosmology part of the course.

Suffice it to say for now that these quantum fluctuations do exist, and their influence on the properties of atoms and things has been measured. They really are real, but you can't measure any specific quantum fluctuation individually because to do so would be a violation of this Heisenberg Uncertainty Principle. They only exist because they themselves are violating the Uncertainty Principle. If you try to make a measurement on them, the product of the energy and the time would have to be bigger than Planck's constant/2π. That's not allowed because they don't exist to allow you to make that kind of a measurement. This is crazy stuff. No one has a good intuitive feel for quantum mechanics, and that's just the way it goes. But it works; it gives predictions that agree with experimental measurements in so many ways.

Back to quantum fluctuations, now near a black hole. Near a black hole, these fluctuations can occur all over the place. They can occur outside the black hole. For example, an electron and a positron can spontaneously appear and then disappear. That can happen within the black hole as well, or maybe it happens outside the black hole, and then both of them go into the black hole and then they disappear; they annihilate. Most of the time, they just spontaneously form, and they annihilate. But occasionally, one of the two particles—and it doesn't matter whether it's the particle or the antiparticle; in this case, the electron or the positron—can go in, leaving the other one outside. That's weird. If the other one is left outside, it doesn't have a partner with which to annihilate, and so—in principle—it could escape. This is amazing; it can escape, carrying with it energy, positive energy.

Where did that energy come from? From our perspective, the particle or antiparticle that went in, went in with negative energy. This is hard stuff; it's not intuitive. Remember, in the black hole case, where we outside think of the singularity as being a point in space. We always talk about it being the center of the black hole. But then I said if you actually enter a black hole, specifically a non-rotating one, the singularity is horizontal in the space-time diagram. That is, it exists over a large region of space, not at a single point. Space and time kind of become reversed inside a black hole. The singularity is not a point in space; it's sort of everywhere in space, and it occurs at some time in your future. Time and space reverse their meaning, as seen from the outside. Similarly, particles can travel in with negative energy, as seen from the outside. That's just the way it works.

The other guy, the one that didn't go into the event horizon, can escape. As a bunch of them escape, all together, they interact, create photons, and annihilate each other, creating even more photons, or they accelerate past each other, emitting photons. The result is that you get a thermal distribution of energies for the resulting photons and the particles that escaped. This is called Hawking radiation. In fact, it is the evaporation of black holes. When I give this lecture in my class at UC Berkeley, I dress up as a black hole, and I've got a little alien hanging from my neck. That alien is being tidally disrupted, or spaghettified, by the tidal forces of the black hole. There's a little thing in the alien that says "Take me to your leader." I have buckets attached to me with celestially appropriate candy—like Mars bars, and Eclipse gum, and Orbit gum, and Starburst candy, and Milky Way, and all that. I have these little buckets attached to me, and I start throwing the candy out to the audience, and that's to illustrate an evaporating black hole. It's very memorable to the students, so they actually get that part of the exam questions right.

This black hole evaporation is almost like a tunneling process; that's another way you can think of it. Again, I could ram myself against the wall and eventually make it through. You can think of these black-hole evaporation processes as, in a sense, particles being created inside the event horizon and then tunneling their way through. That's another way of thinking about it. The mathematics is very complex, but there are various more or less intuitive ways of trying to think about it. I spoke at length about the particle creation and annihilation process, the quantum fluctuations. But another way of thinking about it is quantum tunneling or something. In any case, you get this evaporation.

Outside the event horizon, there are all these particles and photons coming out. The black hole is losing energy in the process, and the temperature associated with the black hole is given by Planck's constant times the speed of light^3, divided by the quantity $16\pi^2$ times a thing called Boltzmann's constant, K, times G, Newton's constant of gravity, times M, the mass of the black hole. The important part here is that the temperature is proportional to the reciprocal, or the inverse, of the mass of the black hole. As the black hole evaporates, its mass decreases. The temperature rises because it's the reciprocal of the mass. The luminosity of a black body is proportional to R^2T^4, if you've got a spherical one. Remember, $L = 4(\pi)R^2T^4$? So L is proportional to R^2T^4; r is proportional to M; t is proportional to 1/M. In the end, L is proportional to $1/M^2$, if you do the math. As the mass decreases, the temperature of the black hole increases. The rate of evaporation increases, and as the mass goes to zero, the evaporation rate goes to infinity. That is, the thing, in the end, explodes.

For any stellar-mass black holes, or supermassive black holes in the middles of galaxies, the rate at which they evaporate is utterly negligible. They accrete material from their surroundings much, much more quickly. This evaporation is only at a high rate for black holes that are tiny, maybe a billionth of the mass of the Earth or less. Hawking has suggested that such tiny black holes do form shortly after the birth of the Universe, due to density variations shortly after the Big Bang. If that's the case, then all of the ones that are born with less than 10^{15} grams—that's about the mass of a big mountain—have already evaporated away. There's been enough time, 14 billion years, for them to evaporate away.

The ones that are 10^{15} grams in mass initially, having an event horizon roughly the size of a proton, are right at this time evaporating furiously away. Since the rate of evaporation approaches infinity as the mass approaches zero, you end up with these bangs. Most of the photons that should come out should be gamma-ray photons. If miniature black holes exist from the birth of the Universe, then Hawking says some of them should be evaporating right now, and should be giving rise to gigantic bursts of gamma rays, high-energy photons. Have we ever seen any gamma-ray bursts that might be indicative of the evaporation of miniature black holes? Stay tuned, and I'll tell you in the next lecture.

Enigmatic Gamma-Ray Bursts

Lecture 65

"We do think that gamma-ray bursts are linked with the formation of black holes—not miniature ones, but stellar-mass black holes. … These gamma-ray bursts, or GRBs, are among the hottest topics of modern astrophysics."

Celestial gamma-ray bursts (GRBs) have been seen, and their presence is one of the most exciting and intriguing areas of study in the field of astrophysics. GRBs were first detected in the late 1960s when the U.S. Air Force launched the Vela spy satellites to monitor Soviet compliance with the Nuclear Test Ban Treaty. They did not find any violations of the treaty, but they did find peculiar celestial bursts of gamma-ray light. The bursts appeared in random parts of the sky, and when we look at curves of the brightness of the gamma rays versus time, we find that no two of them are alike. Some of the curves are spiky, while others have a smoother distribution; some have only two spikes, and some have many. Nevertheless, there are two main types of GRBs: long-duration and short-duration. The long-duration GRBs can last as long as several hundred seconds (average: about 20 seconds) and tend to emit fewer very-high-energy gamma rays. The short-duration GRBs last for less than 2–3 seconds (average: a few tenths of a second) and tend to emit more very-high-energy gamma rays.

For many decades, we didn't know what produced GRBs, though there was a plethora of hypotheses. Between 1973 and 1992, theoretical astrophysicists published about 120 distinct hypotheses in an attempt to explain the physical nature of GRBs. Could they be explosions on neutron stars or comets hitting such stars? Could they be annihilations of large quantities of matter and antimatter? Unfortunately for Stephen Hawking, the hypothesis that they are exploding miniature black holes was not viable. If it were, then GRBs would not show spikes in brightening and fading (and there were other problems, as well). A major impediment to the understanding of the physical nature of GRBs was the fact that we didn't know their distance. Are they close to Earth or billions of light years away? Are they halo objects bound to our Galaxy? Between 1991 and 2000, NASA's orbiting Compton Gamma

Ray Observatory detected and mapped the approximate positions of 2700 GRBs. The most significant finding was that GRBs occur with a completely uniform distribution. No area of the sky has a concentration of GRBs that is statistically greater than any other area. This fact strongly suggested that GRBs occur at cosmological distances—very, very far away.

However, even with the Compton Gamma Ray Observatory, we still could not precisely pinpoint the locations of GRBs in the sky. In 1996, a huge breakthrough occurred with the launch of BeppoSAX, an x-ray satellite capable of taking images of the sky where bursts occurred at lower photon energies—x-ray energies rather than the high-energy gamma rays. The satellite could use the x-ray image to pinpoint where a GRB occurred. With these precise positions, optical astronomers could then search the sky for fading optical afterglows corresponding to GRBs. It was found that long-duration GRBs appear to be associated with galaxies, indicating that they really are far away and that, undeniably, their luminosity must be incredibly great.

NASA

The Compton Gamma Ray Observatory.

Further satellite launches, as well as ground-based telescopes, have been able to gather still more information about GRBs. Thus far, we have found definitive optical counterparts primarily for long-duration GRBs. There are only a few known counterparts for short-duration bursts because our satellites generally don't have enough time to determine an accurate position for the short-duration bursts. Studies of the radiation from GRBs show that the bursts must be ejected in jets in a highly beamed fashion, like the narrow beams of lasers. Some kind of driving force creates a burst of relativistic particles—those traveling near the speed of light—which are ejected along two oppositely directed axes. The particles hit each other, causing internal shocks. The collisions produce gamma rays; then, sometime later, those energetic clumps of particles hit clouds of external gas to produce radio waves, optical radiation, and x-rays. The external shocks produce the afterglow at other wavelengths, and internal shocks produce gamma rays.

Long-duration bursts occur in galaxies forming enormous numbers of massive stars; thus, the jet mechanism should be associated with massive stars. The **collapsar model** describes this process. A collapsing massive star can form a jet along its axis of rotation. This jet of radiation and particles pummels its way through the star, bursting through the surface and creating two oppositely directed beams, or jets. If one of the jets happens to be pointing toward our line of sight, it appears very bright. If neither of them points our way, we don't see the GRB, but we may see a relatively normal supernova. Such bursts probably work best in stripped, core-collapse supernovae—for example, those that don't have much or any hydrogen or helium envelopes. Massive stars with an iron core and an outside layer of helium (which produce Type Ib supernovae) and those with an iron core and an outside layer of carbon and oxygen (which produce Type Ic supernovae) don't have as much material in their envelopes, so the jets can pummel through much more easily than in the case of hydrogen-rich massive stars. So far, the model suggests that at least some long-duration GRBs arise from core-collapse supernovae. Further studies are needed to verify this theory. ■

"In the 20 years between 1973 and 1992, 118 distinct hypotheses had been published by theoretical astrophysicists trying to explain the physical nature of gamma-ray bursts."

Important Term

collapsar model: Model proposed for some types of gamma-ray bursts, wherein a rotating, massive star collapses and forms two highly focused beams (jets) of particles and light.

Suggested Reading

Katz, *The Biggest Bangs: The Mystery of Gamma-Ray Bursts, the Most Violent Explosions in the Universe.*

Pasachoff and Filippenko, *The Cosmos: Astronomy in the New Millennium*, 3rd ed.

Schilling, *Flash! The Hunt for the Biggest Explosions in the Universe.*

Wheeler, *Cosmic Catastrophes: Supernovae, Gamma-Ray Bursts, and Adventures in Hyperspace.*

Questions to Consider

1. What would we expect the gamma-ray light curve of a GRB to look like if GRBs were the evaporation of miniature black holes according to Stephen Hawking's hypothesis?

2. Why did the essentially uniform (isotropic) distribution of GRBs in the sky found by the Compton Gamma Ray Observatory support the hypothesis that GRBs are at cosmological distances?

3. Assume that the Andromeda Galaxy (M31, 2.4 million light years away) is very similar to our Milky Way Galaxy. If GRBs were associated with a very extended, spherical halo of our Galaxy, do you think there should be a non-uniformity across the sky in the observed distribution of GRBs, given enough data points?

Enigmatic Gamma-Ray Bursts

Lecture 65—Transcript

In discussing the evaporation of miniature low-mass black holes, like the mass of a mountain or something, I said that a burst of gamma rays would be emitted, a burst of radiation just as the thing's mass is going to zero. Interestingly, celestial gamma-ray bursts, or GRBs for short, have indeed been detected. Unfortunately for Stephen Hawking, they are very unlikely to be evaporating black holes. They just don't have the right properties, and I'll describe the properties of gamma-ray bursts in a few minutes. However, we do think that gamma-ray bursts are linked with the formation of black holes—not miniature ones, but stellar-mass black holes, it turns out. These gamma-ray bursts, or GRBs, are among the hottest topics of modern astrophysics.

There's just a flurry of activity going on right now in research on GRBs. You can see, in fact, that even Bill Clinton was interested in this question. In his science and technology policy address, he said, "…there are so many more questions yet to be answered…And so I wonder…Are we alone in the Universe? What causes gamma-ray bursts? What makes up the missing mass of the Universe? What's in those black holes, anyway?" He covered a lot of the important questions, as I had mentioned in a previous lecture. Gamma-ray bursts were on his mind during this science and technology policy address in the year 2000.

The history of gamma-ray bursts, of celestial ones, starts in the late 1960s, when several spy satellites—called the Vela spy satellites—were launched by the U.S. Air Force to monitor Soviet compliance with the Nuclear Test Ban Treaty. They wanted to see whether the Soviets were essentially exploding nuclear weapons, which create burst of gamma rays in the process. They did not find any violations of the treaty, but what they did find were these celestial sources that went bang in the night, or in the day—it doesn't matter—bursts of gamma rays over a very short interval of time, just a few seconds up to a few hundred seconds. It was really weird because no one knew what could produce such celestial bursts. Here's a picture of one of the Vela satellites looking around, looking for violations of the Test Ban Treaty.

Instead, it made this amazing discovery, which launched a whole new area of astrophysics.

Here's sort of an animation of what you might see with the Vela satellite. There goes a burst of gamma rays, and then a few days later, there's another burst, and another burst. They appear in random parts of the sky. They didn't appear to be concentrated in any single region, although the Vela satellites only found 73 of these things. You might say that, with such a small number, there might not have been any apparent concentration toward any part of the sky. When you look at the brightness of the gamma rays versus time, the so-called gamma-ray light curve of a GRB, you find that no two of them are the same. They're like fingerprints; none of them are the same. Here's a bunch of them, where on the vertical axis we're plotting the number of photons, or counts, detected per second in units of one thousand, versus time in seconds. You can see that some of them are very spiky. Some have a more smooth distribution. Some consist of only two spikes, some of many spikes. The joke goes when you've seen one GRB, you've seen one GRB. You haven't seen them all because they're all different.

Nevertheless, there are two main types of GRBs: the so-called long-duration and short-duration GRBs. The long-duration GRBs are represented by this typical example. They don't last long on a cosmic time scale, just a few tens of seconds or a few hundred seconds, but that's long compared to another class, the short-duration ones, which last only a second or two, or less. Look at this one. It just spiked up and down in much less than a second, whereas this one had many spikes over about 100 seconds. Those are the long and short kind. Here, you can also see one that lasted a few seconds, four or five seconds. Some astronomers thought that that was yet a third class of GRBs. But really, now that we've found many, many, many GRBs, we see that there's a distribution of times for the short ones, and there's a distribution of times for the long ones.

There's a valley that's right around a couple of seconds. We simply say that the ones that are shorter than two or three seconds' duration are the short ones, and the ones longer than two or three seconds' duration are the long ones. But any specific GRB of duration of two or three seconds, you don't really know whether it's fundamentally the long-burst variety or the short-burst variety. They have some other differences as well. The short ones tend to emit more really high-energy gamma rays than the long ones. There are other differences, but in some cases, you can't really tell which one you're looking at.

These gamma-ray bursts appeared with no particular concentration in the sky. Let me show you a map again of the Earth. This is the whole Earth, and an analogous map of the sky would look like this. You can see the plane of the Milky Way Galaxy going across the center of the map here. There's the halo above and below it, and the bulge region in the middle. What we're saying is that the GRBs found by the Vela satellites kind of appeared everywhere randomly in the sky. If these things had been, for example, surface explosions on neutron stars, you might expect that neutron stars accreting material might go through a nova process, kind of like a white dwarf accreting material. That was one postulated model for gamma-ray bursts.

If they're neutron stars, you would have expected their distribution to resemble that of the Milky Way Galaxy—that is, especially of the massive stars in the Milky Way Galaxy because massive stars tend to produce neutron stars at the end of their lives, and they would be concentrated toward the plane. Some neutron stars escape, and so they're seen up here, but most are along the plane. This is, indeed, the distribution of pulsars that I showed in an earlier lecture. The fact that they were distributed more uniformly, more isotropically, as it is called, immediately showed that these things are not basically neutron star novae. What could they be with such an isotropic distribution? One possibility was that they are very nearby, just a few light years away. If they're all just a few light years away, you might expect a spherical or an isotropic distribution around you.

But in that case, why don't we see the ones that are farther away? By that, I mean all the gamma-ray bursts that the Vela satellites detected were easily detected. They were easily bright enough to be seen. There could have been fainter ones that would have been barely bright enough to be seen, but the Vela satellites didn't see them. When you see the gamma-ray burst, it's really bright. So, if there's a spherical distribution of weird objects around us, why does it end at a couple of light years? Why aren't there ones farther away, in which case we'd see fainter gamma-ray bursts. That possibility was ruled out rather quickly.

Another possibility was that these things are enormously distant, billions of light years away. The Universe extends in all directions for billions of light years, so if they're uniformly distributed in the Universe, but mostly very, very far away from us—because they're rare, let's say—then you'd see this spherical distribution, or an isotropic distribution is another way of putting it. In that case, if they're billions of light years away, the intrinsic power, or luminosity, of these gamma-ray bursts must be truly stupendous—because they're bright, and if they're, in addition, at such gargantuan distances, that means they must be intrinsically very, very powerful. No one knew of a way to come up with this stupendous amount of power.

Or, people said, look at the distribution of our Milky Way Galaxy. What if they are halo objects? What if they're way out here, but basically bound to our galaxy? That would produce, to a first approximation, a spherical or isotropic distribution. But there are problems with that model because, as I will tell you in more detail later, our Sun is about one-half to two-thirds of the way out from the center of our galaxy toward some ill-defined edge. We're out here somewhere, and there's more halo volume in these directions than in the opposite directions. Therefore, you'd expect more GRBs in these directions over here, where there's more volume for the halo. Then astronomers said well, what if they are a population of an even bigger halo, sort of 10 times bigger than the halo drawn in this picture here? In that case, we're nearly in the center of that distribution, and you might expect it to be isotropic.

The problem with that is that, if you make the halo too big, then you would expect the Andromeda Galaxy, if it has a similar halo, to have a similar number of gamma-ray bursts, and you might expect to start seeing a concentration of them toward the Andromeda Galaxy. If our galaxy is typical, and the Andromeda Galaxy is kind of like ours, and they have these vast halos of gamma-ray bursts, you should see Andromeda's population of gamma-ray bursts; yet there was no apparent concentration of gamma-ray bursts toward Andromeda. There was a big debate. Are they halo objects? Are they at these so-called cosmological distances, billions of light years? It's very difficult to tell. We really didn't know for many years.

Well, this was great for theorists because they could come up with all sorts of models, physical mechanisms, for what gamma-ray bursts might be. They were essentially unconstrained by the data because we didn't really know how far away they are. In fact, in the 20 years between 1973 and 1992, 118 distinct hypotheses had been published by theoretical astrophysicists trying to explain the physical nature of gamma-ray bursts. For example, they might be matter/antimatter collisions. You have balls of matter and antimatter. Who knows where they came from, but they would produce gamma rays. Or maybe they're comets hitting the surfaces of neutron stars. Neutron stars have a high surface gravity. There's a lot of pull there, so when a comet hits the surface, it releases a lot of energy, and that tends to come out in energetic gamma rays—or maybe they're a new class of exploding stars or something like that.

The theorists were just running wild. Unfortunately for Stephen Hawking, the theory that they are miniature black holes exploding was not a viable theory for very long, despite the fact that there were not that many observations. There were some, and the observations pretty much don't agree with his hypothesis of exploding black holes, miniature black holes. The basic reason is that you saw that these gamma-ray bursts are kind of spiky; they brighten and fade, and brighten and fade. Evaporating miniature black holes shouldn't do that. It should just go zoom; it explodes at the end, and there's an exponential sort of ramp-up. It goes bang, like that, and it shouldn't do all these little wild oscillations in brightness. There were other problems as well. So, although one of the 118 hypotheses was that these are the long-sought miniature black holes, unfortunately that's not what they are. That's the way it goes.

Great progress was made in this area by the Compton Gamma-Ray Observatory, one of NASA's four great orbiting observatories. There's the Hubble Space Telescope, the Spitzer Space Telescope, the Compton Gamma-Ray Observatory, and then there is the Chandra X-ray Observatory. These are fantastically important telescopes. In particular, the Compton Gamma-Ray Observatory lasted from 1991 to the year 2000, and it detected a whole bunch of gamma-ray bursts. Here's an artist's impression of the Compton Gamma-Ray Observatory. You can see a number of gizmos here; these are the different detectors that it had. It was deployed by the Space Shuttle, and then the various instruments began to collect data.

In particular, there was an instrument called the Burst and Transient Source Experiment. You can see that there are actually eight little devices at the eight corners of this rectangular box. These devices scan the sky in eight octants. They essentially monitor most of the sky at any given time, this BATSE thing. Here, two of the eight are shown. This BATSE device then detected a whole bunch of gamma-ray bursts. You can see one going off right there. In the sky above the Milky Way, in this particular case, you can see a burst, and you can plot the brightness, or the photons per second, versus time and see this thing brightening and fading. That's what BATSE did, and it found 2,700 of these things during the nine or 10 years of operation. When you plot the distribution of these nearly 3,000 gamma-ray bursts in the sky, you can see that the distribution is completely uniform; it's isotropic. No area of the sky has a concentration of gamma-ray bursts that is statistically greater than any other area.

You might say, "Well, there's a hole right there, and there's a clump right here." But if you throw darts at the wall in a room, you will get statistically significant looking clumps and areas where there weren't any darts, even if you just throw blindly. You have to actually look at the statistics and figure out whether this is a physical clumping or not. It turns out this is consistent with random throwing of darts onto the wall. There's no clear clumping in any part of the sky. The different colors are used to signify different energies for the gamma-ray bursts—how bright they were, how intense they were.

These and other data suggested that gamma-ray bursts are, in fact, at cosmological distances—very, very far away. They couldn't really be in an extended halo of our galaxy because we still live sort of toward one side, and you'd see more of them in one region than the other. Moreover, if Andromeda is similar to our galaxy, we should have started seeing gamma-ray bursts corresponding to Andromeda, which is only 2 million light years away. If our halo is nearly that big, and we're seeing gamma-ray bursts in our halo, then we should see them in Andromeda as well. Those are comparable distances. We didn't see a distribution that was concentrate toward Andromeda, and so most people started reluctantly saying that these things are billions of light years away. But that was a conclusion that didn't make people feel good because the energetics of these objects was still unfathomable. Moreover, we didn't have any other way of studying these objects because no counterparts to gamma-ray bursts at other wavelengths had ever been found.

The BATSE device on the Gamma Ray Observatory told you roughly where a gamma-ray burst occurred in the sky, but it had an uncertainty that was pretty large, five degrees or so—a diameter kind of like that of an outstretched hand. You couldn't tell exactly where these things were. People were trying to find optical, and x-ray, and radio counterparts, or afterglows, to gamma-ray bursts in order to study them at other wavelengths and learn more about them. But there was nothing to be seen because these instruments wouldn't tell us precisely where to look, and so people were frustrated. In 1996, a huge breakthrough occurred when an x-ray satellite known as BeppoSAX was launched. It had the capability of not only detecting gamma-ray bursts, but also of getting an image of that part of the sky where the burst occurred at lower photon energies, x-ray energies rather than just the high-energy gamma rays. It was able to look and pinpoint where the gamma-ray burst occurred.

Lo and behold, it found x-ray counterparts to the gamma-ray bursts. Here's a visual picture of one of them. These are optical photons, but they're used to represent x-ray photons. Here, you could finally tell, with considerable accuracy, where the gamma-ray burst occurred. With these precise positions provided by the BeppoSAX satellite, optical astronomers could then look at the sky and look for fading optical afterglows corresponding to the gamma-ray bursts. Sure enough, they were found. Here is a bright object—it looks dark in this picture because this is a negative image, not a positive image. The sky should be black, and the stars should be white. But here—on February 28, 1997—there's a bright thing. This was the same day as BeppoSAX detected a gamma-ray burst at this position and found an x-ray afterglow. Then, a week or so later, that bright optical thing had faded, and there was nothing left there. There was a little fuzzy thing next to it. That fuzzy thing is a galaxy. Spectra of the galaxy showed that it was really far away. I'll tell you soon how we deduced those things; we're getting to galaxies in the next part of the course.

This optical transient was associated with a distant galaxy. Other optical transients were found as well. Here's one for GRB 971214. They're named after the year in which they were found, the month, and the day—so this was December 14, 1997. Here is a Keck pair of images, taken shortly after the GRB occurred, and then some time later, and it had faded. Being associated with galaxies in all of the cases where we could see an object next to the gamma-ray burst meant that these things really are far away and that, undeniably, their power or luminosity must be incredibly large. Some of these things are unbelievably powerful. There was one found in 1999, a gamma-ray burst that alerted astronomers here on the ground, and they studied that burst. But there were also robotic telescopes that were scanning the sky, and happened to catch the optical afterglow of that gamma-ray burst.

This telescope in particular, called ROTSE, consists of four wide-angle cameras that are sort of staring at the sky all the time, taking pictures, multiple exposures. They caught the optical afterglow of the gamma-ray burst without even having been notified that a gamma-ray burst occurred. It was in their data already because they were taking data over a wide swath of the sky. Here, you can see this optical afterglow on January 23, 1999. You can see that in a previous image taken in 1994, there was nothing there. This optical afterglow was so bright that, if you had been looking at the right part of the sky through a good pair of binoculars, you would have seen it. It was ninth magnitude, a factor of 16 or so fainter than can be seen with the naked eye. With a good pair of binoculars, you could see it; yet it turns out that this object is billions of light years away, roughly 10 billion light years away. That is amazing. When the Hubble Space Telescope looked at the afterglow, it found that it is in a fuzzy thing whose spectrum showed that it's a distant galaxy.

Indeed, gamma-ray bursts are cosmological; they occur in very distant galaxies. That was the big breakthrough of the BeppoSAX satellite. But BeppoSAX only lasted a few years, and we wanted to get lots of these things. It found a few dozen, but we wanted hundreds in order to study their properties overall. Some are different from others—we already know that from their light curves—so you need to study an ensemble of objects. Some astronomers said, "Let's launch satellites that will specifically find these things, and then alert (in real time) ground-based telescopes to the presence of the gamma-ray bursts, so that those ground-based telescopes can right away, maybe even robotically, turn their gaze toward the right part of the sky and capture a possible optical afterglow."

There was an experiment called the High Energy Transient Explorer, HETE; there were two of them. HETE-I failed after launch, but HETE-II was out there, and it found gamma-ray bursts and measured their positions. Then it transmitted those positions, through radio signals, to telescopes on the ground—to one main station on the ground, which then distributed the data, or the position of the gamma-ray bursts, to telescopes throughout the Earth. The way that this HETE thing was able to do that was that, during its orbit, it is always above a receiving station on the ground. There's a network of radio telescopes on the ground that could capture the signals from the HETE-II satellite and transmit those signals to a central processing unit, which would then tell all the ground-based radio telescopes what's going on.

Here you have HETE flying above all of these radio receivers on the ground. There was one in the Galapagos and Hiva Oa; that's in the Marcasis Islands; on Maui and all over the place. This HETE thing was great, and it alerted robotic telescopes, like my telescope at Lick Observatory, KAIT, the telescope that Weidong Li and I, and my entire team, run primarily to search for supernovae. But we want to get into the gamma-ray burst game as well. Why not? It's an exciting game. This HETE satellite would send signals that our computer program would interpret correctly, and those signals would tell us the position of the gamma-ray burst.

The telescope would then automatically slew its position to the right part of the sky and start taking photographs repeatedly, even as we slept. We don't need to be present because HETE tells the computer the coordinates, and the computer is programmed to respond. Here's one of our best successes. There's the optical afterglow of a gamma-ray burst, and some time later, you can see it has faded. These things fade away quite quickly. This was a gamma-ray burst that was first seen on December 11, 2002. You can see that we took so many snapshots of it that we could monitor its light curve (or brightness) on the vertical axis, versus time. This was—at that time, in December 2002—the best-ever sampled light curve of the optical afterglow of a GRB. I'm very proud of our group and of our telescope for getting this light curve.

The only kind of gamma-ray bursts for which optical counterparts had been found through the early 2000s were the so-called long-duration gamma-ray bursts. We had never actually found a counterpart to a short-duration burst. That's because these satellites don't have enough time to determine an accurate position for the short-duration bursts because they don't last very long. But at least HETE and these other satellites had found a lot of long-duration bursts, had pinpointed their positions, and allowed us to study them. It turns out that studies of the radiation from these bursts showed that the radiation must be coming out in jets, in a highly beamed fashion, like a laser. What emerged was a so-called fireball model, where for some reason, jets of relativistic particles—particles traveling at nearly the speed of light—are ejected along two axes. Then they interact with material and cause a lot of radiation to come out.

Here's the idea. There's some engine here, which might be very small, creating a burst of energetic particles, which hit each other—there are many such bursts—causing internal shocks because these various bursts of particles slam into each other. The collisions produce a lot of gamma rays; that's the gamma-ray burst. Then some time later, those energetic clumps of particles hit clouds of gas, the interstellar medium, and those collisions produce radio waves, and optical radiation, and x-rays. It's this so-called external shock that produces the afterglow at other wavelengths, and an internal shock that produces the gamma rays themselves. What was found was that these long-duration bursts occur in galaxies forming lots and lots of massive stars—here's an example of one—so this jet mechanism must be something associated with massive stars.

Something called the collapsar model was developed by Stan Woosley and his group. He basically said that, in some cases, the collapse of a massive star can form a jet along the axis of rotation of that star, and that jet can burst through the star. If the jet is oriented along our line of sight, then we will see it glowing brightly. If it's pointing the wrong way, then we won't see it. It's kind of like a pulsar pointing the wrong way. Here's what could happen, then. You've got a massive star blowing off a wind of material. But then near the end of its life, that star decides to collapse. It's collapsing, like the first stage of a supernova, but it's also rotating. Along the axis of rotation, there's this jet of radiation and particles, which pummels its way through the star, bursting through the surface and creating two oppositely directed beams or jets that—if they happen to be pointing our way—appear really, really bright. If they're not pointing our way, then we won't see the gamma-ray burst. We might just see a normal supernova or something like that.

Here's one of these jets pummeling its way through the star, in a computer simulation. These are very complex calculations that are done. It looks like, in some cases, the jet can pummel its way out through the star, and then these accelerated particles shine a tremendous amount. They emit a lot of gamma rays, collide with other particles, and create this great burst. A snapped picture might look something like this. To demonstrate what might be happening, you can try this at home; I won't do it here in this nice classroom. You can take a yogurt container of some yogurt, but just make sure it's one of the containers that is narrower at the top than at the bottom. It's hollow here; I don't have anything inside. If you were to drop this thing very carefully in a vertical way and let it hit the ground, then you would see the yogurt go bursting out in a jet because it's being confined by the sides of the container. It's a dramatic thing to do, but just don't do it near any expensive furniture that can't be cleaned, or near first-of-their-type paintings or something like that, because it is very messy.

This is sort of like the supernova bounce demo, where the tennis ball was dropped on top of a falling basketball, and it got ejected. In a similar way, if you have a rotating star, then the sides can constrain the material to flow more or less along the axis of rotation. This works best, probably, if you have a stripped core-collapse supernova—for example, one that doesn't have much or any hydrogen or helium. So, it only has the carbon and oxygen layers, and the oxygen, neon, and magnesium, and the iron core. In that case, there's not as much material for the jet to pummel through, so it's able to get out more easily. The collapsar model suggests that at least some long-duration gamma-ray bursts are actually supernovae, and we see them along the jet axis of ejection. There's a perfect test for this model. We should go out, look at long-duration gamma-ray bursts, and see whether they indeed are, or are not, in some cases associated with supernovae.

Birth Cries of Black Holes

Lecture 66

"Why do we prefer a stripped star for the massive progenitor of a GRB? It's simply because the more material that has been lost prior to the implosion and explosion, the easier it is for this jet of particles to pummel its way through the remaining material and actually get out."

We continue our discussion of long-duration GRBs, beginning with a recap from the previous lecture. The leading hypothesis—the collapsar model—for long-duration GRBs is that they arise from massive, stripped, core-collapse supernovae that implode, forming a jet of high-energy particles along the axis of the star's rotation. We think that GRBs are more likely emitted from stars whose hydrogen and helium envelopes have been stripped away by stellar winds or through transfer of material to a companion star. In addition, these collapsing stars are likely to rotate rapidly, and rotation creates a natural axis along which to funnel jet-like ejecta. Moreover, rotation creates an accumulation of material in the equatorial plane, forcing the jets to shoot along the path of least resistance, the rotation axis.

Why would some massive stars produce jets—which we detect as GRBs—when they implode while others don't? One leading hypothesis states that this is the same mechanism as the core collapse previously described for Types II, Ib, and Ic supernovae. In this case, however, the star collapses not just to a smaller neutron star but, instead, to a black hole. Thus, it's possible that GRBs signal the birth of black holes. If the material collapses to a radius smaller than 10 kilometers (the average size of a neutron star), a greater release of gravitational energy will occur. As the material collides, it emits radiation in the form of photons, as well as high-speed charged particles. If there is more rapid rotation than normal, a high-speed jet can sometimes form. Material that isn't ejected could collapse into the emerging black hole to form an even bigger black hole, or it could explode. On the other hand, if a neutron star formed during the collapse, a flood of neutrinos could push out the remaining material in the normal way of a supernova, or some other effect might push out the remaining material. Thus, the prediction is that at least some long-duration GRBs are also associated with visible supernovae. A GRB afterglow fades with time, but perhaps a week later, supernova light becomes visible, dominating the declining afterglow of the GRB itself.

In 1998, a peculiar, low-luminosity GRB (GRB 980425, found on April 25) was detected that happened to be spatially consistent with the position of an optical supernova (SN 1998bw); the two observations were also consistent in time. The supernova occurred about a week after the GRB, consistent with the time it takes for supernova light to brighten. In this particular case, there was no GRB optical afterglow associated with the burst of gamma rays. The absence of the GRB optical afterglow made the supernova especially obvious in this case. Despite misgivings about the general connection between normal GRBs and stripped-envelope supernovae, the properties of the low-luminosity GRB 980425 and its associated SN 1998bw supported the collapsar model. The supernova was not just a highly stripped Type Ic but perhaps an even more highly stripped variety in which even part of the carbon-oxygen layer was absent.

An artist's depiction of a supernova.

In 2003, another GRB and an associated supernova occurred, and a comparison of its spectrum with that of SN 1998bw showed great similarity between the two events. The supernova was again clearly of the stripped-envelope, core-collapse variety. Because this GRB was more typical, more luminous, than GRB 980425—including a normal optical afterglow—this sealed the case in support of the collapsar model. But whether such events really do signal the birth of a black hole, rather than a neutron star, we still are not certain. Further satellite explorations have observed many GRBs. In particular, the Swift satellite, launched in 2004, has detectors that scan the sky in search of GRBs. It also has x-ray and UV optical telescopes that quickly take pictures of the part of the sky in which GRBs occur and find x-ray and optical afterglows possibly associated with GRBs. Swift has confirmed that some low-luminosity GRBs emit much of their energy at x-ray wavelengths rather than in gamma rays. At least a subset of these might be associated with the formation of a neutron star rather than a black hole.

So far, we've discussed properties of long-duration GRBs, but what about short-duration bursts, which are only about one-fifth or one-sixth as frequent as long bursts? Swift has discovered several short-duration GRBs, for which it also detected an optical counterpart. We have verified that these short-duration GRBs often come from distant galaxies having only old stars. The data also reveal that there is no supernova light associated with those particular afterglows, suggesting that rather than a massive star undergoing core collapse and exploding, the event may be two neutron stars merging to form a black hole. If true, this would indicate that short-duration GRBs are also the birth cries of black holes. Some short-duration GRBs might arise from the merger of a neutron star with a black hole, as opposed to the merger of two neutron stars. This would indicate a growth spurt of the black hole because it accretes a neutron star. At least some short-duration GRBs could arise from *magnetars*, or starquakes, during which the surfaces of neutron stars with exceptionally high magnetic fields (1000 times stronger than those of normal pulsars) redistribute their crusts and magnetic fields. In December 2005, a magnetar burst produced enough gamma rays to ionize part of Earth's atmosphere. This magnetar was atypical in that it occurred in our own Galaxy, yet it was as bright as a GRB.

"In December of 2005, there was a magnetar burst that produced enough gamma rays to actually affect Earth's atmosphere."

Should we worry about GRBs? Could gamma rays affect Earth's atmosphere so much that life on Earth would be greatly endangered? We think that GRBs are rare enough that they don't produce mass extinctions more than once every billion years or so. It's unlikely that there have been more than one or two mass extinctions on Earth from GRBs. In particular, new evidence suggests that at least long-duration GRBs tend to occur in galaxies with low abundances of heavy elements compared to our own Galaxy. The fact that our Galaxy is abundant with heavy elements decreases the chances that a GRB will extinguish us.

How can we be sure that GRBs are produced by the collapse of a star to form a black hole, or the merging of two neutron stars forming a black hole, or some other energetic process? One way to verify the claim, and to test general relativity, would be to detect *gravitational waves*, ripples in the fabric of space-time created during the formation of a black hole. When a massive star collapses, gravitational waves are emitted. Likewise, when two neutron stars orbit each other and merge, gravitational waves are emitted. But these waves are extremely difficult to detect because they're very weak. Such waves have interesting signatures: They first stretch an object along one axis, while squeezing it along the other axis. Then, they stretch the object along the second axis and squeeze it along the first. Gravitational waves are so weak, however, that they would cause a 1-meter-long rod to change size by an amount only equal to 10^{-6} of the diameter of a proton! Joseph Weber first tried to detect this movement, and though he wasn't successful, his methods paved the way for future attempts.

The basic idea for detecting tiny movements is based on the concept of oscillating an object at its natural *resonant frequency*, thus amplifying the effect, similar to pushing someone on a swing at just the right time to increase his or her amplitude (the height reached). Hence, to attempt the detection of gravitational waves from merging neutron stars, physicists have built large, L-shaped vacuum tubes through which laser beams pass. The two such gravity-wave detection facilities in the United States (in Washington and Louisiana) are called, collectively, LIGO, the Laser Interferometer Gravitational-Wave Observatory. The laser beams measure the distance between mirrors mounted on hanging masses at the ends of the tubes and at their intersection point. If a gravitational wave passes through, the space in one tube contracts, while in the other, it expands. LIGO has not yet detected any gravitational waves; however, it is an important step in the development of better instruments for the future. ■

Important Term

interferometer: Two or more telescopes used together to produce high-resolution images.

Suggested Reading

Katz, *The Biggest Bangs: The Mystery of Gamma-Ray Bursts, the Most Violent Explosions in the Universe.*

Pasachoff and Filippenko, *The Cosmos: Astronomy in the New Millennium*, 3rd ed.

Schilling, *Flash! The Hunt for the Biggest Explosions in the Universe.*

Thorne, *Black Holes and Time Warps: Einstein's Outrageous Legacy.*

Questions to Consider

1. If GRBs are beamed, are the energy requirements per GRB smaller than if isotropic emission (i.e., uniform across the sky) is assumed? Is the number of GRBs that we detect per galaxy affected by the beaming?

2. Why was identification of the optical afterglows of short-duration GRBs of critical importance to the interpretation of such GRBs?

3. Given that we feel the effects of gravity every day (it is a very familiar force), why do you think gravitational waves are so difficult to detect?

Birth Cries of Black Holes

Lecture 66—Transcript

We have seen that, roughly once per day, somewhere in the sky there's a tremendous burst of high-energy photons, gamma rays lasting just a few seconds or a few hundred seconds, just coming from some random spot, a different spot each time, each day roughly. These so-called GRBs (gamma ray bursts) have a distribution of duration time, but generally there are two groups, those with durations of less than two seconds or so, and those with durations of more than two seconds. It really is what's called a bimodal distribution, as is shown here. There's this group of short-duration GRBs, centered on perhaps .01- or .02-second duration, and then the long-duration GRBs, centered on something like 20-second duration.

As I mentioned last time, the leading hypothesis—at least for the long-duration GRBs—is that they come from massive stripped stars that implode, forming a relativistic jet of high-energy particles along the axis of rotation of this rotating star. If the jet happens to be pointing in our direction, then we see it as this bright flash. Here's this imploding star with an axis of rotation, along which relativistically accelerated particles—that is, particles going at 90%, 95%, even 99% of the speed of light—are zooming along, emitting radiation. If that jet is pointing at us, it looks really, really bright, but it doesn't last very long. We think that these objects come from the so-called stripped massive stars—not stars that have a thick hydrogen envelope, such as the one at the left here, but rather stars in which at least the hydrogen envelope is gone, and maybe even both the hydrogen and helium envelopes. These are the massive stars that, either through winds of their own, as this one here shows, or through transfer of material to a companion star, as in this artist's illustration, have lost their outer envelopes of material, leaving a more or less bare core—not all the way down to the iron, but at least down to the carbon and oxygen layer, something like that.

Why do we prefer a stripped star for the massive progenitor of a GRB? It's simply because the more material that has been lost prior to the implosion and explosion, the easier it is for this jet of particles to pummel its way through the remaining material and actually get out. If you look at this jet trying to get out, if there's a lot of material through which it has to blast its way, it'll lose energy. It'll slow down while trying to get out, and it'll be less likely to get out. If there's not much material, then it can make its way out more easily. Ideally, there'd be no material at all blocking its path, so that the high-speed charged particles could zoom out very, very easily. But we're not sure whether this ideal is actually ever reached in practice. We think that there's always at least a little bit of material through which it has to go, but the jet has so much energy that it's able to blast its way through.

Why do we want these objects to be rapidly rotating? That's part of this so-called "collapsar model," that the star not only implodes, but is also rapidly rotating. If you have rapid rotation, then you get a natural ejection axis, a natural axis along which to funnel these high-speed charged particles. Moreover, with rotation you get an accumulation of material in the equatorial plane, and that material blocks any jet from going along the equator, and forces it to go along the path of least resistance—that is, the axis. If you've got sort of a doughnut of material in the equatorial plane, the high-speed charged particles are unlikely to go in that direction; rather, they will go along the path of least resistance, which is this axis of the jet.

Then, you might ask, "Why should some massive stars, when they implode—when their cores implode—produce jets, and others not produce jets? What is it that makes the GRBs so energetic and so powerful compared to normal supernovae?" One leading hypothesis is that it's the same mechanism as the core-collapse mechanism I had previously described for Type II and Type Ib and Type Ic supernovae. But in this case, you have not only more rapid rotation than normal, but also collapse not just down to a neutron star, having a radius of 10 kilometers or something like that, but instead a black hole. So, if the material collapses even more than to the size of the neutron star, all the way down to a black hole, you can get a greater release of gravitational energy, gravitational potential energy. When you drop things, they pick up speed—they fall. That's at the expense of their gravitational energy. If something falls into a deeper gravitational field, it picks up more speed. If all these things collide with one another, those collisions can lead to the emergence of radiation, of photons and also of high-speed charged particles.

It's thought that these GRBs may be, literally, the birth cries of black holes. The black hole is formed during the implosion of the star, and somehow—

the details are still not understood—you get the conversion of gravitational energy into the high speeds of the ejected particles going along these two bipolar jets. That's the basic idea. The remaining material could fall all the way in and continue to form an even bigger black hole, or it could explode. Just because you have these two jets coming out from the vicinity of a black hole doesn't mean that the rest of the material absolutely has to collapse down into that black hole as well. If you formed a neutron star along the way during this collapse, for example, then you might have a flood of neutrinos that pushes out the remaining material in the more or less normal way of a supernova, or some other effect might push out the remaining stuff.

So, you could get a supernova explosion that's relatively normal for a core-collapse object, except that it has these two jets. Or you might have all the material collapsing directly down to a black hole and not producing a supernova explosion. The collapsar model comes in these two basic flavors. But the prediction is that, at least in some cases, you not only have the jet, but you have the material in the equatorial plane, which ultimately explodes outwards, forming a more or less normal supernova. The prediction might be that at least some long-duration GRBs are also associated with visible supernovae. If we plot the optical brightness on the vertical axis versus time along the horizontal axis, the GRB afterglow fades with time—that's the energy associated with the GRB and with the jets crashing into circumstellar material, gas around the star. That afterglow dies down with time. But then after a while, maybe a week or two after the GRB, supernova light starts making its appearance because it starts to dominate over the declining afterglow of the GRB itself. You might expect to see bumps in the light curves of GRBs, at least if this hypothesis is correct.

In 1998, there was a GRB, 980425, which happened to be spatially coincident, or at least consistent, with the position of an optical supernova, as shown here. There's the optical picture of a supernova, and we know the GRB went off somewhere in a little circle, sort of surrounding that supernova, and it's at least in position consistent with the supernova, and also consistent in time. In other words, the supernova occurred maybe a week or so after the GRB, and that's consistent with the time it takes for supernova light to brighten. In this particular case, there was no optical afterglow visible to the GRB itself.

In other words, there wasn't any optical radiation associated with the burst of gamma rays themselves. But a week later or so, there was this supernova.

Some physicists said, "Well, this is just a spatial and temporal coincidence. These are two unrelated objects, and they happen to be in about the same place at about the same time." But others said, "This is a really weird supernova. It has some really strange properties, and we think it really was associated with the GRB and supports this collapsar model where, at least in some cases, there's a supernova associated with the GRB." But there was doubt in many people's minds. If you look at that supernova, it was a weird one. It looked like not just a highly stripped Type Ic, but perhaps an even more highly stripped variety, where even part of the carbon-oxygen layer was gone, beginning to reveal the next layer in, the oxygen-neon-magnesium layer. This supernova, Supernova 1998bw, associated with the GRB, or at least possibly associated with it. It looked weird, but supported the hypothesis that the progenitor stars, and at least some collapsed stars, are these highly stripped stars.

Nevertheless, as I said, there was doubt, so people waited for another example. In 2003, there was GRB 030329 on March 29, 2003. Associated with it was the optical light of a supernova, and a spectrum—brightness versus wavelength from violet to red—of the putative supernova associated with GRB 030329 looked very much like the spectrum of that supernova 1998bw, which had been supposedly associated with GRB 980425. The spectrum is noisier—that's what all these little up and down undulations are—but the big undulations, the broad ones, are very similar to those in supernova 1998bw. This pretty much sealed the case in the minds of many astronomers because GRB 030329 was a much more typical GRB. It was more luminous than the one in 1998; it had a normal optical afterglow. In other words, it had all the characteristics of normal GRBs, and it had a visible supernova associated with it. What more could you want? That was just really great, many of us thought.

By 2003 or so, the collapsar model was reasonably well established, and people thought that this may really signal the birth of a black hole. We wanted more such cases, and so the Swift satellite was sent up in 2004. This satellite was a great example of a NASA mission that didn't cost very much—it was about $250 million. It was special-purpose, and it just exceeded expectations. It found lots of GRBs and did lots of cool things. Here's the lift-off of the rocket that launched the Swift satellite into an orbit high above Earth's atmosphere. In a 96-lecture astronomy course highlighting so many wonderful NASA results from NASA telescopes, I've got to show at least one launch of one of these telescopes. There it is: a beautiful, successful launch in November of 2004. The Swift satellite, since then, has been finding a GRB roughly every three days.

It's a very good satellite because it has not only detectors that scan much of the sky, finding the GRB bursts—that's what this BAT thing is, a burst and transient sort of a detector—but it also has an X-ray telescope and a UV optical telescope, which can quickly take pictures of the part of the sky from which the GRB occurred and find an X-ray and optical afterglow, possibly associated with the GRB. Here you can see the BAT device detected a GRB. The telescope then slued quickly over toward the direction of the GRB, and the X-ray part of the telescope, and the UV optical part of this whole spacecraft, then started snapping pictures of that part of the sky. In this way, they provide very good coordinates and pictures of optical afterglows of GRBs. Here is sort of the field of view of the X-ray telescope. It said that somewhere in there the GRB occurred, and in fact, the UV optical telescope might even show the afterglow sitting in this galaxy right there. It's been a great satellite, and most recently, in February of 2006, it found yet another case of a rather weak GRB, with which there was clearly associated a visible optical supernova. That supernova, as well, had the characteristics of a stripped star. However, this particular weak GRB may have been associated with the formation of a neutron star instead of a black hole.

I've been talking so far about the long-duration GRBs. What about the population of short-duration bursts, which are only about one-fifth or one-sixth as frequent as the long duration bursts, but seem to be a class of their own? They have shorter time scales. They tend to emit more high-energy gamma rays, and so on. Swift has discovered several short-duration GRBs, for which it also found a detected optical counterpart. Finally, we see something glowing at other wavelengths, corresponding to the short-duration GRBs. We didn't have that until 2004 or 2005. What has been found is that these short bursts seem to be in galaxies containing only old stars.

Here's one case of the Swift X-ray afterglow being somewhere in this little circle here. Everywhere in that circle are only old stars associated with the halo of this galaxy here. This particular afterglow didn't show any optical radiation. Since then, there's been a little bit of optical radiation, and we have absolutely verified that these short-duration GRBs come from distant galaxies having only old stars.

If you look at the position at which the short-duration GRB occurred, not only is it consistent with there being a population of old stars there, but there's been no supernova light associated with those particular afterglows. There's been no bump in the light curve, as in the long-duration GRBs. This suggests that it's not a massive star undergoing core-collapse and the rest of it exploding. Rather, maybe it's two neutron stars merging together to form a black hole during their final merging process. If that's the case, then short-duration GRBs are also the birth cries of black holes. You might have two neutron stars orbiting each other closely like this, and they spiral together. As they merge, some still-mysterious process produces these two oppositely directed jets.

Here's an animation of what might be going on; two neutron stars closely orbiting one another, like the binary pulsar I've discussed before. They spiral in toward each other, and during the final merger process, a tremendous amount of energy is emitted. Boom, like this! Much of it is directed along two oppositely directed jets. In that particular animation, we're going the wrong way, by the way. They should have been going along the orbital axis of the two stars, and instead they went off in some cockeyed direction. That was a minor mess-up by the animator, but that's the way it goes. There's also the possibility that some of the short-duration GRBs are mergers of neutron stars with black holes, not mergers of neutron stars with other neutron stars. This would mean that, instead of being the birth cry of a black hole, it would be sort of a growth spurt of the black hole because it accretes a neutron star.

Here is an animation that shows this process. There's a neutron star going around the black hole. It gets tidally disrupted when it gets close enough to it, and then it accretes onto the black hole. While it's still outside the black hole, these two oppositely directed jets go zooming out. There's one other possibility, and that is that at least a small fraction of the short-duration

GRBs might be magnetars, these starquakes on the surfaces of neutron stars having inordinately high magnetic fields—a quadrillion Gauss, a million billion Gauss. I discussed the magnetars a few lectures ago, and here's the animation of what one might look like. There's the spinning neutron star. There's a starquake, a redistribution of the crust and the magnetic field, and it emits this burst, and then basically keeps on spinning; it survives. It's still a neutron star that's intact. It just had a redistribution of the crustal material and the magnetic field. This can give rise to a large amount of energy. In December of 2005, there was a magnetar burst that produced enough gamma rays to actually affect Earth's atmosphere. Here they are coming out, and they travel many, many light years, and they come toward the Earth's atmosphere, and they actually ionize part of the Earth's atmosphere. That was a really hefty blast of a neutron star, a magnetar in our own galaxy, yet it was as bright as a GRB, and much of the radiation came out in the form of gamma rays. This was a GRB, just not the typical kind. The typical kinds come from more-distant galaxies. But you could see even this one, had it been in a fairly nearby galaxy. Probably at least a small fraction of the GRBs are not the total destruction of a neutron star, forming a black hole, but rather one of these magnetar starquakes.

This blast that hit the Earth's atmosphere raises an interesting question. Should we worry about GRBs? Could the gamma rays affect Earth's atmosphere so much, like by destroying the ozone layer, that life on Earth would be greatly endangered? Of if we're really close to a GRB, within a few hundred light years, could the gamma rays just sterilize all the planets surrounding a star that's near a GRB? Indeed, in *Sky & Telescope* magazine, there was this headline "Gamma Ray s of Doom." If there's a GRB sufficiently nearby, could it really hurt us? Could GRBs have caused at least some of the mass extinctions over the history of our planet? We think that they're rare enough that they don't produce mass extinctions more than once every billion years or so. It's unlikely that there have been too many mass extinctions here on Earth as a result of these GRBs. In particular, the extinction of the dinosaurs was almost certainly due to the collision of a comet or an asteroid with the Earth, not a GRB, despite what this dinosaur here is thinking, looking up at the GRB and calculating the total energy emitted by it. I'm not sure dinosaurs could do that.

We think that these events are rare enough that probably, at most, maybe one mass extinction in the last billion or two years was due to a GRB. In particular, there's new evidence suggesting that at least the long-duration GRBs tend to occur in galaxies having fairly low abundances of heavy elements compared to our own galaxy, in which 2% or 1% of the elements are heavy elements. The observed GRBs tend to occur in these galaxies that seem to have a lower abundance of heavy elements, as though there's something about the heavy element composition of massive stars that makes it more likely to form a GRB if there's not many heavy elements in the atmosphere of the star. Maybe the collapse occurs more easily, or maybe the winds are able to get rid of the atmosphere of the star or something like that. Given that our galaxy has quite a high heavy-element abundance, that further decreases the chances that our own galaxy will have a GRB that will extinguish us. I think we have to worry a lot more about comet and asteroid collisions for humans, for the existence of humans; not to speak of our own social diseases like war, and pestilence, and all that.

How can we be sure that GRBs are produced by the collapse of a star to form a black hole, or the merging of two neutron stars forming a black hole, or some other energetic process like that associated with either the birth or the growth of a black hole? We'd like to be sure of that. One way of verifying this claim, and also testing general relativity, would be to detect gravitational waves, ripples in the fabric of space-time created during the formation of the black hole. When you have the collapse of a massive star, and especially when you have a very non-spherical situation like two neutron stars orbiting each other merging together, you get emission of gravitational waves. Here are the warpings of space produced by two neutron stars. As they orbit one another, they produce these waves of gravity, gravitational waves, ripples in the fabric of space-time. In this animation, which I've shown before, you can see these ripples going out.

This animation happens to have been made for two white dwarfs that merged to form either a bigger white dwarf or a neutron star, but you could just as well have two neutron stars coming together like this, emitting gravitational waves more and more quickly, until there's a tremendous burst of them, just flowing out tremendously as these two things merge together in their final moments. There you have a rapidly spinning remaining neutron star, in this case, or black hole. You get a large amount of gravitational waves coming out in a situation like this, resulting in the formation of a black hole. That would be a wonderful signature to see. But gravitational waves are extremely difficult to detect. They're very weak. They have a rather interesting signature.

If you take a gravitational wave, and it's coming along like this, it first stretches an object along one axis and squeezes it along the perpendicular axis. Then it stretches it along that perpendicular axis and squeezes it along the first one. They go like this; it's that kind of a thing, but only ever so slightly. Gravitational waves are so weak that we expect a one-meter-long rod to only change size by an amount equal to a millionth (10^{-6}) of the size of a proton. A one-meter rod won't oscillate very much, and yet we'd like to detect this. The first attempts were made by Joseph Weber, a pioneer in such studies, where he made a cylindrical aluminum bar two meters long, and he tried to detect these small fluctuations in length of only 10^{-6} of the diameter of the proton. He didn't succeed, but his methods paved the way for future methods.

Many of them rely on an interesting idea, where if you oscillate an object at its natural resonant frequency, you can amplify the effect and either detect it with a detector or even hear it. Your ears are a detector. An example is if you take a wine glass and rub your wet finger on the edge, if you do it just the right way, you can sometimes hear a rather loud audible effect. Let's try it; there it goes. Try this at home. You can get it to work quite well if you change the pressure and angle of your finger. Some orientations work really, really well.

Joseph Weber chose a frequency of 1,000 Hz—1,000 cycles per second—to be the resonant frequency of his rod. He was hoping to amplify this minute effect by having the rod oscillate at this resonant frequency. By resonant frequency, I kind of mean if you're pushing someone on a swing, if you're pushing them in phase, the amplitude builds up, right? Whereas if you push them at the wrong time—when they're coming back, you push them forward, but they're only midway through their swing or something like that—you won't build up the amplitude. You want to resonate with something, and you can build up the amplitude. He paved the way. He didn't detect anything, but he got people excited about the prospect of detecting gravitational waves.

In the past decade or so, LIGO (the Laser Interferometer Gravitational-Wave Observatory) was built. There are two facilities—one in Washington, one in Louisiana—where there's an L-shaped structure about 4 kilometers long on each arm. Inside that structure, there's a tube with a vacuum inside. Along the tube, laser light can go. It goes through a thing called a beam splitter, sending half of the laser light toward one mass over here, and the other half toward another mass over there. There are mirrors attached to these two masses, which reflect the light back. If you focus your attention on the light reflected back from mass 1, part of it then gets split by the beam splitter toward a detector. The other part goes back toward the laser, where another mirror can actually send it back, it turns out. The light from mass 2 also bounces off, goes through the beam splitter, hits the detector—or part of it can go back to the laser and bounce off of yet another mirror.

Now, by setting up the system so that, at least initially, the two laser beams that hit the photodetector destructively interfere, giving you zero light—that is, they're out of phase. One laser beam is going up, down, up, down. The other one is going down, up, down, up, like this. You set the experiment up so that, if there's no gravitational wave, the two paths differ in such a way that you get exact destructive interference of the waves, giving you zero light at the detector. Now suppose a gravitational wave comes through. It makes one arm slightly longer, the other one slightly shorter. Now the two waves coming into the detector will be not so perfectly out of phase, producing a small signal. Here I have them 5% not quite perfectly out of phase, and then 10%, and then 25%. You can see that the more in phase they are, the more light you will see. What you can try to detect are changes in the lengths of the tubes as a result of this gravitational wave coming by and producing a signal in the resulting light measured at the detector. That's the idea.

LIGO has not yet measured any gravitational waves. Again, they're very, very weak, and a 4-kilometer-length arm is not really long enough to produce a measurable effect for anything other than the most massive neutron stars and black holes emerging very, very close to us. The LIGO scientists didn't even really expect to detect anything. But again, they're paving the way for more advanced techniques, hoping to show that, with even better experimental set-ups, they will some day find the distinctive signature of merging neutron stars, or the collapsing core of a massive star, signaling the birth of a black hole and verifying one of the major predictions of general relativity: the existence of these ripples in the fabric of space-time carrying gravitational wave energy. That will be a fantastic discovery once it's made.

Our Home—The Milky Way Galaxy

Lecture 67

"We'll start with our own home, the Milky Way Galaxy, a grand structure—a spiral galaxy about 100,000 light years in diameter, and only a couple of thousand light years thick—containing several hundred billions of stars."

In this fifth and final unit of the second major part of the course, we will examine the contents of the Universe on large scales, beginning with a look at our own Milky Way Galaxy. The Milky Way Galaxy is spiral, about 80,000 (though perhaps 100,000) light years in diameter, and a few thousand light years thick. It contains several hundred billion stars, all gravitationally bound together and orbiting the central part of the Galaxy. Our Sun is one-half to two-thirds of the way out from the center of the Galaxy, orbiting in a near-circular fashion at about 200 kilometers per second. Being about 24,000 light years from the center of the Galaxy, our Sun takes about 250 million years to complete one full orbit around the center. Given that the Sun is roughly 4.5 billion years old, it has thus far made about 18 orbits around the center of the Galaxy. The Milky Way has a nucleus and a central bulge of stars from which a bar-like structure emerges. The bar's ends have two major arms that break into two or more other arms. The four main arms of our Galaxy are Norma, Sagittarius, Orion, and Perseus. The Sun is on the inside edge of the Orion arm. Star clusters and nebulae tend to congregate in the arms, with many nebulae and clusters in the Sagittarius arm toward the center of the Galaxy.

Star clusters arise from giant clouds of gas and dust, like the Orion Nebula. They are gravitationally unstable and, thus, begin to contract, fragmenting into smaller sub-units and eventually forming stars, as we discussed in a previous lecture. From Earth's position near the outskirts of the Galaxy, we can look toward the center and see the bulge. If we look to the sides, we see a disk of stars growing fainter farther away from the bulge. When we look along the plane of the Galaxy, we can see a multitude of stars. When we look at other angles through the disk of the Galaxy, we don't see as many stars. This effect is what produces the "Milky Way" in the sky, as discussed in Lecture 5. The ecliptic plane of our Solar System is tilted by about 60 degrees relative to the Galactic plane. In addition, Earth's axis is tilted 23.5 degrees relative to the rotation axis of the Solar System. During the northern-hemisphere summer, we can see views toward the center of our Galaxy. During the northern-hemisphere winter, we look in a direction opposite the center. Because of the combined tilting effects, the center of our Galaxy is nearly overhead from the southern hemisphere during its winter, enhancing our view from that vantage point.

Now we look at nebulae, clouds of gas and dust. The Orion Nebula, about 1500 light years away in the sword of the constellation Orion, is a region where stars have been forming for the past few million years. Peering into its depths, we can witness the process of star formation as it occurs. Some nebulae, in particular *spiral nebulae*, are actually distant galaxies. We will consider them further in Lecture 69. The spiral arms of our Galaxy and other spiral galaxies contain most of the nebulae, which glow from massive young stars forming within them. These are called *emission nebulae*. The gas is ionized by ultraviolet radiation from the newly forming, hot, massive stars, making the surrounding clouds of gas glow. When atoms are ionized, such as hydrogen, an electron is liberated from the hydrogen atom. If the electron recombines with a free proton, it can emit a photon of light and, subsequently, jump to still lower energy levels, continuing to emit light. An electron can also collide with another electron already bound in an atom, bumping it up to a higher energy level. When that electron subsequently jumps to a lower level, it emits light. Previously, in Lecture 51, we considered both processes when discussing the glowing cloud of gas ejected by dying stars (planetary nebulae), but here, we concentrate on nebulae from which new stars are forming.

Photons can ionize the atoms in their vicinity, but beyond some distance, there are not enough photons to ionize atoms, so they remain neutral. Ionized hydrogen is called HII, while neutral hydrogen is called HI. If sufficiently energetic ultraviolet photons are present, they can produce ionized helium, though this happens only around very hot stars. We can produce wonderful photos of nebulae and their various colors by using special filters that capture the different emission lines produced by a glowing nebula. *Reflection nebulae* shine because visible photons from nearby stars reflect off small particles of matter, or space dust, which is usually mixed in with gas. Because the reflection process works best for blue light, reflection nebulae tend to glow blue. Reflection nebulae don't glow from ionization but, rather, from light bouncing off particles. Their color is blue for essentially the same reason our sky appears blue; as the light filters through our atmosphere, the blue, green, and violet photons are reflected more easily than the red and orange ones. *Absorption* (or *dark*) *nebulae* are dense and dusty, with so much material that Earth's view of their light is blocked. These are the regions in which stars are currently forming. All three types of nebulae can occur together. Not only can we view visible wavelengths of the light they emit, but using infrared and radio telescopes, we can also see new stars forming in these nebulae, especially the dark nebulae.

"The Milky Way is what you get when you look along the plane of our galaxy."

Diffuse clouds of gas and dust between the stars form part of the *interstellar medium* (ISM), much of which is low in density. However, some regions have dense clouds of gas, which can become gravitationally unstable and collapse to form new stars. The densities of particles in these clouds can be so high, up to 1 million particles/cm^3, that molecules begin to form. These dense clouds span regions up to a few hundred light years across and are where giant clusters of thousands of stars are formed. We know that our Sun and planets formed from this ISM, whose composition is gradually changing as a result of the heavy elements ejected into the cosmos by supernovae. The realization that we formed from such structures, chemically enriched by previous generations of massive stars, was a monumental step in our understanding of our place in the cosmos and our origins in this vast Universe. ■

Suggested Reading

Croswell, *The Alchemy of the Heavens: Searching for Meaning in the Milky Way*.

Ferris, *Coming of Age in the Milky Way*.

Henbest and Couper, *The Guide to the Galaxy*.

Pasachoff and Filippenko, *The Cosmos: Astronomy in the New Millennium*, 3rd ed.

Verschuur, *The Invisible Universe Revealed: The Story of Radio Astronomy*.

Questions to Consider

1. How would the Milky Way appear if the Sun were closer to the edge of our Galaxy?

2. If the Sun is 8 kiloparsecs (about 24,000 light years) from the center of our Galaxy and it orbits with a speed of 200 kilometers per second, show that the Sun's orbital period is about 250 million years. (Assume that the orbit is circular.)

3. Compare (a) absorption (dark) nebulae, (b) reflection nebulae, and (c) emission nebulae.

4. Describe the relation of hot stars to H I (neutral hydrogen) and H II (ionized hydrogen) regions.

Our Home—The Milky Way Galaxy

Lecture 67—Transcript

I now move on to the fifth and final unit of the second major part of the course, the Contents of the Universe. In the first four units, I discussed the Solar System, other planetary systems, the many types of stars and their lives, and black holes. Now in this fifth unit, I'll go on and discuss galaxies, the larger structures in which all of these other components exist. We'll start with our own home, the Milky Way Galaxy, a grand structure—a spiral galaxy about 100,000 light years in diameter, and only a couple of thousand light years thick—containing several hundred billions of stars, all gravitationally bound together, going around in circles or ellipses around the central part of a galaxy. Seen from the outside, our galaxy might look something like this: a spiral with the Sun just one of these hundreds of billions of stars, perhaps one-half to two-thirds of the way out from the center.

We can't see our Milky Way's structure very clearly because we live inside it. It's sort of like a mouse inside a maze. It's hard to see the overall structure of the maze if you're snooping around inside. If you're looking at it from a low angle, you begin to see the structure of the maze. But if you look at it from a high angle, then you can see it much more clearly. We shouldn't fault the mouse for having a hard time making its way to the cheese. We say "Oh, go that way," but for the mouse, it's hard to tell which way to go. In a similar way, we don't really know in detail the structure of our own Milky Way Galaxy because we live within it like this. We know the structure of other galaxies much better.

The Sun orbits in a nearly circular orbit, slightly elliptical, around the center of the galaxy with a speed of around 200 kilometers per second. At its distance from the center of our galaxy, about 24,000 light years, it takes it 250 million years to complete one full orbit around the center of the galaxy. Given that the Sun is about 4.5 billion years old, this means that it has completed, since its birth, about 18 orbits around the center, so it could consider itself a young adult in terms of galactic years. Of course, the Sun is about halfway through its main sequence lifetime, and that's a much more fundamental measure of the ages of stars—how old they are intrinsic to themselves, not how many times they've gone around the galaxy. Anyway, the Sun is a young adult, basically, in galactic years.

If we look at what we think our galaxy looks like, it's something like this. This is just sort of an artist's representation of what our galaxy is like, as seen face-on from the outside. It has a nucleus and a central bulge of stars, and then a bar-like structure like this, from the ends of which emerge two major arms that then break up into two or more other arms. Our Sun is about one-half to two-thirds of the way out from the center to the ill-defined edge of the galaxy, and it's on the inside edge of what's called the Orion Arm, a spur from a bigger arm known as the Sagittarius Arm. The Sagittarius Arm is toward the center of our galaxy. In the opposite direction, there's the Perseus Arm, and we can see some clusters of stars there as well, but we see many more nebulae and clusters in the Sagittarius Arm, toward the center of our galaxy. The clusters and nebulae tend to congregate in the arms. There's another major arm known as the Norma Arm, but this is still kind of a vague picture because we don't really have a detailed view of our galaxy.

If we look closely at what's in those arms, we see nebulae and clusters. Here's a star cluster in or near a spiral arm, and here's another one. These clusters, as I've discussed before—and will do even more in today's lecture—come out of the formation process intrinsic to a giant cloud of gas and dust, like the Orion Nebula. These clouds are gravitationally unstable, and they start contracting. Then those contracting clouds fragment into smaller sub-units, which then turn into stars. But the clusters form from these nebulae, and they form rather quickly. Since the nebulae are in the spiral arms, new stars are formed in the spiral arms as well. An edge-on view of our Milky Way looks something like this. This is an actual photograph, or a collection of photographs, taken at infrared wavelengths. From our position near the outskirts of our galaxy, we look toward the center and see this bulge. If we look off to the sides, we see a disk of stars getting fainter and fainter as you go farther and farther away from the bulge. This is sort of an all-sky view of our galaxy.

A schematic representation is shown here, with the thin disk of our galaxy, maybe 1,000 or 2,000 light years thick, and perhaps 100,000 light years across. In the very center, there's the nucleus, and then a bulge of rather old stars. There's also a halo containing old stars and globular clusters. The disk contains both old stars and younger stars. In particular, the young stars are formed in the spiral arms. The globular clusters are a beautiful sight. Here's a fantastic photograph of one of them: giant clusters of hundreds of thousands, or even millions, of stars gravitationally bound together, orbiting as units the center of our galaxy. You can take quite good photographs of these things even through relatively small, amateur-sized telescopes. Here's a beautiful photograph taken through an amateur telescope, and here's another one.

They're just really great objects; although if you look through a telescope, you rarely can see the individual stars. They kind of blur together because they're faint, and there are many of them, but in a photograph, you can discern them. You get great views with the Hubble Space Telescope, where you can go all the way down into the central regions of globular clusters and see individual stars. Here in this photograph, they're a little bit burnt out, but in other stretches of this photograph, you can see stars all the way down to the center because you don't have the turbulent effects of the Earth's atmosphere.

The band of light stretching across the sky, which we call the Milky Way, I've already described in Lecture 5. It's got this splotchy distribution, dark regions like this and bright regions, concentrations of stars, clouds of illuminated gas, clouds of dark gas and dust, a very splotchy thing. Basically, as I've said before, the Milky Way is what you get when you look along the plane of our galaxy, where you see a multitude of stars along the plane—whereas perpendicular to the plane, or at other odd angles, you're just not looking through regions that have as many stars. That's what the plane of the Milky Way is caused by. We're looking through the disk of our galaxy.

Our own Solar System, the ecliptic plane, is tilted by about 60 degrees relative to the galactic plane. During the summer months, as viewed from the Northern Hemisphere, we are actually crossing the galactic plane, and the center of our Milky Way is in this direction. During the Northern Hemisphere winter months, at night we're actually looking toward the direction opposite to the direction of the center of our galaxy. Recall that our own axis of rotation is tilted relative to the vertical. Here's the axis of rotation of our whole Solar System, and Earth's axis is tilted by 23.5 degrees. In the summer, in our

Northern Hemisphere summer, the Milky Way Galaxy's center is off to the right here, to my right, and the Sun is there.

Here's the night sky, looking out this way. You can see that the center of our galaxy is quite low above the horizon, as seen from the Northern Hemisphere. But from the Southern Hemisphere, the center of our galaxy is high above the horizon. In fact, it passes nearly overhead. So, you get a better view of the Milky Way from the Southern Hemisphere of the Earth during the Northern Hemisphere summer, or hence the Southern Hemisphere winter. You don't get such a great view from the Northern Hemisphere, although you still can get quite a nice view if you have dark skies. If you really look up at the sky when it's really dark and away from the city lights, and the Moon is down, you get a totally glorious view of the Milky Way. The central part of our galaxy is over here in this direction. I described how, when I observed in Chile, far from any city lights, I just felt like I was immersed in the Milky Way, with the whole thing forming this giant structure above me, like an inverted bowl. The celestial sphere is this inverted bowl with this giant structure hovering above me. It was just a marvelous feeling. It made me feel like I'm part of the Milky Way Galaxy in a way that I'd never felt before.

If you take a pair of binoculars and sweep across the Milky Way, you'll see lots of these nebulae: dark nebulae, bright nebulae, clusters, and things like that. In the Southern Hemisphere, you can see Alpha and Beta Centauri, which are pretty much right smack in the Milky Way Galaxy, near the Southern Cross, which within it has a region called the Coal Sack because it looks so dark relative to the surrounding regions that have lots of bright stars. There, there's a bunch of dust blocking our view of the more-distant stars. Looking over toward the constellations Sagittarius and Scorpius, you can see many such dark clouds, and also some brighter regions, and then some glowing clouds of gas where you have hot stars ionizing the gas. In nice, deep photographs of the Milky Way, you can see all these magnificent structures. I urge you to just take a good pair of binoculars, go to a dark site, and explore the Milky Way. It'll just be a fantastic experience.

I should mention that not all the nebulae are clouds of gas within our galaxy. Some nebulae—in particular, nebulae called spiral nebulae, like the ones shown here—turn out to be galaxies of their own, far from our own galaxy. The nature of these spiral nebulae was controversial about a century ago, but now we know that they are other galaxies. These other galaxies cannot be seen in the plane of our own Milky Way Galaxy very easily because of all the dust and gas blocking the view through the plane. You tend to see the spiral nebulae, the other galaxies, perpendicular to the plane of the Milky Way, or at pretty steep angles to the plane of the Milky Way. These nebulae can be stunningly beautiful.

The Orion Nebula is just a great object, 1,500 light years away in the sword of the constellation Orion. There's the belt; there are the feet, the shoulders, the head, and then the sword. Let's zoom in on it. It's this region of star formation where stars have been formed for the past few million years, and we can see these young stars. Moreover, peering down into the depths of the Orion Nebula, we can witness the process of star formation as it's going on right now. With many different telescopes, we've achieved great views of the central regions of nebulae like this, showing in detail the process of star formation.

If we look at a spiral galaxy in general and talk about these spiral arms, which have the nebulae within them, the arms glow largely because of the massive young stars that form within them and that emit a lot of light, which then ionize and cause to glow the surrounding clouds of gas. These are the emission nebulae. The gas is ionized by ultraviolet radiation from these hot, massive stars that form in these giant nebulae. Here with the Orion Nebula, you see in particular four central, very massive, very hot stars that are emitting a flood of ultraviolet radiation. That ultraviolet radiation can ionize atoms—in particular, I show hydrogen here—kicking the electron away. When that electron finds another free proton, it can recombine, emitting a photon, emitting a quantum of light, and then cascade down to lower and lower energy levels, emitting yet more light. That's the process by which these nebulae glow. Here we have the electron recombining with the proton and emitting a photon. Then it cascades down in one or more jumps to lower energy levels, again emitting photons. That's one of the processes by which these things glow.

The other process I've already described is when an electron kicks another electron already bound in an atom up to a higher energy level. It gives some of its kinetic energy, its energy of motion, to the electron that's bound in an atom, kicks it up to a higher energy level, and then from that higher energy level, the electron cascades down and emits light. I discussed this when I discussed planetary nebulae, like the one we're zooming in on here, the Cat's Eye Nebula. It's a wonderful structure, glowing because of the ultraviolet radiation emitted, in this case by the dying star in the middle of this nebula. The process by which the nebula glows is the same as in the emission nebulae I've just described. But in the case of the emission nebulae I just described, it's young, new, massive stars forming the flood of ultraviolet photons that ionize the gas, rather than an old, dying star producing the ultraviolet photons.

Here's another planetary nebula where you've got this old, dying star, whose core is becoming exposed, and the hot core emits ultraviolet photons and ionizes the cloud of gas. I had discussed that in Lecture 51, but the important thing to keep in mind is that, though the glowing process is the same in a nebula where new stars are forming, in this case the new, massive, hot stars are the things that produce the ultraviolet photons, in contrast to the planetary nebulae—where it was an old, dying star whose exposed core is really hot and produces the ultraviolet photons. There's a slight distinction there.

If you then look at a schematic of the gas around a hot, massive star that emits a lot of ultraviolet photons, you see that those photons ionize the atoms in its vicinity, out to sort of a spherical region like this. Beyond some distance, there are not enough photons to ionize the atoms, and so they remain neutral. Here's a region of neutral atoms. Surrounding the star there's a region of ionized atoms. In particular, when hydrogen is ionized, it's referred to as HII. Neutral hydrogen is referred to as HI. Here's an HII region surrounding a hot, massive, young star. Here's another region that has a concentration of hydrogen atoms. It's a cloud, a nebula, but it hasn't yet formed a star that's emitting ultraviolet photons. Therefore, in this nebula, the gas is still neutral; it is an HI region. In some cases, there is enough ultraviolet photons that they can even eject one of the electrons from a helium atom, producing ionized helium as well, although it takes a lot of energy to ionize helium, and most stars don't produce ultraviolet photons having so much energy. Instead, they just produce ultraviolet photons with enough energy to ionize the hydrogen.

Within this HII region, then, these different processes that I just described can occur: recombination of an electron with a proton, or collisional excitation and then cascading downwards. In either case, you get a spectrum of emission lines. Plotting apparent brightness versus wavelength or color, you see a continuum produced by the hot stars of the nebula, and then you see these spikes, these emission lines, produced by the ionized gases—in this case, doubly ionized oxygen—or by neutral hydrogen. Here's the Hβ line; there's the Hα line, due to electrons cascading down the different levels that are allowed in the hydrogen atom. You get this spiky structure. From analysis of the spikes, you can determine things like the chemical composition of the nebula, the temperature of the star producing the ionizing photons, and things like that.

These nebulae then glow at these various colors, these various emission lines. If you isolate the light from the various emission lines and produce several different photographs through filters, passing light from one emission line versus another, and then combine them, you can get these incredible photographs through this tri-color process. For example, you combine three different pictures, isolating light from three emission lines at, say, blue, green, and red wavelengths, and then you combine them and get these fantastic pictures like the ones produced by Richard Crisp and other amateur astronomers who do this process. It's a fantastic process because they get these great pictures. Earlier, in fact, I illustrated how you get these pictures for an object like the Crab Nebula—where, again, you have an emission-line spectrum. If you isolate, or filter through, the light from each of these strong emission lines, you get a different picture depending on which emission line you are imaging. When you combine the pictures, you get this fantastic structure. I just love those photographs. Here's a picture of the Orion Nebula, taken using this tri-color process with several different filters.

Those are the emission nebulae. They glow because the ultraviolet photons of hot, massive stars ionize the gas and essentially cause it to fluoresce. There's another type of nebula, a reflection nebula, where you have visible photons being reflected off of small particles of matter. We call them dust; it's just fine particulate matter. This dust tends to be mixed in with gas wherever you have nebulae. Where you have gas, you also have dust. Some places have more dust than others, but in any case, if you have the dust, it can reflect the light. The reflection process works best for blue light, so the reflection nebulae, as they're called, tend to be blue like the nebula surrounding the stars of the Pleiades, the Seven Sisters. That nebula is a reflection nebula. It's not glowing because these stars have ionized the gas. These stars aren't hot enough to ionize gas. They don't emit enough ultraviolet photons, but they do emit visible light photons, which then reflect off of the dust, causing the surrounding nebulosity to glow.

The light that is reflected is predominantly blue for the same reason that our sky is blue. I had already discussed with you how sunlight going through the atmosphere gets reflected by molecules, and also by dust particles, in our atmosphere. The blues, greens, and violets get reflected more easily than the red and orange colors. The red and orange photons go streaming through, whereas the blues tend to get reflected, and they reach our eyes. Away from the Sun, the sky looks blue. Of course, the sunset looks red because the sunlight is filtering through so much of the atmosphere that most of the violet, blue, and green photons have either been reflected away or absorbed by gas and smog in our atmosphere, leaving the orange and red colors to filter through to your eyes.

We had discussed that before, and here we see a celestial example where the blue reflection nebulae are blue for essentially the same reason as the daytime blue sky. Here you have a bright star that's not emitting much ultraviolet light, but it is emitting optical light, which reflects off of the dust particles, causing this blue glow. In this nebula here, the Trifid, there are some hot stars that are ionizing the clouds of gas, and those ionized clouds of gas then recombine, and you get Hα photons, which tend to be red. That's what causes this reddish glow here. You also see some very dark regions. Those are dense, dusty nebulae called dark nebulae (or absorption nebulae), where there's so much material that it blocks the light that's behind that object, as seen by us, and doesn't let that light come through, so those look dark in relation to the very bright surroundings.

Here's a good example of a reflection nebula surrounded by a dark nebula. This reflection nebula glows blue, and it looks kind of like the North American continent. Here's a big version of Florida, and there's the Caribbean and Mexico. Then here's the California, Oregon, and Washington coast, and up here is Canada. That's called the North American Nebula—kind of a cool thing. There are usually all three types of nebulae. Where you find one, you tend to find others as well. In particular, here you have the Horsehead Nebula, where there's a reflection nebula around a bright star emitting visible photons down here. Over here there's some hot stars that are ionizing the clouds of gas and causing them to glow through the various processes that I described.

But then here, in the shape of a horse's head, is a particularly dark and dense cloud of gas that blocks the background light. There are not that many stars visible within it as well because it's just too dense for light to pass through. Here's another example of a reflection nebula next to this bright star, some emission nebulae there, and some dark nebulae here. In some cases, you can't even see the star that's ionizing the emission nebula because it might be behind this dark nebula. Here's another example of a reflection nebula surrounding a dark nebula, and then there are some emission nebulae over here. And then the Eagle Nebula is a fantastic object, where the dark nebulae are so dense that we know that stars are forming down in there because we've seen them with infrared and radio wavelengths.

If we zoom down into the Eagle Nebula, you can see the grandeur, the majesty of our galaxy, and of all these stars and the glowing clouds of gas and dust from which the stars formed. We go down, down into this nebula, peering ever closer into the heart of what's going on in there. Although it might be hidden at visible wavelengths, if you look at the same region in the infrared, or at radio wavelengths, you can see new stars forming. Here's another example, the Trifid Nebula, zooming in on this thing from far away. As we come closer and closer, we can see the glowing clouds of gas and the reflection nebula produced by a bright star shining photons off of dust particles. Down in the core of the dark cloud of the Trifid Nebula, there are some new stars that recently formed, as well as stars that are still forming.

Here's a schematic of what's happening in these nebulae. In the densest parts, new stars form. They emit ultraviolet light that ionizes at least a certain region around that cluster of hot stars. That's then the HII region that emits the emission lines. Surrounding that is a region of neutral gas that doesn't produce such emission lines. If you have a very dark, dense nebula, the ultraviolet photons don't penetrate it very much, and neither do other photons of optical or ultraviolet wavelengths. But radio and infrared photons might penetrate it and allow you to see what's going on inside. Then out here, there might be a reflection nebula, where optical photons made it through this ionized cloud and got to the reflection nebula. Then dust in the reflection nebula reflected those visible light photons. In this photograph, you can see once again the Horsehead Nebula, a dark nebula surrounded by the emission nebula with various other reflection nebulae in its vicinity. This whole structure is just below the belt of Orion.

These diffuse clouds of gas and dust between the stars are part of what we call the interstellar medium (ISM). Much of the ISM has a very low density and is very diffuse. But in some regions, you have these denser clouds of gas, which can become gravitationally unstable and collapse to form new stars. The densest of the clouds form molecules. Here you can see the central part of what's called a giant molecular cloud. The densities of the particles are so high—up to a million particles per cubic centimeter—that molecules can begin to form. In those densest regions, new stars are forming as well. Here we're seeing just the central few light years, half a parsec or so, of this vast thing, the Orion Nebula, which we've looked at before.

Here's the core of it, where you have these densities of a million particles per cubic centimeter—still trillions of times less dense than the atmosphere of the Earth at sea level. This is better than any vacuum we've ever produced on Earth. But for interstellar space, this is dense. The typical density is one particle per cubic centimeter. Loose clouds might be 100 or 1,000 particles or atoms per cubic centimeter, but these dense molecular clouds contain 100,000, up to a million, particles per cubic centimeter, and they span regions up to a few hundred light years across. These are the regions that form giant clusters of thousands of stars. If we zoom in now on one of these regions, the Cone Nebula, we can see, deep inside, stars forming right before our very eyes.

By that, I mean that we can't see the process overnight or anything, but by looking at these objects with different telescopes, and looking at nebulae having different ages—like this Cone Nebula and other nebulae, like the Eagle Nebula that we can zoom in on—we can see the process of star formation at different stages of development. It's kind of like looking at the evolution of stars. You look at different stars, and you can see them at different stages of their development. In a similar way, zooming into the Eagle Nebula, we can see stars in different stages of their development. We see this whole process of star formation, in that sense, going on right before our very eyes. That's not to say that one particular star formed before our very eyes, but the whole collection of them in different stages of development formed, and we can see those different stages of formation.

The realization, then, that our Sun and planets formed from this ISM, whose composition is gradually changing as a result of the heavy elements ejected into the cosmos by supernovae—the realization that we formed from structures like this, chemically enriched by previous generations of massive stars, was a monumental step in our understanding of our place in the cosmos and our origins in this vast Universe.

Structure of the Milky Way Galaxy

Lecture 68

"The astronomers in the late 19th century who looked at the sky and counted stars in different parts of the sky saw about the same number of stars in all directions of the band of the Milky Way, and they concluded that our Sun is in the center of the Milky Way. ... Harlow Shapley, in 1917, realized that [it is] not."

Based on observations that the naked-eye stars within the band of light called the Milky Way appeared to be roughly evenly distributed in all directions, astronomers once believed that our Solar System is at the center of the Milky Way Galaxy. They also concluded that the Galaxy is relatively small because they couldn't see stars more than a few thousand light years away. Further studies showed these conclusions to be false. Many stars are hard to see, and others appear dim, because dense clouds of gas and especially dust block their light, causing an *extinction*, or *obscuration*, of light similar to the effects of smog. In 1917, **Harlow Shapley** realized that our Solar System is not in the center of the Galaxy. His conclusion was based on the observed distribution of globular star clusters, which are more highly concentrated in one region of the celestial sphere than in the opposite region. Globular clusters contain hundreds of thousands of stars bound to each other, as discussed in a previous lecture. Shapley correctly assumed that these clusters form a roughly spherical distribution around the Galaxy's center. Because of the observed concentration of globular clusters in one region of the celestial sphere, Shapley realized that our Sun is near the edge of the Galaxy. From his measured distances to the globular clusters, he was able to determine the Sun's distance to the center of the Galaxy. It was much larger than previously thought, thus implying that the entire Galaxy was very large. However, he overestimated the size of the Galaxy because he wasn't aware of the obscuration (extinction) by dust.

Obscuration also prevents us from seeing, in detail, the Milky Way's spiral arms. Other galaxies have spiral arms that are readily visible from the outside, such as those of the Whirlpool Galaxy (M51). We can map the spiral arms in our own Galaxy by looking at the distribution of types O and B

"How do we get the motions of stars relative to the Sun? The radial motion, toward us or away from us, is easy to obtain from the Doppler shift."

main-sequence stars, young and massive stars that don't wander far from their nebular birthplaces in spiral arms. Recall that open star clusters also form from nebulae; thus, we can map spiral arms by looking at the distribution of open clusters. We can better map spirals by measuring radio waves emitted by clouds of atomic hydrogen because such waves easily penetrate clouds of gas and dust. In its ground state, hydrogen has one electron in its lowest energy level. However, that level can be split by a *hyperfine transition* into two closely spaced energy levels, which are defined by the energy of the electron and proton when both spin in the same direction versus opposite directions. Because spinning protons and electrons behave like magnets, they naturally want to flip over so that they spin in opposite directions. This is a lower energy state than when both point in the same direction. During this flip, a photon of radio radiation is emitted.

A hydrogen atom with its electron in the lowest energy level can be kicked up to the higher level by a collision with another electron. That upper level is not perfectly stable, though typically, it takes about 11 million years before the electron decides to flip its spin back down to the lower level. But because clouds of gas can be hundreds of light years across, there are enough hydrogen atoms constantly producing radiation that we can map the spiral arms of our Galaxy with this technique. Our Galaxy doesn't appear to have a very orderly structure, as some other spiral galaxies do. Instead, it has clumps and spurs, making its spiral arms look stubby.

How do the stars and clouds of gas and dust move in our Galaxy? Because we know the Sun orbits the center of our Galaxy in roughly a circle, with a speed of 200 kilometers per second (km/s), we can determine the absolute motions of the stars and gas-dust clouds. The *radial motion*, or velocity, of stars (toward us or away from us) is easy to obtain from the Doppler shift; we look at starlight spectra and measure how much the pattern of lines is shifted to redder or bluer wavelengths. The amount of shift gives us the radial velocity. **Transverse velocity**—the physical speed in the plane of the sky—is harder to obtain. We know that stars move relative to one another with time in what are called *proper motions*, an *angular motion* across the sky (measured in seconds of arc per year). Some stars have a greater proper motion than others. If star A is the same distance from us as star B but physically moving faster than star B, star A will have a greater angular motion across the sky than star B. Alternatively, stars A and B could be moving at the same rate, but if star A is much closer to us, it would appear to move more quickly than star B. Thus, distance affects how we perceive stars to move. In addition, our view of a star affects how it appears to move. For example, if the star's motion is largely perpendicular to our line of sight, it appears to move faster. If the star is moving mostly toward or away from us (radial motion), it will appear to move more slowly across the sky.

Our Sun moves around the Milky Way Galaxy at about 200 km/s.

If we know a star's angular motion and its distance, we can determine its transverse velocity, which together with radial velocity, produces what we call *space motion*, or *space velocity*. If we measure the space velocity of stars in the vicinity of the Sun, we see much random motion; some stars move away from us and others move toward us. However, if we measure the space velocity for many stars, we observe an average, or mean, motion. Stars that are the same distance as our Sun from the center of our Galaxy appear to move with us at about 200 km/s, so their relative speed is essentially zero. Stars closer to the center of our Galaxy than the Sun appear to overtake us. Though they move at about the same physical speed, they have a smaller circumference to travel. Stars that are farther from the center than the Sun (but not too far) move at about the same physical speed as the Sun. But because they have a larger circumference to travel, they appear to lag behind the Sun. Thus, the stars appear to move at different speeds relative to us; this is called *differential motion*. We can map this *relative motion* at even larger distances in our Galaxy by looking at the radio radiation emitted by clouds of hydrogen gas.

What happens when we plot the speed of the stars and clouds of gas in our Galaxy as a function of distance from the center? This gives us the **rotation curve** of our Galaxy. Our Sun (and Solar System) moves with a speed of about 200 km/s, as does most of the rest of the Galaxy, on average. In the inner 1000 parsecs or so, the speed gradually rises from roughly zero near the center to about 200 km/s. Here, we temporarily ignore the very central parsec, where speeds turn out to be much higher because of the presence of a supermassive black hole. At distances of about 1000 pc to more than 20,000 pc, the speed is roughly constant, 200 km/s. We say that the rotation curve is "flat." Other spiral galaxies show a similarly flat rotation curve in general. But if the closer-in stars have a smaller orbital distance to traverse than the farther-out stars, why don't the spiral arms of galaxies eventually become tightly wound up? It turns out that the arms are actually *density waves*, a compression of gas where new stars form and then gradually move away. The regions of compressed gas rotate around the center of our Galaxy at a different speed than that of typical stars. Therefore, at any given time, the spiral arms consist of different material than at some other time. Further, the orbits of stars are often elliptical because of perturbations from other galaxies. The ellipses rotate—or precess—at different rates, causing an apparent clumping, which forms the spiral structure. As we said, these clumps contain massive clouds of gas and dust from which new stars form.

If we look at the rotation curve of the Milky Way Galaxy, it appears as if the stars at the very center have a speed of zero. However, on closer inspection, we see that the actual speed here is enormously high because of a black hole in the center of our Galaxy. The Milky Way's central black hole is no ordinary one. From the movement of stars around it, we have deduced that its mass is 3.6 million solar masses confined within a region smaller than 45 AU. So far, the center of our Galaxy provides the most compelling evidence for the existence of supermassive black holes, which we will see are a common characteristic of the centers of galaxies. ■

Name to Know

Shapley, Harlow (1885–1972). American astronomer; correctly deduced that the Sun is not at the center of the Milky Way Galaxy and that the Galaxy is larger than previously believed. Incorrectly concluded that the "spiral nebulae" are within the Milky Way, but most of his reasoning was logically sound.

Important Terms

rotation curve: A graph of the speed of rotation versus distance from the center of a rotating object, such as a galaxy.

transverse velocity: The speed of an object across the plane of the sky (perpendicular to the line of sight).

Suggested Reading

Croswell, *The Alchemy of the Heavens: Searching for Meaning in the Milky Way.*

Ferris, *Coming of Age in the Milky Way.*

Henbest and Couper, *The Guide to the Galaxy.*

Melia, *The Black Hole at the Center of Our Galaxy.*

Pasachoff and Filippenko, *The Cosmos: Astronomy in the New Millennium*, 3rd ed.

Verschuur, *The Invisible Universe Revealed: The Story of Radio Astronomy*.

Questions to Consider

1. If the band of light called the Milky Way stretched only halfway around the sky, forming a semicircle rather than a circle, what would you conclude about the Sun's location in our Galaxy?

2. Can Shapley's conclusion regarding the Sun's location in our Galaxy be considered an extension of the Copernican revolution?

3. While driving at night, you almost instinctively judge the distance of an approaching car by looking at the apparent brightness of its headlights and using the inverse-square law of light. Is your answer correct if you don't account for fog along the way?

Structure of the Milky Way Galaxy

Lecture 68—Transcript

I have discussed how the space between the stars is filled with gas and dust. Gas is just little atoms and molecules, and dust is fine particulate matter. In general, this interstellar medium (ISM) is splotchy. There are denser regions, and there are less-dense regions. Stars form out of the densest clumps. The gas and dust, though the site of star formation, does present a problem for us because it blocks our view of various parts of our Galaxy. It's like smog, and you can't see through it very easily, and so you can't see stars that are behind dense clumps. Even if you can see the stars, they appear fainter than they would have otherwise appeared. If you look at the sky with the naked eye and start counting stars, and plot their positions around you, you will see that there's about the same number of bright stars in all directions of the band of light we call the Milky Way.

That is, you can't see a huge concentration toward the center of our Galaxy because, in fact, the center of our Galaxy is blocked or obscured by clouds of gas and dust between us and the center of our Galaxy. This is a process called extinction, or obscuration, of light. The astronomers at the late 19th century who looked at the sky and counted stars in different parts of the sky saw about the same number of stars in all directions of the band of the Milky Way, and they concluded that our Sun is in the center of the Milky Way. Moreover, they concluded that the Milky Way Galaxy isn't very big because you couldn't see stars that were more than a few thousand light years away. That's the size of the Milky Way, and we're at the center.

This extinction, or obscuration, of the starlight is similar to smog. I did my graduate studies at Caltech in Pasadena, and in the late '70s and early '80s, LA was a much smoggier place than it is now because now there are better emission controls on cars and things like that. But from Caltech, the San Gabriel Mountains are just a few miles away. On some days, you could see them clearly; on other days, they were barely visible. On some days, I'm sad to say, you couldn't see them at all because of the smog, not just fog or clouds. You can see this effect when you look through fog as well. Through thick fog, you can't see anything. Through thin fog, parts of the road here, for example, are visible, but they're clearly obscured. You can't see them as easily, and the parts that you can see are dimmer than they would otherwise have been.

Harlow Shapley, in 1917, realized that we're not in the center of our Galaxy. He saw that the distribution of globular star clusters had a higher concentration in one part of the celestial sphere than in the opposite part. Because the globular clusters were these grand structures containing hundreds of thousands of stars bound to each other, he assumed that the globular cluster system is centered on the center of our Galaxy. This was a bold assumption and not particularly well motivated by other data, but he just thought that this system of globular clusters probably, in a sense, is a backbone for our Galaxy. He was right in his gut feeling. From the distribution of the globular clusters—from their distances and their positions in the sky—he realized that our Sun is out near the edge of our Galaxy, and that our Galaxy is much bigger than had been thought before.

Here is the schematic that he drew. Our Galaxy has this concentration of globular clusters surrounding the center. Because there were more globular clusters in these general directions of the sky than in the opposite directions, he concluded that our Sun is offset by a considerable distance from the center of our Galaxy. From the distances of the globular clusters, he concluded that the center of our Galaxy is very far away. I think he concluded that it's something like 100,000 light years away or more. In fact, it turns out he overestimated the size of our Galaxy. We now know that our Sun is only 24,000 or 25,000 light years from the center.

That overestimate resulted from the fact that he didn't properly take into account the dimming, or obscuration, of even the globular clusters by this interstellar gas and dust. Its effects were not known yet. They became known about a dozen years later, from the work of Trumpler. Shapley didn't know about the dimming effect, and therefore, looking at the brightnesses of the globular clusters and figuring out their distances from their apparent brightnesses, and comparing with the known luminosity of nearby ones, he overestimated their distance, and hence the distance to the center of our Galaxy. Nevertheless, he was clearly right that our Galaxy is much bigger than previous astronomers had thought, and that our Sun is not in the center.

Now let's look at our Milky Way Galaxy, taking a deeper view. If you take a long exposure photograph like this one, then you begin to see the faint stars, and then you can see that, indeed, there is a concentration of stars toward the center of our Galaxy compared with other regions of the Milky Way. But this isn't so readily visible when you're looking only at naked-eye stars because you only see out a couple of thousand light years or so, and you can't see to the center of our Galaxy. But in a deep photograph like this, you can see much more distant, fainter stars, and their concentration toward the center starts becoming apparent. It becomes even more apparent if you look at our galaxy with infrared wavelengths or radio wavelengths, where you can penetrate the clouds of gas and dust and see most of the stars. In that case, you can readily tell that we are offset toward the edge of our Galaxy.

The interstellar extinction, or obscuration, that you can see when you look at these clouds of gas in the Milky Way Galaxy prevent us from seeing in detail the spiral arms. I discussed this in the previous lecture, that we can't really tell the structure of the maze in which we're located. Other galaxies have spiral arms that are readily visible because we see them from the outside. Especially for a face-on galaxy like this, there's very little gas and dust blocking our view of the spiral arms. If that galaxy were edge-on, then the gas and dust within that galaxy would block our view of its spiral arms. But here, for a face-on galaxy, they're easy to see. Here's a beautiful galaxy, the Whirlpool Galaxy, where the spiral arms are once again easy to see.

How do you map the spiral arms in our own Galaxy? As I said, massive stars form in the arms because there are nebulae—clouds of gas and dust—in the arms. You can look for Type O and Type B main-sequence stars—young, massive stars that have not had much of a chance to wander away from their birthplaces because they simply don't live a long time. You look at how bright they appear to be, figure out their distance (taking into account interstellar extinction and all that), and you can kind of tell where the arms are, based on the distribution of Type O and Type B main-sequence stars. But they're rare, so you can't really tell the detailed structure of our arms. You can also find the distances of star clusters. Since clusters form in the arms and don't have that much time to move away in general, especially if they're young clusters, you can map out the spiral arms by looking at the distribution of clusters in our Galaxy.

An even better way to map the distribution of the arms is to use radio astronomy techniques and look at the light emitted by cold clouds of atomic hydrogen. Those clouds of atomic hydrogen emit radio waves, which find themselves in essentially a transparent galaxy. They don't care about the clouds of gas and dust that obscure optical and ultraviolet light. Radio waves just bend around those particles, just as on a smoggy or a foggy day, you can hear your radio stations in your car even though you might not be able to see more than 100 yards in front of you. Radio waves pass through space even if it has clouds of gas and dust. The line of neutral hydrogen is produced by the following process: You have hydrogen in its ground state, where the electron is in its lowest energy level compared to the higher energy levels out here.

But it turns out that that lowest energy level actually is split by what's called a hyperfine transition into two very closely spaced energy levels. They are defined by the energy of an electron and a proton when both are spinning in the same direction, as shown here, as compared with when the electron and proton are spinning in opposite directions. Protons and electrons are fundamental particles, or nearly fundamental particles, that have this property called spin. The spin makes them behave kind of like a magnet. They have this orientation effect—where with two magnets that are aligned parallel to each other like this, there's no tendency for the two magnets to attract each other; in fact, they repel one another. But if I now flip one of them over, they attract each other like this. This is a lower energy state than when they are both pointing in the same direction.

It's natural for the electron and proton to want to flip over in such a way that their spins are opposite. During this flip, a photon of radio radiation, having a wavelength of just 21 centimeters—well into the radio part of the spectrum—gets emitted. So, cold clouds of hydrogen can emit this radiation. But you might ask, "How do the atoms get in such a state where both the electron and the proton are oriented with their spins in the same direction in the first place—that is, how do they get into the higher energy state?" If you have cold hydrogen atoms sitting around, and occasionally they collide with other particles, those collisions can flip the electron to be in the higher energy state relative to the proton. That is, you flip them in such a way that the magnets are aligned. After something like typically 10 million or 11 million years, the electron decides to flip over.

In other words, it's not too unhappy in this higher energy state because it's not a very much higher energy state. It's only higher by the energy equivalent of one 21-centimeter radio photon, and that's not a very much higher energy, so the electron doesn't mind being there. But after 10 million or 11 million years, it decides okay, on average that's about how much time I wait before I flip over to the lower energy state and emit this photon. They do it only very rarely, but if you have clouds that are hundreds of light years across, there's enough hydrogen atoms that enough of them are producing this radiation at any given time, that you can see that radiation coming from a clump of gas, a cloud. There are special techniques for determining how far away that cloud is, but I simply don't have time to go into them.

In this way, you can map the distribution of clouds of gas in our Galaxy and see whether it has a nice, orderly spiral structure like some of the other spiral galaxies I showed. The result is that you don't see a very orderly spiral structure. You see some hints of some spiral arms, but also blobby distribution with lots of spurs and things sticking out. It could be that our Galaxy doesn't have a grand design spiral structure. It might have more short, stubby arms. Or this could partly be due to uncertainties in the distances that are determined for these clouds. It turns out to be quite tricky to determine these distances after all. In any case, our Galaxy definitely does have at least some stubby arms, and maybe a few bigger ones, off of which these stubby ones emerge.

How do the clouds and stars in our Galaxy move? What are their trajectories? What are their orbits? We can study their relative motion—that is, their motions relative to the Sun. If we know that the Sun is orbiting the center of our Galaxy in roughly a circle, or a little bit of an ellipse, with a speed of 200 kilometers per second, and if we know those relative motions, then we can determine the absolute motions of the stars and clouds of gas in our Galaxy. How do we get the motions of stars relative to the Sun? The radial motion, toward us or away from us, is easy to obtain from the Doppler shift. It's this business where you take the spectrum of the star, see how much the pattern of lines is shifted to the red or to the blue wavelengths. From the amount of that shift, you get the speed of the star. So, the radial velocity is easy to get.

The transverse velocity—that is, the speed in the plane of the sky—is harder to get. We see that stars are moving over the years ever so slightly relative to one another in the sky. Those are called proper motions. If you look at the proper motion of the fastest-moving known star, Barnard's star, you can see that, from 1997 to 2004, it moved a considerable distance relative to the more-stationary, more-distant, less-rapidly moving stars in its vicinity. If you look at the Big Dipper, and you ask yourself what did its shape look like 100,000 years ago, and what will it look like 100,000 years from now, and in the meantime; we can look at the proper motions of all the stars in that part of the sky and see how the Big Dipper changes in appearance from 100,000 B.C. to 100,000 A.D. Let's run through that again. It starts at 100,000 B.C.; here it is about now. Then 100,000 years from now, it'll have a very different shape indeed.

You can see that the other stars are moving as well. They're moving every which way because all stars have some amount of proper motion, and some stars have more than others. Why might a star have a bigger angular motion across the sky than another star? Well, maybe it's physically moving faster. If a star is at the same distance as another star, but physically moving faster, it'll have a greater angular motion across the sky. Or maybe they're moving with the same speed, but one of them is much closer to you than the other. Then the one that's closer will appear to move in angular motion across the sky more quickly.

You've seen this. When you're near an airport, you see the airplanes taking off, and they go zooming across the sky, from horizon to horizon. At takeoff, they're not moving as quickly as they are once they're nicely high up in the sky. Yet high up in the sky, you see an airplane, and it's sort of crawling across the sky; the angular motion is small. Clearly, distance is an important effect. Finally, it's important whether most of the star's motion is across your line of sight, like this, or toward you, radial. For a given physical speed, clearly the angular motion will be greater if more of it is perpendicular to your line of sight than if it's along your line of sight. All these factors come into play when getting the proper motion, or angular speed, of a star across the sky.

If you measure the angular motion, and you also determine the distance using triangulation, or trigonometric parallax, or spectroscopic parallax, or other techniques, then you can get the physical motion in kilometers per second. That is, if you know the angular motion, and you know the distance, then you can figure out the physical speed. Similarly for an airplane—if you know its angular motion, and you know its distance, you can figure out its physical speed. That physical speed in the plane of the sky is called transverse motion, as opposed to the radial velocity. Those two velocities together give you the space motion, or the space velocity. Here it's illustrated. Here's a star, which moves in a certain amount of time over to here, radially along the direction to the Sun, and also moves transverse to the direction to the Sun—that is, in the plane of the sky by a certain amount. Its total space motion is this distance divided by a certain amount of time, which gives you a space velocity. That space velocity is the combination of the radial velocity and the transverse velocity.

When you measure the space velocity of a bunch of stars in the vicinity of the Sun, you find that there is a fair amount of random motion, sort of chaotic. Some stars are moving toward us and in the same direction of the sky; others are moving away. But when you do this for a bunch of stars, you find that there's a sort of an average motion, a mean motion. What you find is that, if here's the Sun, then stars that are the same distance from the center of the Galaxy, which is down here somewhere, are not moving at all toward us or away from us. That means they must be moving with us at about 200 kilometers per second. On the other hand, stars that are closer toward the center of our Galaxy than our Sun are seen to be overtaking us. They're moving at about the same speed, but they have a smaller circumference to go, and so they're passing us up, it turns out.

Conversely, stars that are not too far away, but farther from the center than our Sun is, they turn out to be moving at about the same speed as our Sun. But because they have a bigger circumference to go through, they are lagging behind the Sun, as seen by us. So, everything is rotating in this direction, from left to right, but the stars closer in are passing us up, and the stars farther out are lagging behind. This is known as differential rotation. You can also get the relative motion by looking at the radio radiation emitted by clouds of hydrogen gas throughout our Galaxy. They're doing the same sort of thing; there's a differential rotation. If you then look at the picture of our Galaxy and how the stars and clouds of gas are moving, here we are out here at 24,000 light years. We're moving at about 200 kilometers per second, but most parts of our Galaxy are also moving at about 200 kilometers per second. There's a little dip here at 180, and a slightly higher velocity at 230, closer to the center. Then even closer to the center, the stars and clouds of gas are moving more slowly. But over much of the Galaxy, everything is orbiting at about 200 kilometers per second.

If you plot the speed of rotation of our Galaxy versus the distance from the center, close in to the center, the stars are moving more slowly than 200 kilometers per second. But out at 1,000 parsecs and beyond, the speed is roughly constant, at about 200 kilometers per second. The rotation curve of our galaxy is said to be flat. This is true for many other spiral galaxies it turns out. You measure how quickly they're rotating as a function of distance from the center, and they seem to have this flat rotation curve. It's a generic property of spiral galaxies, but it has strong implications for the formation of spiral arms in the first place. If you ask yourself why don't the arms wrap up quickly, like the cream in a coffee cup, which you put into the coffee cup, and then you swirl it around, and the cream makes a spiral pattern that wraps up quickly. So, too, should the arms of our Galaxy.

Look, if all the stars and clouds of gas are moving at about 200 kilometers per second, but the ones closer in to the center of the Galaxy have a smaller orbital distance, a smaller circumference for the circle, they will get in front of the ones that are farther away. Similarly, the ones that are really far away will lag behind the ones that are in the middle. So, very quickly, what you have is this stretching out of an initially radial spoke—let's say you started out that way—and that radial spoke quickly winds up and forms spiral arms that are too tightly wound compared with the typical ones that we think are in our Galaxy, and the ones we definitely see in other galaxies. If the whole thing is rotating at 200 kilometers per second, why don't the arms get very tightly wound up?

The solution is that the arms are actually what are called density waves. The arms don't consist of a bunch of stars that forever remain in those arms. Rather, the arms are a compression of gas where new stars form and then gradually move out of that region. The compression of gas rotates around the center of our Galaxy at a different speed than the speed of the typical stars that are wandering around the center of our Galaxy. Therefore, you have this pattern, the spiral density waves, which are moving through the disk at a different speed than the orbital speed of the stars. At any given time, the spiral arms consist of different material at one time than at some earlier or future time. Let me give you an example.

Suppose there's a roadblock because people are fixing potholes in a four-lane highway. You're up in a helicopter, looking at the traffic flow. Cars are zooming past, but they have to clump up a little bit in order to fit into two lanes here—whereas before they fit comfortably into four. But if there are not many cars, they don't actually have to slow down. Let's say there are no rubberneckers, and they just keep on going at the same speed; they're not slowing down to look at the action. All these cars are zipping through, but meanwhile the people fixing the potholes are making slow progress, and they gradually move along the road, fixing the potholes along the way. Meanwhile, all these cars are zipping past at a much higher speed. What you would see from a helicopter is a clump of cars that gradually moves along the highway at a slow speed, and a much slower speed than the speed of individual cars that went zipping through that bottleneck.

In a similar way, then, in our Galaxy, what we think is happening is that the orbits of stars are elliptical because of a perturbation from other galaxies, like the Magellanic Clouds, little dwarf galaxies that orbit our own. They cause the stellar orbits to be not quite circular, rather elliptical, and the ellipses precess (or rotate) at different rates, causing an apparent clumping up, and the clump has a spiral structure. Those clumps are where there are massive clouds of gas and dust, from which new stars form. Then those new, massive stars light up the surrounding nebulae, producing the beautiful arms that we see. We see the arms defined by young, massive stars, and the ionized clouds of gas from which they form, and dark clouds of gas next to those ionized gas clouds. But that clumping of clouds, forming a spiral pattern, moves through the disk at a different rate than the stars do, just as the cars moved past the bottleneck at a different rate than the bottleneck itself was moving along. It's kind of a neat thing.

We see that our own Galaxy has these companions, which may be hassling it, which may be perturbing it, to form spiral arms in this way. Indeed, the best so-called "grand design spiral galaxies" like this one here, the Whirlpool, almost always have a companion, which might be gravitationally perturbing gas in the disk, causing the spiral structure to form in the first place. You might think, looking at this velocity curve, that the speed of stars in the very center is zero. It looks like it's zero, right? But that's because we haven't taken a very close look at the very central region. If you take a magnifying glass—or better yet, a telescope—and zoom in on the central region of our Galaxy, you find actually that the speeds are enormously high. That's because there's a supermassive black hole in the center of our Galaxy, having about 3.6 million times the mass of our Sun.

Let's look at the center of our Galaxy. There it is, toward the region of Sagittarius and Scorpius; it's in this green box. With the technique of adaptive optics on big telescopes—like the Keck Telescope or the Very Large Telescope that the Europeans own in Chile—you can get very clear views of those stars. Andrea Ghez (at UCLA) and Reinhard Genzel (in Germany) have been, for years, mapping the motions of those stars in the central part of our Galaxy. Here's a picture taken with adaptive optics or a related technique in 1994, then in 1996, and then in 2000. You can see that the stars have moved. In fact, if you plot their positions, you can see that (zoom, zoom, zoom) they're orbiting around something. Let's look at it again (zoom, zoom, zoom). These stars are moving very quickly around some object marked with a red cross here. Look at that thing. That one went so close to whatever is there pulling on it. It's fantastic.

Now, if you plot these motions of the stars, and you look at where they are—this star started here, and then over the years it moved to there. This one started here, and over the years it moved to there, and so on. If you look at a few of these in particular, Star-2 and Star-16, you find that Star-2, for example, passes 124 AU from some sort of a central object that's pulling on it. Star-16 passed only 45 AU from that same object. At the time of closest approach, Star-2 is moving at 5,000 kilometers per second and Star-16 was moving at 12,000 kilometers per second. From those numbers, you can figure out that there's a common single object right smack in the middle there, marked by this black cross, causing these stars to have such incredible trajectories. The mass at that central position is 3.6 million solar masses, and it's confined within a region smaller than 45 AU because we see one of these stars coming that close to that spot in the sky.

There is nothing that that could be other than a black hole. Whatever other ideas people have at this point, they are far more exotic than even black holes. The center of our Galaxy provides, at this time, by far the most compelling evidence for the existence of not just any old black holes, but the supermassive variety, which we will see are a common characteristic of spiral and other galaxies.

Other Galaxies—"Island Universes"

Lecture 69

For a long time, astronomers thought that our Milky Way Galaxy was the only galaxy in existence—essentially, the entire Universe. But by the early 20th century, there were good reasons to believe that other galaxies, or "island universes," existed.

Some nebulae are clearly members of our own Galaxy because of their distance from us. The Orion Nebula, for example, is only 1500 light years away, and we can easily see the individual stars that light it up. But the nature of other so-called spiral nebulae was much more controversial because individual stars could not be seen within them and their overall structure differed from that of gaseous nebulae, such as the Orion Nebula.

In 1920, a famous debate took place between Heber Curtis and Harlow Shapley, two major astronomers of the time, on the subject of the "Scale of the Universe." If our Galaxy is all that exists, then the Universe is pretty big but not gargantuan. On the other hand, if spiral nebulae are actually galaxies in their own right, then the Universe is far more vast. Curtis correctly believed that spiral nebulae were separate entities, far beyond the outskirts of our Milky Way, but part of his reasoning was flawed. Shapley incorrectly believed that spiral nebulae were clouds of gas within our Galaxy, but his reasoning was sound. His conclusion was partly based on an erroneous measurement of the apparent rotation of one spiral nebula and the false assumption that the Andromeda Nebula was closer to Earth than it really is.

In the mid-1920s, the famous astronomer Edwin Hubble resolved this controversy through his recognition of a certain type of star—a Cepheid variable (Lecture 45)—in spiral nebulae. Recall that Cepheid variables are evolved supergiant stars that grow bigger and smaller with time and, hence, brighter and fainter; these oscillation periods can last from 1 to 100 days. More luminous Cepheid variables have longer periods than less luminous ones. From this period-luminosity relationship, we can compare the average apparent brightness of a Cepheid variable of unknown distance with the known average luminosity of a Cepheid variable having the same period to determine the distance of the first Cepheid variable and, hence, the distance of its associated galaxy or nebula. Hubble noticed that Cepheid variables in the Andromeda Nebula looked faint and correctly concluded that they must be incredibly distant. Furthermore, given the apparent angular size of the Andromeda Nebula (it spans several degrees in the sky), Hubble deduced that it was a huge system containing billions of stars—a galaxy. Today, we can see anywhere from 50 to 100 billion galaxies with our most powerful telescopes.

Hubble classified two main types of galaxies: *spiral galaxies*, like our own Milky Way, and *elliptical galaxies*. There were also several additional types of galaxies. Spiral galaxies differ from one another in shape. Some have fairly open arms and fairly small central regions (bulges); others have larger bulges and more tightly wound arms. Some have bars through their centers with arms protruding from the ends. Generic spiral galaxies have a disk containing arms. New stars form continuously out of nebulae located primarily within those arms, then migrate away. The disks have a mixture of old and young stars, but in particular, the arms contain young stars. Old stars are present almost exclusively in the halo and in the bulge, the central part of the galaxy. The vast halo around a spiral galaxy tends to contain almost entirely old stars.

Elliptical galaxies appear roughly spherical or elliptical (elongated). They have no disk or arms, and there is very little gas and dust. Further, they have no nebulae and, hence, very little active star formation. Presumably, large stars in these galaxies existed long ago and died as supernovae or GRBs. The largest elliptical galaxies (300,000 light years across) are more than three times the size of the Milky Way and contain up to 10 trillion stars. The Milky Way, on the other hand, has only a few hundred billion stars. Most elliptical galaxies are relatively small (5000 to 10,000 light years across), with perhaps only 1 million to 10 million stars.

A third class of galaxies, *irregular galaxies*, is relatively rare in the local parts of today's Universe. But as we look far away and back in time, we see more of these irregular galaxies. Nearby examples are the Large Magellanic Cloud and the Small Magellanic Cloud, satellite galaxies of the Milky Way Galaxy. There are also *peculiar galaxies*, which differ from irregular ones in that they have fairly distinct spirals or elliptical shapes but also an unusual feature. Some spiral galaxies have rings or tails, both of which may be caused by gravitational interactions with neighboring galaxies. Some elliptical galaxies have lots of gas and dust, unlike the typical ones.

One final class is a cross between spiral and elliptical galaxies, called *S0 galaxies* (pronounced "ess-zero"), or *lenticular galaxies*. They have a bulge and a disk like a spiral galaxy, but the disk doesn't have much gas and dust. There are no arms, and there's very little evidence for the formation of new, massive stars, either in the recent past or in the present.

NASA, ESA, and The Hubble Heritage Team (STScI/AURA)-ESA/Hubble Collaboration

The barred spiral galaxy NGC 1672, imaged with the Hubble Space Telescope.

Edwin Hubble tried to arrange the different types of galaxies in a "tuning fork" diagram: Ellipticals were arranged on the handle of the tuning fork, while normal spirals and barred spirals were arranged on one tine each. Hubble thought that there might even be an evolutionary progression from ellipticals to spirals or vice versa. We now think, in general, that no such evolutionary progression exists, although some spirals do merge to eventually form ellipticals. A single lone spiral will not later turn into an elliptical, nor will an elliptical, on its own, turn into a spiral. In this sense, then, galaxies don't undergo an evolutionary sequence.

Gravity binds galaxies together. Indeed, they are formed out of primordial clouds of gas that collapse, creating many galaxies in about the same place. Orbiting our Milky Way Galaxy are the two Magellanic Clouds and six or seven smaller companions, dwarf galaxies that are harder to see. Our nearest big neighbor, the Andromeda Galaxy, also has two main companions and half a dozen smaller ones. We live not just in a binary system with the Andromeda Galaxy but with several dozen smaller galaxies in a loose cluster known as the **Local Group**, which has a diameter of a few million light years. Galaxies also occur in large clusters. The Virgo cluster spans about 15 degrees in the sky, contains more than 1000 galaxies, and is 60 million light years away. The Coma cluster is about 300 million light years away and contains more than 10,000 galaxies. Clusters appear to form even larger superclusters, spanning 50 to 100 million light years and separated from each other by vast empty regions called *voids*.

Astronomers have come a long way since the days when they thought that ours was the only galaxy. We now know that there are up to 100 billion galaxies out to 14 billion light years from us—a giant leap in our understanding of our cosmic origins and our place in the Universe. ■

Important Term

Local Group: The roughly three dozen galaxies, including the Milky Way, that form a small cluster.

Suggested Reading

Christianson, *Edwin Hubble: Mariner of the Nebulae.*

Dressler, *Voyage to the Great Attractor: Exploring Intergalactic Space.*

Hirshfeld, *Parallax: The Race to Measure the Cosmos.*

Lightman and Brawer, *Origins: The Lives and Worlds of Modern Cosmologists.*

Malin, *View of the Universe.*

Pasachoff and Filippenko, *The Cosmos: Astronomy in the New Millennium*, 3rd ed.

Petersen and Brandt, *Visions of the Cosmos.*

Waller and Hodge, *Galaxies and the Cosmic Frontier.*

Questions to Consider

1. Explain why Cepheid variable stars are so useful for determining the distances of galaxies.

2. Why can't we determine the distances to galaxies using the geometric method of trigonometric parallax (triangulation), as we do for stars?

3. How is the discovery of other galaxies an extension of the Copernican revolution?

4. The sense of rotation of galaxies is determined spectroscopically. How might this be done, in practice?

Other Galaxies—"Island Universes"

Lecture 69—Transcript

We have seen that the Milky Way Galaxy, our home, is a vast system of stars, containing perhaps a few hundred billion stars, spanning 100,000 light years in diameter—a huge place of which our Sun is just one of a myriad of stars. But is this the entire Universe? Is our Milky Way Galaxy the entire Universe, or is there something more out there? For a long time, astronomers thought that the Milky Way is all that there is. But by the early 20th century, there were good reasons to think that there were, in fact, other galaxies, other "island universes," as they were called. I will now discuss the evidence.

Some nebulae, like the Orion Nebula, are clearly members of our own Galaxy. You can see the stars within them causing the gas to glow. One can use a number of techniques to measure the distances of those stars, and they are clearly within our Galaxy. The Orion Nebula, in particular, is about 1,500 light years away. But the nature of other nebulae, the so-called "spiral nebulae," was much more controversial. Here are a couple of examples of spiral nebulae as they appeared perhaps through the telescopes being used by early 20th-century astronomers. They didn't quite look like Orion because they had more of a structure, a spiral shape to them like this. Here's one in particular that definitely looked like a spiral.

These things were more controversial because it's conceivable that they are vast systems of stars, gravitationally bound together, beyond our Milky Way. Or maybe they're just a special, peculiar-looking nebula in our own Galaxy—like the Orion Nebula, but with a different shape. But you couldn't see very clearly any massive, ionizing stars within those things, so people began to doubt that they are clouds of gas in our own Galaxy. In 1920, there was a famous debate between Heber Curtis and Harlow Shapley, two major astronomers of the time. The debate was called "The Scale of the Universe" because really, that's what was at stake. If our Galaxy is all that there is, then the Universe is pretty big, but not gargantuan. Whereas if these little spiral nebulae are actually systems of stars of their own, galaxies in their own right, then the Universe is far, far more vast than just our Galaxy.

Shapley used excellent reasoning to conclude that, in fact, our Galaxy is all that there is. The trouble is that he had faulty data and incomplete knowledge. With those faulty data and incomplete knowledge, he came to the wrong conclusion. We now know that the spiral nebulae are other galaxies in their own right. That's the conclusion that Curtis came to, but with perhaps somewhat weaker reasoning. One of the things that Shapley believed was that Adriaan Van Maanen, an astronomer, had measured the apparent angular rotation of at least one spiral nebula, M101, shown here. He thought he could see it actually rotating. You could calculate the entire rotation period of that galaxy, and it turned out to be only about 100,000 years.

Therefore, Shapley reasoned, if this thing is really millions of light years away, but it completes one rotation in 100,000 years—yet it's a vast object; it's huge because it appears pretty big, and it's millions of light years away, so it must be physically very large. To complete one rotation in 100,000 years would mean that its rotation speed would be highly relativistic, close to the speed of light, and it would be just torn apart, flung apart, by the rapid rotation. Shapley concluded that this thing must be nearby. For a given size, an angular size in the sky, its physical size would be much smaller than if it were millions of light years away, so a given angular rotation would translate into a smaller physical speed, and it wouldn't be torn apart. Excellent reasoning, but it turns out that von Maanen's observation of the apparent angular rotation was simply wrong. There is no measurable rotation over a human lifetime. So, that's the breaks.

Another argument that Shapley used was that there was an apparent nova, a new star, in the Andromeda Nebula. If you compared its brightness with the apparent brightness of novae in our own Galaxy, you would conclude that, in fact, the Andromeda Nebula wasn't very far away. However, it turns out that that nova was not a nova of the normal sort; it was a supernova. It was a much, much more luminous object. Had Shapley known about such objects, he would have concluded that Andromeda is really very, very, very far away in order to make this very luminous object appear so faint. Curtis, in fact, said, "Well, maybe there are two classes of novae. Maybe there are really luminous ones and not so luminous ones." He said, "Maybe this one is the really luminous variety that's far away," but there wasn't any real evidence for that at that time. Now we look at galaxies—here's one taken through an amateur-sized telescope—and it shows a supernova, SN2006X. Had one assumed that that's a regular nova, one would have concluded that this object is quite nearby, not at distances of millions of light years.

In the mid-1920s, the famous astronomer Edwin Hubble, after whom the Hubble Space Telescope was named, made a major breakthrough in resolving this controversy. He used the 100-inch telescope at Mount Wilson Observatory, known as the Hooker Telescope, to observe in great detail some of these spiral nebulae—in particular, the Andromeda Nebula. Within the Andromeda Nebula, he recognized a certain type of star. We've already discussed them; they're called the Cepheid variable stars. You will recall that Cepheid variables brighten very quickly, then fade more slowly, and their total period of brightening and fading is, of order, a few days or up to about 100 days. They have this very distinctive light curve, or graph of brightening and fading versus time.

As we discussed in Lecture 45, Cepheid variables are evolved supergiant stars that actually grow bigger and smaller with time, and hence brighter and fainter. They have these periods of 1 to 100 days. The more luminous Cepheid variables are the ones with the longer periods than the less-luminous ones. Indeed, Henrietta Leavitt—shown here on the right next to Annie Jump Cannon, another famous early 20th-century astronomer—figured out this period-luminosity relationship, Leavitt did, by looking at the Large and Small Magellanic Clouds, where the stars are all at about the same distance from us, and so the Cepheid variables that appear brighter are also the ones that are more luminous than the ones that appear fainter. All the stars are at the same distance, so you don't have to worry about the distance.

She saw that the ones that appear brighter have the longer period than the ones that appear fainter, so she determined this so-called period-luminosity relationship, where you plot the luminosity on the vertical axis—that's similar to absolute magnitude, going up to –8, so these are very luminous stars, actually. Our own Sun has an absolute magnitude of only about +5 or so. You see the less luminous ones have short periods, and the luminous ones have long periods. From this period-luminosity relationship, one can look at the average brightness of a Cepheid variable of unknown distance and compare it with the known luminosity of a Cepheid variable having the same period. In this way, comparing the luminosity with the apparent brightness, you can determine the distance of that Cepheid variable, and hence the distance of the galaxy or nebula in which it's located. You know the luminosity by having measured the period and by comparing with this graph of Cepheids having known luminosity and known period.

Hubble noticed that the Cepheid variables in the Andromeda Nebula, and a few others, looked really faint. They're very, very faint on average. He could tell that they're faint not because there's smog, gas and dust blocking his view—because they didn't look anomalously red; they looked the normal color, but just really faint. He concluded that, to look so faint, they must be incredibly distant, unbelievably distant almost. This Andromeda Nebula, he found, was way outside the confines of our own Milky Way Galaxy. Given its apparent angular size—it actually spans a couple of degrees in the sky—he deduced that the Andromeda Nebula is this huge system, presumably containing billions of stars. Indeed, on the clearest nights and the sharpest scene, the smallest amount of turbulence in the atmosphere, he could actually see (in his best photographs) some of the individual stars of the Andromeda Nebula.

This wasn't a nebula; this was a galaxy, a huge stellar system in its own right, an "island universe," as Immanuel Kant had said in 1755—although at that time, there was really no evidence that these nebulae were galaxies outside our own Milky Way Galaxy. Here's one of these galaxies, now—not just a little smudge of gas and dust in our own Galaxy, but a giant collection of gravitationally bound stars far outside our Galaxy. If you look at deep pictures of the sky, you see galaxies all over the place. In fact, one can estimate that you can see something like 50 billion to 100 billion galaxies with today's most powerful telescopes. This was an enormous expansion not of the Universe itself, of course, but of the human perception of the Universe. Galaxies are the fundamental building blocks of the Universe, just as stars are the building blocks of galaxies. It is fitting to put galaxies on T-shirts, given their great importance. In fact, I have lots of T-shirts that have galaxies on them, especially this beautiful Whirlpool Galaxy here.

It turns out there are two main classes of galaxies that Edwin Hubble classified. Having discovered that these nebulae are systems outside our own Galaxy, he started to systematically classify them. He came up with two main classes. One is the spirals that sort of precipitated this whole debate. The spirals are like the Milky Way Galaxy in general form, but they have differences in shape depending on which galaxy you look at. Here's a schematic showing a bunch of spiral galaxies, and you notice a few things. Some of them have fairly open arms like this one, and a fairly small central region. Others have a bigger central region and more tightly wound up arms. Some have a bar going through the center, which is a relatively straight thing like this, and then the arms hanging out off of the ends of the bar. Then some look different, presumably because they're seen from the side. Here's one of these galaxies seen from the side rather than face-on. Others are seen at some intermediate angle.

Let's take a look at some of these things. Here's a beautiful galaxy, one of my favorites, in a picture taken from the ground. Here's another one, the Whirlpool Galaxy, which has a companion gravitationally interacting with it. I mentioned before that it is the interaction of companions with galaxies that sometimes help to produce the grand spiral arms seen in some of these galaxies. Here's another one, a nice galaxy with a couple of background galaxies here, some bright background galaxies. If you look closely, you can see lots and lots of fainter, more-distant, smaller-looking galaxies all over this photograph. Galaxies are just all over the place: along the line of sight to nearby bright ones. You can find little galaxies in the background all over the place. That's a pretty one right there.

Here's one of the edge-on galaxies. You can see, in fact, some dust blocking our view of many of the stars in the disk, or plane, of this galaxy. That effect is beautifully seen in this example, where there's a lot of gas and dust along the plane of this so-called Sombrero Galaxy, blocking our view of many of the distant stars there. I like that galaxy a lot. These are generically the spiral galaxies, and they have a disk with arms. New stars are forming pretty continuously out of nebulae, clouds of gas and dust, primarily within those arms. Then they migrate away from the arms. The disk has a mixture of old and young stars, but in particular the arms contain the young stars because that's where the nebulae are able to form the arms.

Old stars are present almost exclusively in the halo and in the bulge, the central part here that you can see especially in the side view of spiral galaxies. The central bulge, and then this vast halo around the galaxy, tends to have almost entirely old stars—no new, young stars. Occasionally, there might be a young star found in the bulge region, but almost never. If you look at these galaxies, you can classify their shapes. Often, the arms come out of the nucleus. But as I mentioned earlier, in some of the cases, the arms come off of the end of a relatively straight-looking bar. These are a variant known as the barred spiral galaxies. There's an example. Here's another example; I like this one a lot. There's this bar that's relatively straight, and then the arms come off of the end of the bar.

Usually there are two arms. Sometimes they split up into four or more, but often spiral galaxies have two main arms. Our own Galaxy is thought to look something like this. This is an idealized view of our own Galaxy, as I mentioned in a previous lecture. The actual maps of clouds of gas and their distribution in our own Galaxy aren't nearly as regular-looking. But when we piece things together, we think that our Galaxy may look something like this. It's got two major arms, which then split up into a larger number of more minor arms. Those are the spiral galaxies.

Now let's move on to the next major class that Hubble defined, the elliptical galaxies. Here's a schematic of what they look like. They're kind of boring compared to spiral galaxies. They either look roughly circular or spherical, presumably, or elongated to varying degrees. They don't have a disk; they don't have arms. They have very little gas and dust in general, and no nebulae, so there's very little active star formation going on right now in elliptical galaxies. Indeed, you look at the stars, and you can see that they are old stars. There are very few young, massive stars in elliptical galaxies. Presumably, they existed long ago, but they've all died out as supernovae or whatever, gamma-ray bursts. What are left over are the low-mass, long-lived stars.

The biggest elliptical galaxies are huge. They could be more than three times the size of the Milky Way, perhaps 300,000 light years across and containing up to 10 trillion stars. Our own Milky Way has, perhaps, a few hundred billion, something like that, so the biggest elliptical galaxies are giants. But most elliptical galaxies are actually relatively small, and indeed, they're called dwarfs. They have perhaps a million to 10 million stars, and they might be 5,000 to 10,000 light years across; perhaps 6,000 light years is a nice little average. Here are some elliptical galaxies. There's one, and you can actually see the individual stars in this galaxy. They're pretty boring-looking—no nebulae, no arms, kind of just blah. But that's not to say they're not important; they are. Here's an amazing one. This is one of these very massive elliptical galaxies. This one actually is in the center of a big cluster called the Virgo Cluster. In its center, as I'll discuss later, we have evidence for a supermassive black hole, 3 billion times the mass of our Sun, nearly 1,000 times more massive than the supermassive black hole in our own Milky Way Galaxy. So, those are the elliptical galaxies.

There's another class of galaxies called the irregular galaxies, and they are a relatively minor number in today's Universe right now, nearby. But as we look far away and back in time, we will find that there are progressively more of these irregular galaxies. An example nearby is the Large Magellanic Cloud, which orbits our own Galaxy, and the Small Magellanic Cloud. You can see that these things have no distinct spiral or elliptical shape. They're sort of more irregular. They've got blotchy appearances, and they've got, in general, a fair amount of gas and dust from which new stars are forming. Those are the irregular galaxies.

There are also peculiar galaxies, and they differ from the irregulars in that the peculiar ones usually do have a fairly distinct spiral or elliptical shape, but they have one or more peculiarities associated with them. Here's an example of a ring galaxy. It's sort of got a nuclear region here, which looks more like a spiral galaxy. It's got these little arms coming out to a ring, way out here. This peculiar shape is thought to have been caused by a gravitational interaction with two neighboring galaxies here. Here's another peculiar galaxy. It's a spiral galaxy, but it has a very large amount of gas and dust, especially on one side, blocking our view of a lot of the stars. That's kind of a weird one. It's a beautiful galaxy. It's called the Black Eye Galaxy.

Here's an elliptical galaxy that differs from normal ellipticals in that it has a lot of gas and dust, especially going through a plane right here. Usually, elliptical galaxies don't have much gas and dust, so that's a weird one. For a lot of years, astronomers thought that this was actually two colliding galaxies, but we now think that's not the case. Maybe there's a galaxy that was eaten by an elliptical galaxy, and that's what caused this region of gas and dust. Maybe a spiral galaxy got consumed by this elliptical galaxy. But it wasn't two ellipticals that crashed together because two ellipticals wouldn't have had gas and dust, and so that doesn't explain the formation of that structure in the middle there. Here's another elliptical galaxy with, once again, a lot of gas and dust, this time in a more chaotic distribution than in the previous galaxy. That's a weird one.

Here's one that looks basically like an edge-on spiral galaxy, but there's a plume of material coming out of it. Here's a deeper view. There's all this gas coming out from the plane of the galaxy toward either direction here. We think what's going on there is that there's a huge burst of star formation going on in this galaxy, and all the winds from the massive stars, and the supernovae, are blowing gas out of this galaxy perpendicular to its plane because that's the path of least resistance. Here's a pair of galaxies called "The Mice," and we think that they are a result of a gravitational interaction between two spiral galaxies, as is shown here in this computer animation. We stop the animation for a minute and show that it looks a lot like the observed Mice. Then we can let the computer go forwards in time and predict what The Mice will look like billions of years in the future. We think they will merge into a single galaxy like this one, maybe not with that detailed shape, but something like it. Those are the peculiar galaxies.

There's one final class, and that's sort of a cross between a spiral and an elliptical. It's called an S0 galaxy, sometimes called a lenticular galaxy. It's got a bulge and a disk like a spiral galaxy, but in that disk there's not much gas and dust. There are no arms, and there's very little evidence for the formation of new, massive stars—either in the recent past or right now. Here's an example of an S0 galaxy. This one is seen edge-on, and you can see the big bulge and the disk here, and there's a little bit of gas and dust. You can see that narrow, little black stripe there, which is a little bit of gas and dust in this galaxy, but certainly not as much as in typical spiral galaxies. This is sort of a cross between an elliptical galaxy and a spiral galaxy.

Edwin Hubble tried to put these galaxies into cubbyholes, into pigeonholes, that he could then arrange in certain ways. In particular, he had this so-called tuning-fork diagram, where he had the ellipticals arranged at the left, from the ones that look roughly circular—he called those E0s—to the ones that look more elongated; those are the E5s. He even had some E7s. Then he had the S0 or lenticular galaxies, which were sort of a cross between a spiral and an elliptical. Then on one branch of this fork, he had the normal spirals: Sa, Sb, Sc; and then the barred spirals: SBa, SBb, SBc. He thought there might even be an evolutionary progression from ellipticals to spirals, or maybe the other way around. We now think, in general, there isn't such an evolutionary progression, although we will see later on that some spirals do merge together, forming ellipticals eventually. A single lone spiral will not later on turn into an elliptical, nor will an elliptical, on its own, turn into a spiral, so it's not an evolutionary sequence in that sense.

Going back to the tuning fork, here, you can see that the Sa's, Sb's, and Sc's are arranged in order of decreasing size of the central bulge region; the Sa's having the biggest bulge, Sb's a smaller bulge, and Sc's an even smaller bulge. Moreover, this sequence of decreasing bulge size also seems to correspond to a greater and greater openness of the arms. See, the Sc's have less-tightly wound arms in general than the Sa's. Moreover, the Sc's have more gas and dust and more active star formation than the Sa's, so in that sense, the Sa's are closer to the ellipticals and the S0's, in that they don't have as much star formation. This was a nice way to classify these things and then try to understand how they form and evolve. The Sb's, the barred spirals, have the same general sequence. From a to b to c, you have a decreasing bulge size, a more open set of arms, and more gas and dust from which new stars are forming.

Galaxies tend to be social creatures—not because they think about it or anything, but because gravity binds them together. Indeed, they are formed out of primordial clouds of gas that collapse and form many galaxies in about the same place. Our own Milky Way has several prominent companions—the Magellanic Clouds, shown here. And our Milky Way actually has six or seven other smaller companions, which are harder to see. They're dwarf galaxies like this one, the Leo I dwarf. You can actually see the individual stars in it very easily. It's very nearby and very distended like this, and it's hard to recognize these little dwarf galaxies because they sort of merge with all the other stars that you see in the sky. Our nearest big neighbor, the Andromeda Galaxy, also has two main companions. There's one of them, and there's the other one. It, too, has another half-dozen smaller companions.

If we take a view from our Earth, past the Chandra X-Ray Observatory, past Venus, past Mercury, past the Sun, we're going to go out and see, now, sort of in a Powers of 10 fashion, progressively bigger views of the Universe, all the way out to the "Local Group" of galaxies—the galaxies nearby to our own. Now we're passing by nearby stars and a nebula, and now we're going to pass through a cluster of stars in our own Galaxy. Then we'll pass through a bunch of other stars and nebulae, and now we're going to go a little bit above the plane of our Galaxy, so we can see it more clearly. It's still nearly edge-on in this view, but we're sort of above the plane. Going farther and farther out, we will start seeing additional galaxies—our companions and other members of our so-called Local Group of galaxies, consisting of us, the Milky Way, and our immediate companions; the Andromeda Galaxy and its immediate companions; and then a few other galaxies thrown in for good measure. This small, loose cluster is called the Local Group, and it contains about three-dozen galaxies or so.

We live not just in a binary system with the Andromeda Galaxy, but with lots of other smaller galaxies thrown in, in a loose cluster known as the Local Group, having a diameter of a few million light years. There are other similar small groups of galaxies like this one here, consisting of four main galaxies and some other smaller ones. Here's another group of galaxies. Galaxies tend to come in groups. Then there are the bigger clusters. The Virgo Cluster of galaxies, which spans about 15 degrees in the sky, contains about 1,000 galaxies. It's a much bigger cluster than our own little Local Group. It's about 60 million light years away. This is the nearest big cluster, and only a few of its galaxies are shown here. Then there's the Coma Cluster of galaxies, about 300 million light years away, containing over 10,000 galaxies. Nearly every blob you see in this photograph is a galaxy in the Coma Cluster, not a star in our own Galaxy. There's a star in our own Galaxy, and there are a few others here, but all these other blobs are galaxies in a cluster. Here's a really striking cluster of galaxies, containing several tens of thousands of galaxies gravitationally bound together.

The clusters are an amazing set of structures, containing vast numbers of galaxies. But they, too, appear to be clustered in even bigger configurations called superclusters, spanning 50 million to 100 million light years. If you plot our position down here at the point of this pie diagram, and you plot a certain region of the sky—here's an arc across the sky—and look at the distances of various galaxies in these directions here—distance is along this axis. Some direction in the sky is along that axis there. You can see that each of these dots is a galaxy, or a cluster of galaxies, and they themselves, then, are clustered in superclusters, separated from each other by voids—vast, empty regions.

If you look at three-dimensional projections of certain regions in the sky, and sort of rotate them like this, you can see the clusters upon clusters, which are themselves grouped in superclusters. This is one region of the sky toward a structure known as the Great Wall. Here's another region of the sky, toward a structure known as the Perseus-Pisces supercluster. You can see all those dots are galaxies and clusters of galaxies, and they tend to congregate in larger structures separated by relatively empty voids. As I'll discuss later, supercomputer calculations can now reproduce, at least approximately, the observed structure in our Universe—clusters of galaxies and superclusters arranged along bubbles, or filamentary structures, surrounding relatively empty regions that we call voids.

Astronomers have come a long way since the days when they thought that our Galaxy was the only galaxy in the Universe, the whole Universe itself. We now know that there are hundreds of billions of galaxies, spanning billions of light years of space. This has been a giant leap in our understanding of our cosmic origins and of our place in the Universe.

The Dark Side of Matter

Lecture 70

"There's now strong evidence that most of the mass in galaxies and clusters of galaxies may be dark matter—material that gravitationally binds the galaxies and clusters of galaxies together, but does not emit any significant amount of electromagnetic radiation, and so it can't be seen."

Until a few decades ago, astronomers thought that galaxies were composed primarily of stars. There is now strong evidence that most of the mass of galaxies may be dark matter. Dark matter is material that gravitationally binds galaxies and clusters of galaxies, yet it does not emit any significant amount of electromagnetic radiation, so it can't be seen. The first evidence for this came in the 1930s when Fritz Zwicky saw that galaxies within clusters move incredibly quickly. Given that these clusters are unlikely to be chance, fleeting groupings, there must be some additional material (beyond that of visible stars) gravitationally binding them together; otherwise, the galaxies would fly apart. Though Zwicky's theory was largely ignored, additional evidence came later through studies of the rotation curves (orbital speed versus distance from the center) for spiral galaxies.

Let's take a closer look at rotation curves. The rotation curve of our Solar System shows that distant planets move more slowly than nearby planets. Plotting these speeds (v) on a graph against their distance from the Sun (R) produces a curve that can be described as speed proportional to 1 divided by the square root of distance: $v \infty 1/\sqrt{R}$. This inverse-square-root law occurs when a single central mass dominates the gravity of a structure, as our Sun does in our Solar System. Because the Sun's mass is so great, our planets' masses are negligible in comparison. Recall from Lecture 68 that most of the stars in the Milky Way Galaxy orbit at about 200 km/s. In the central regions of the Galaxy, the stars orbit more slowly, but in the very center—near the supermassive black hole—stars orbit very quickly. The rotation curve of our Galaxy, as we've seen, levels off at a distance of about 1000 parsecs. However, the rotation curve for the Solar System is much different—it declines with distance from the Sun according to an inverse-square-root law.

The Sun's mass accounts for the rotation curve of our Solar System, but in the Milky Way Galaxy as a whole, there is no single dominant mass; the supermassive black hole in the center (roughly 4 million solar masses) dominates only within the central few light years and isn't big enough to dominate the cumulative mass of all the stars in our Galaxy. Most of the mass in our Galaxy is distributed throughout. There is enough mass in the bulge and inner disk to account for the observed speeds of the stars, but farther out into the disk, there are too few stars to account for the observed rotation curve. When we calculate the mass of the Milky Way Galaxy out to a certain distance, we get $M = v^2R/G$, in which M is mass within an enclosed circle of radius R, v is orbital speed, and G is Newton's gravitational constant. Only stars within the orbit of the Sun (or whatever other star we're considering) count; stars outside that orbit don't affect the orbital speed as long as they are uniformly distributed. This calculation is derived using Newton's version of Kepler's third law: $(M + m)P^2 = (4\pi^2/G)R^3$. The equation $M = v^2R/G$ implies that the mass enclosed within progressively larger orbits grows in direct proportion to the radius of the orbit. Twice as far out encloses twice as much mass; four times as far out encloses four times as much mass. But farther from the Galaxy's center, we notice that the observed speeds are as much as twice that of the expected speeds based on visible matter only.

"Where is most of the mass in our Galaxy? It's distributed over the Galaxy."

Clearly, there are not enough stars at large distances from the Galaxy's center to account for differences between the observed and expected rotation curves. Therefore, there must be other matter influencing the stars that orbit in the outer parts of our Galaxy. This material must be dark matter, gravitationally influencing the stars. Most dark matter occurs in the halo surrounding our Galaxy, not in the disk or the bulge. We think that the **galactic halo** extends far out, even beyond the globular clusters. This is because we see very distant stars moving quickly, far beyond the region where there is much visible matter. Dark matter increasingly dominates farther away from our Galaxy's center.

Vera Rubin was the first to recognize that the rotation curves of most other spiral galaxies show this same general trend. If we plot the observed rotation speed against distance from the center of a spiral galaxy, we notice that one side rotates away from us and the other side rotates toward us, with a roughly flat rotation curve. Rubin concluded that there is extra, invisible mass in the galaxies, though few other astronomers took her seriously at the time. Historically, dark matter was called the "missing mass," though this term is not favored anymore because the mass is not missing; rather, it is faint or invisible. Elliptical galaxies also show evidence of dark matter, as do pairs of galaxies that interact with speeds too great to have been produced by only the visible stars alone. If it weren't for dark matter, the hot gases in galaxy clusters would quickly evaporate away from the clusters.

Gravitational lensing also provides evidence for dark matter. Recall that a foreground object can bend, or lens, the light of a background object. Likewise, dark matter lenses the light of distant galaxies and clusters. A galaxy lensed by a foreground cluster of galaxies tends to produce arcs. The more galaxies being lensed, the more arcs are produced. Studying the brightness and distribution of the arcs can tell us more about dark matter. Clusters of galaxies are 90% dark matter, some of which may be visible because it emits x-rays, being quite hot. Indeed, recently, we've discovered that perhaps 10% of dark matter actually glows at x-ray wavelengths.

If 10% of dark matter is visible at x-ray wavelengths, but the other 90% cannot be detected in any form of electromagnetic radiation, what could it be? One theory is that this invisible dark matter consists of **massive compact halo objects (MACHOs)**, astrophysical objects in the halos of galaxies. MACHOs include stellar-mass black holes—not supermassive black holes—as well as old and dim white dwarfs or neutron stars, brown dwarfs, and free-floating planets. We can find MACHOs by studying their gravitational influence on light. For example, if one were to move between Earth and our line of sight to a star in the Magellanic Clouds, it would bend the star's light, causing the star to brighten and fade for a time. The amount by which the star brightens at blue wavelengths should be the same as the amount by which it brightens at red wavelengths, because the warping of space is not dependent on the wavelength of electromagnetic radiation. Through observations of this effect, we've found brown dwarfs, free-floating planets, and even some solitary black holes, especially toward the bulge of our Galaxy.

Unfortunately, astronomers have determined that most of the lensing objects were not in the halo of our Galaxy. In fact, at most, 20% of the dark matter in the halo of our Galaxy consists of MACHOs. What could the other 80% of dark matter be? It could be golf-ball-sized rocks, but it is difficult to see how such objects could form. Moreover, other studies show that very little dark matter could consist of objects containing normal neutrons and protons—that is, normal matter. Dark matter must be predominantly in the form of odd little particles known as **weakly interacting massive particles** (**WIMPs**), predicted to have been produced shortly after the Big Bang. If these particles still exist in great abundance, they could account for most of the dark matter in our Universe. ■

Name to Know

Rubin, Vera (1928–). American astronomer; was the first to observationally show that the rotation curves of most spiral galaxies imply the presence of considerable amounts of dark matter. She also obtained early evidence for large-scale *peculiar motions* of galaxies relative to the smooth expansion of the Universe.

Important Terms

dark matter: Invisible matter that dominates the mass of the Universe.

halo (galactic): The region that extends far above and below the plane of the galaxy.

massive compact halo objects (**MACHOs**): Brown dwarfs, white dwarfs, and similar objects that could account for some of the dark matter of the Universe.

weakly interacting massive particles (**WIMPs**): Theorized to make up the dark matter of the Universe.

Suggested Reading

Dressler, *Voyage to the Great Attractor: Exploring Intergalactic Space.*

Lightman and Brawer, *Origins: The Lives and Worlds of Modern Cosmologists.*

Pasachoff and Filippenko, *The Cosmos: Astronomy in the New Millennium*, 3rd ed.

Rees and Gribbin, *The Stuff of the Universe: Dark Matter, Mankind and the Coincidences of Cosmology.*

Rubin, *Bright Galaxies, Dark Matters.*

Waller and Hodge, *Galaxies and the Cosmic Frontier.*

Questions to Consider

1. Show that the mass of the Milky Way Galaxy within a radius of 10,000 light years from the center is about 3×10^{10} solar masses. (Assume a rotation speed of 200 km/s at that radius.)

2. To explain the flat rotation curves of spiral galaxies, one might argue that gravity does not behave according to the inverse-square law at large distances. For this to work, would gravity have to be stronger or weaker than in the standard description? (It turns out that there is evidence against this hypothesis, but it is nevertheless useful to consider.)

3. Can the discovery that most of the matter in the Universe might be dark be considered an extension of the Copernican revolution?

4. Are you bothered by the notion that most of the matter in the Universe might be dark, detectable only through its gravitational influence? Can you think of any alternative explanations for the data?

The Dark Side of Matter

Lecture 70—Transcript

Galaxies, the building blocks of the universe, are composed primarily of stars, or so people thought until a few decades ago. There's now strong evidence that most of the mass in galaxies and clusters of galaxies may be dark matter—material that gravitationally binds the galaxies and clusters of galaxies together, but does not emit any significant amount of electromagnetic radiation, and so it can't be seen. The first evidence for this came actually over half a century ago, when Fritz Zwicky studied clusters of galaxies, such as this one, and figured out that the galaxies within them are moving around at incredibly rapid speeds—so fast that these clusters of galaxies would go flying apart unless there were extra material gravitationally binding them together, more material than corresponding to the visible stars and galaxies alone.

Here's Fritz Zwicky again, who figured out that massive stars collapse and then explode, producing neutron stars. He, in the 1930s, realized that clusters of galaxies couldn't remain gravitationally bound unless there was a lot of additional material. About 80% to 90% of the material might be dark, invisible, not seen, at least with the optical telescopes that were being used back then. Nor can that material be seen easily with other kinds of electromagnetic radiation. He measured the Coma cluster of galaxies in particular, 300 million light years away. From the spectra of the galaxies in the Coma cluster, he got their radial velocities, their motions toward us and away from us, and he assumed that the transverse velocities are of similar size, so he could get the space velocities. He figured out that wow, this thing would get ripped apart.

Could it be that these galaxies are just passing through the night and happen to be close together? There are so many clusters of galaxies in the sky that probably they are physically bound objects, rather than chance superpositions of a bunch of galaxies passing through the night. Another indication of dark matter came many decades later. Zwicky was largely ignored in the '30s and beyond for what he said because he was this crazy guy that didn't make much sense in many other discussions, so why should people believe him? He was ignored. But then, decades later, there was this additional evidence that came about as a result of studies of the rotation curves, or orbital speed versus distance from the center, for spiral galaxies, starting with our own Milky Way.

If we look at the rotation curve of the Solar System—that is, the orbital speed versus distance—you can see here in this schematic that Mercury moves with a speed of 50 kilometers per second; Venus, 35; Earth, 30; and Mars, 25 kilometers per second. Jupiter, Saturn, Uranus, and Neptune move more slowly still. If you plot the orbital speed, or the speed of revolution, on the vertical axis, versus the distance from the Sun—in this case, in astronomical units—on the horizontal axis, you can get the positions of the planets on a curve, here, and this curve follows a 1/square root of the distance mathematical relationship. Here's Venus up here, and then Earth, Saturn, and Jupiter are down there. The curve connecting those observed data points is mathematically represented by something proportional to 1/the square root of the distance.

You can get this inverse square root law if you have a single central mass dominating the gravity of a structure like the Solar System. In the case of the Solar System, the Sun is very massive; it's 1,000 times as massive as Jupiter; it's 330,000 times as massive as the Earth. All the planets have negligible mass compared to the Sun, so gravitationally, the Sun does dominate the Solar System. If we go back to Newton's version of Kepler's third law, and we ignore the mass of the planets in comparison with the mass of the Sun, then we have that the orbital period of a planet^2 is proportional to its distance or semimajor axis^3. The constant of proportionality is $4\pi^2/G$ and over the mass of the Sun.

If the orbit of a planet is roughly circular, and that's roughly correct, then the circumference of the orbit, $2\pi R$—that is, the distance traveled in one period—must equal the orbital speed, V, multiplied by time—that is, the time it takes a planet to orbit the Sun. So $2\pi R = VP$. That's just distance = speed x time. If we now divide by the speed, we get that the orbital period is $2\pi R/V$. Plugging this expression for the orbital period back into Kepler's third law, you can solve for the mass of the Sun in terms of the distance of a planet, the orbital speed^2, and G, Newton's constant of gravity. Then rearranging, you find that the orbital speed is just the square root of G times the mass of the Sun/distance of the planet. This is proportional to 1/the square root of the planet's distance. Distant planets move more slowly than nearby planets, according to a law that says that their speed is proportional to the inverse of the square root of their distance. So, you get the curve that I had before that nicely fits the points.

Now let's consider the Milky Way Galaxy. Viewing it edge-on like this, we can get the rotation curve. If we schematically look at the orbital speeds of the stars—I showed this a few lectures ago. I showed that most of the stars are orbiting with a speed of around 200 kilometers per second, except in the very center where they're orbiting more slowly. Of course, at the very, very center, where the supermassive black hole is, they're orbiting very quickly, but that's not shown here. Basically, most of them are orbiting at 200 kilometers per second. If you plot the rotation curve of the Milky Way Galaxy, the speed of rotation is on the vertical axis. The distance, in thousands of parsecs—where a parsec is 3.26 light years—is given along the horizontal axis. The Sun is up here, about 7 to 7.5 kiloparsecs away from the center, and it's moving with a speed of 200 kilometers per second. Indeed, much of the galaxy is rotating with that speed.

That rotation curve is in stark contrast to the rotation curve that I just showed for the Solar System, where you have a 1/the square root of distance-type law instead of a flat rotation curve. See, this one declines with distance instead of remaining flat. In the case of the Solar System, we understand the rotation curve as being due to the gravitational influence of a single dominant body, the Sun. Clearly, in the case of the Milky Way, this isn't the case. There isn't a single dominant mass. The Milky Way's mass must be more distributed. You might say, "Well, I just told you about a supermassive black hole in the middle of our galaxy, having a mass of almost 4 million suns. Doesn't that dominate the dynamics of our galaxy?" No, it doesn't, because 4 million solar masses, though large, is small in comparison with the mass of all the stars.

Suppose you consider just the first 4 million stars out of hundreds of billions. Four million out of 4 billion is 1/10 of 1%. The central black hole in our galaxy has a dominant influence over the stars in its immediate vicinity, but out where the Sun is—even 1,000 parsecs away, or even a couple hundred parsecs away from the center of the galaxy—it's nothing. The central supermassive black hole is just negligible. So, where is most of the mass in our galaxy? It's distributed over the galaxy. Indeed, in the central region, you can see this bulge, and you can measure how many stars are there. There's enough mass in the bulge to account for the observed speeds of the stars. It turns out that, as you go farther out into the disk of our galaxy, there become too few stars to account for the observed rotation.

We can calculate the mass of the Milky Way Galaxy out to a certain distance, R, by using, once again, Kepler's laws, or Newton's version of them. It turns out that the mass enclosed within a circle of radius R from the center of our Milky Way Galaxy is equal to the orbital speed, v, squared, multiplied by the distance from the center, divided by G. You can figure this out just by using Newton's laws of motion and universal gravitation. You have to take into account the fact that only the stars interior to the orbit of the Sun, or whatever star you're considering, matter. The ones outside that orbit don't matter as long as they are uniformly distributed. This can be proven. Newton proved it, and I would just assume it for now.

Let's take a look at how this is derived. Let's go back to Newton's form of Kepler's third law; $(M + m)P^2 = (4\pi^2/G)R^3$. It's the same sort of calculation we just did. If the mass of the orbiting star, m, is negligible compared to the mass of the galaxy within its orbit, M, then you can ignore m in comparison with M, and you get $Mp^2 = 4\pi^2/GR^3$. Again, for a circular orbit, the circumference, $2\pi R$, is equal to the speed, v, multiplied by the time, T, the orbital period—in this case, denoted by P. Solving for P and substituting, you get that the enclosed mass within that orbit is equal to v^2R/G. If you go to the flat rotation curve, and you ask yourself what does $M = v^2R/G$ imply, it implies that the mass enclosed within progressively bigger orbits grows linearly—that is, in direct proportion to—the radius of the orbit. So, you go twice as far out, and you enclose twice as much mass. At four times as far out, you enclose four times as much mass. Indeed, if you look at the plot of the rotation speed of our galaxy versus distance from the center in thousands of parsecs, here is the observed plot. Again, it's flat.

You go twice as far out as our Sun's distance—go out to, say, 15 kiloparsecs instead of 7.5—and you'd have twice as much mass. If you go four times as far out—say, to 30 kiloparsecs—you have four times as much mass. But if you look at what would be expected from the stars alone, you see that, in fact, the expected orbital speed of not just the Sun, but of stars farther out, is smaller than the observed speed. Here, for the Sun, the orbital speed should have been about 150 kilometers per second if the only mass influencing the Sun's motion were the mass associated with the visible stars within its orbit. It should have a speed of 150 kilometers per second, but the observed speed is 200. There are not enough visible stars within the Sun's orbit to account for the motion of the Sun. There must be dark stuff there.

The problem gets worse as you go farther out. You see, as you go farther out from the center of our galaxy, the difference between the observed rotation curve and the expected rotation curve, based on the visible matter only, increases; this discrepancy increases more, and more, and more. As you go out way far from the center of our galaxy, the observed speeds are twice as big as those anticipated from just the observed material. Clearly, there are not enough stars at the large distances from the center. You can actually see this nicely in this photograph here, where you go far out, and the number of stars starts declining. It's very faint out there, in the outer outskirts of our galaxy, yet the rotation curve is still remaining flat. There's a lot of matter influencing the few stars that are orbiting way out in the outskirts of our galaxy.

This material must be dark matter, the stuff that's causing the extra motion of the observed stars. It's sometimes called the missing mass. Old textbooks, in particular, called it the missing mass, and that's a misnomer. The mass is there; it's the light that's missing. We don't see what's out there, but clearly there's something out there because it's gravitationally influencing the stars that are out there. Moreover, through more detailed studies of the motions of stars, you can tell that most of that dark matter can't be in the disk or plane of the galaxy. It must be in a vast halo surrounding our galaxy. If you look at a schematic of our galaxy, here's the plane, and here's the bulge. Here's this halo of globular clusters. We think that the dark matter is out here where the globular clusters are, but not within the globular clusters, more uniformly distributed.

Moreover, we think that the halo extends far out, even beyond the globular clusters—because even way out here, where there's only a few stars in the disk of our galaxy, you can tell that they're moving very rapidly, 200 kilometers per second, and yet that's so far out that, out there, you don't even see globular clusters anymore. You don't see any visible matter, hardly, in our galaxy. So, there must be dark matter way out there whose presence increasingly dominates the farther out you go. Remember, I said that in the very central regions of our galaxy, there's enough material to account for the orbital speeds of stars? That's fine, but as you go farther and farther out, there's a greater and greater mismatch, and so the dark matter dominates more and more.

The rotation curves of other spiral galaxies show this same general trend. Andromeda rotates, but it rotates more quickly than it should if the rotation was caused by the visible stars alone. Here's the rotation curve measured for the Andromeda Galaxy. You see it remains flat way out here, beyond the regions where the galaxy essentially ends. You might say how is the rotation measured out here. There are a few stars and clouds of gas out here. You concentrate on those few stars or clouds of gas, and you can measure their rate of rotation. But clearly, there's not enough visible stuff out here to account for the magnitude of that rotation. Most of the stars are down in here, but most of the mass is out in a vast halo. Other spiral galaxies show the same thing.

Here's an edge-on spiral galaxy, and you can use a spectrum to plot the observed rotation speed as a function of r. Here it is; the rotation speed is flat out to large distances, and then wiggles around a little bit more. On the opposite side of the galaxy, it's again, roughly speaking, flat. It wiggles around, but basically this plot is showing us that one side of the galaxy, this side, is rotating away from us, and the other side is rotating toward us. This is a spinning disk with one side rotating toward and one side away. But the rotation curve, aside from bumps and wiggles, basically is flat.

Vera Rubin was the first to realize this and the really important implications of this observation. She said there's a lot of dark matter in the outskirts of galaxies in particular. Less of it closer in, but especially far out, there's all this dark matter causing the rotation curves to remain flat. Kind of like

Zwicky decades before her, she was ignored or largely forgotten for a few years. Then people began to notice, and now she's quite celebrated. Zwicky was ignored for decades when he found this same sort of evidence in clusters of galaxies. Now, dark matter, and trying to figure out what it is, is one of the hot topics in astrophysics. Everyone hails the discoverers of these rotation curves, and Zwicky's work, and all that as great stuff. But they weren't properly recognized at the time they did this work.

There's evidence for dark matter not just in spiral galaxies, but in elliptical galaxies as well. You can measure the motion of stars in a big elliptical galaxy like this one, M87. They're moving around too quickly compared with the amount of gravity expected from the visible stars. If you look at pairs of galaxies like this, they are interacting with speeds that are too big to have been produced by only the visible stars alone—that is, there's not enough gravity—or little groups like this. The galaxies are moving around too quickly to be under the gravitational influence of the stars alone. There must be dark matter. I had already mentioned that, from the work of Zwicky long ago, we already knew that clusters of galaxies are, indeed, dominated by dark matter.

Nowadays, we can tell that there's a lot of hot gas in clusters of galaxies, and that hot gas would quickly evaporate away if there weren't a lot of matter holding it in. When you look at a cluster of galaxies like this at X-ray wavelengths, you see that it's immersed in a glow, this orange glow. Let's zoom in on some of the individual galaxies of this cluster. There they are; they contain visible matter. But as we zoom out, more and more of the cluster is being dominated by dark matter, and you can tell because this orange glow is hot, X-ray-emitting gas. That gas would not remain gravitationally bound to this cluster of galaxies if there were only the gravitational influence of the visible stars. There has got to be an extra pull holding that hot, X-ray-emitting gas in; otherwise, it would all spread out and be lost by the cluster of galaxies.

Gravitational lensing also provides a lot of evidence for dark matter. You will recall that, if you have two objects collinear—that is, along the same line of sight—then the foreground object can bend or lens, or focus, the light of the background object. If things are perfectly aligned, you can get an Einstein ring. This focusing is caused essentially by the warping of space. That's a prediction of general relativity that we've already encountered. In a perfect case, you get an Einstein ring like this. But in general, when distant galaxies are being lensed by another galaxy, or a foreground cluster of galaxies, you get a series of arcs. Let me show you a little animation, here, where light from a distant galaxy is being focused, or lensed, by mass, including dark matter, in a foreground cluster. Now, we align things and look at what we see.

In addition to the galaxies that are doing the lensing, you can see these little arcs, these streaks, which are the background galaxy that has been lensed into a bunch of arcs that are all parts of these Einstein rings. A galaxy lensed by a foreground cluster of galaxies tends to produce a bunch of arcs like this. If you have a bunch of galaxies being lensed by a foreground cluster, you get more arcs. Studying the brightness and distribution of those arcs can tell you not only about the luminous matter in the foreground cluster, but also about the dark matter. Here's a visible foreground cluster and a bunch of background galaxies. We can tell that they're farther away, through methods that I'll describe soon. But these are background galaxies arranged in arc-like configurations, centered on this cluster. From the brightnesses and the distributions of these arcs, you can tell how much matter is in the cluster—not just the visible matter, but the dark matter as well.

Here's another cluster, where you can see the foreground galaxies, and then faintly, you can see all these little arc-like structures centered in concentric rings, or parts of rings, around this cluster. From an analysis of those rings, you can tell that this cluster of galaxies, the lensing cluster, is dominated by dark matter, not by visible matter. In fact, 90% of the mass in that cluster is dark. If it weren't there, it wouldn't cause the amount of lensing that we actually observe, so we know it's there. Gravitational lensing provides a beautiful probe of dark matter. Here's a nice little close-up of one of those clusters of galaxies where you can see background galaxies lensed into parts of circles, these little arcs here. Again, that tells us not only about the gravitational properties of the visible matter, but also that there's dark matter there as well.

Clusters of galaxies are dominated by dark matter; 90% of their matter is dark. Some of it may be a little bit visible because it's hot and emits X-rays. Indeed, recently it has been found that perhaps 10% of the dark matter does glow at X-ray wavelengths. It has a temperature of maybe a million degrees, up to 10 million degrees. If you look at the proper wavelengths of the electromagnetic spectrum—in particular, in the X-rays—you can see some of this previously dark form of matter. But that's only 10% of the 90%; most of it does not glow at any wavelengths that we've looked at so far. But if you look at little groups of galaxies like this, for example, you can see now, faintly, X-ray-emitting gas having a temperature of, say, a million degrees. That gas is part of the dark matter that binds together little groups like this one. Indeed, here is an X-ray picture of one of those little groups. You can see the faint glow from the million-degree gas.

Well, 80% or so of the material is dark and hasn't been detected in any form of electromagnetic radiation. What could it be? One general class of models is the so-called machos, MACHOs, massive compact halo objects. These are not clones of Randy Savage, the macho man. MACHOs are an astrophysical object out in the halos of galaxies, not a wrestler or other type of person who likes to be macho. MACHOs include stellar-mass black holes—not the supermassive black holes thought to be in the centers of galaxies. We know about their presence, and they're not flying around in the halos of galaxies. Though they're dark and they're massive, they're not part of the halo, so they're not MACHOs. Stellar-mass black holes could be MACHOs, or old and dim white dwarfs, or old and dim neutron stars—anything that has mass, but doesn't glow very much. Brown dwarfs could be MACHOs; they're hard to see. Free-floating planets could be MACHOs because if they're free-floating, they're not near a star, and so they don't shine very much by reflected light.

You might find these MACHOs by studying their gravitational influence on light. In particular, if you look toward the lines of sight of the Large and Small Magellanic Clouds—they are about 170,000 light years away and 210,000 light years away, respectively—you're looking through a large fraction of our Milky Way's halo when you are looking at the Magellanic Clouds. If there are MACHOs flying around between us and the Clouds, occasionally those MACHOs might go along the line of sight to one star in each of the Large or Small Magellanic Clouds. A few nights later, some other MACHO might pass between us and a star in one of these two Clouds. There are lots and lots of stars in the Large and Small Magellanic Clouds. They form a good background of stars, which we can use to search for MACHOs. Here's a schematic.

There's the Earth in the plane of the Milky Way Galaxy. Here's the halo of the Milky Way Galaxy. The Large and Small Magellanic Clouds are out here, in the outskirts of the halo. If there are a bunch of MACHOs flying around, occasionally a MACHO would be precisely along the line of sight to a star, either in the Large or Small Magellanic Clouds. As that MACHO moves through our line of sight to that particular star, it will bend or focus the light from that star, causing the star to brighten. If you plot the brightness of the star versus time, it'll remain constant for a while, and then right as the MACHO is flying through the field of view, it magnifies or amplifies the light from the star, and then it declines as it moves out of the line of sight. Then the star remains constant again.

This brightening is independent of wavelength. That is, the amount by which the star brightens at blue wavelengths should be the same as the amount by which it brightens at red wavelengths, because the warping of space is not dependent on the wavelength of electromagnetic radiation that you're looking at. This brightening should be independent of wavelength and have this distinctive signature; a rise and then a symmetric fall. You can check whether this method is working by looking at stars toward the bulge of our galaxy. Here, if we go through and look through a star cluster at the central part, or bulge, of our galaxy, we'll be looking at a region where there are lots of stars. This one right here; watch this one right there. See that? It brightened just when a MACHO passed between us and that background star. That MACHO focused the light from that star. We've seen this effect, and we've found some brown dwarfs, and free-floating planets, and even some solitary black holes using this method, especially toward the bulge of our galaxy.

Toward the Large and Small Magellanic Clouds, they've also seen events like this. There's only one chance in 10 million that any particular star will get lensed by a MACHO. But there are tens of millions of stars in the Clouds, so if you watch them long enough, a few of them will brighten. That's been good; astronomers have detected that brightening. Unfortunately, the astronomers doing these studies have determined that most of the lensing objects have been not in the halo of our galaxy, but rather in the Large or Small Magellanic Clouds themselves. In other words, stars in the Clouds can lens each other, just like a MACHO along the line of sight could lens a star. They can lens each other. So, easily 80% of the events that they've observed have been these lensing events caused by stars in the Clouds lensing each other, rather than a MACHO lensing a star in one of the Clouds. We can say that, at most, 20% of the dark matter in the halo of our galaxy consists of MACHOs, just 20%.

What's the rest? Well, maybe it's what I call golf balls, or rocks—little objects the size of golf balls or rocks that exist in great abundance and make up a lot of matter. But it's hard to see how such objects could form. We don't know of any way that rocks, or golf balls, or bricks can just form out in the halo of a galaxy. Moreover, there are additional studies that show that very little of the dark matter could consist of objects containing normal neutrons and protons, normal matter. It turns out that most of the dark matter must be some sort of weird little particle, a weakly interacting massive particle, or a WIMP. So, there are WIMPs, to be compared with MACHOs. WIMPs are predicted to have been produced shortly after the birth of the Universe, the Big Bang. If they're still hanging around in great abundance, they may be what most of the dark matter is in our Universe. As I'll explain later on, there are now searches underway to try to directly detect these weakly interacting massive particles, or WIMPs.

Cosmology—The Really Big Picture

Lecture 71

"What's going to happen to the Universe? What does it consist of? These are among the deepest and most profound questions that humans have ever asked."

We now enter the third major part of this course, the study of the Universe as a whole. In this lecture, we'll talk about cosmic expansion as a tool for determining the distances of faraway galaxies, allowing us to study their evolution. *Cosmology* is the study of our Universe as a whole—its shape, size, structure, composition, age, and future. A fundamental question of our origins is: When did the Universe begin? We now know the answer: about 14 billion years ago. Perhaps a bolder question is: When will the Universe end? Although current data suggest that the Universe will expand forever, we don't yet know whether this will really happen.

Because galaxies are the building blocks of the Universe, it behooves us to understand where they came from and how they evolve. A typical small patch of the sky, roughly the apparent size of a grain of sand held at arm's length, has about 1000 galaxies. If we extrapolate over the whole sky, we conclude that—within the realm of great telescopes, such as the Hubble Space Telescope—there are perhaps 50 to 100 billion galaxies, and that's just in the parts of the Universe that we can see! We now have good reason to believe that the Universe extends far, far beyond the parts that are visible to us. There may even be an infinite number of galaxies. How did they form, and how do they evolve with time? Given the finite speed of light, when we study distant galaxies, we see them as they were in the past. Thus, we can compare them with the galaxies of today, determining how, at least statistically, they evolve.

In seeking and finding answers to these profound cosmological questions, scientists are not trying to dispel the existence of God. Rather, they are attempting to discover the fundamental laws of physics and use them to understand how the Universe works. The goal is to have comprehensive models for the physical properties and behavior of the Universe and its constituent objects. In general, there is no conflict between science and religion, despite popular belief. They have different goals and different operational rules. Scientists don't claim to determine the purpose of the Universe or of humans; those topics belong to theologians and philosophers. Further, scientists do not address questions of moral values or other non-scientific issues. However, as we will see, some conclusions made through the scientific process are not testable (at least, not yet) and, hence, to some extent, remove themselves from the realm of science. This blurs the distinction between religion and science in some cases.

"Because redshift increases with distance, you would conclude that speed of recession increases with distance as well."

Our study of cosmology begins in the 1920s with Vesto Slipher, who noticed that, with a few exceptions, the spectra of galaxies (then known as spiral nebulae) are redshifted to longer wavelengths. Another galaxy might have the same absorption lines as ours, for example, but Slipher noticed that all of them are shifted toward redder hues of the spectrum. If we denote redshift by z, the definition is $z = \Delta\lambda/\lambda_0 = (\lambda - \lambda_0)/\lambda_0$, where λ_0 is the rest wavelength of a given absorption line and λ is its measured wavelength in the galaxy spectrum. Taking this a step further, in 1929, Edwin Hubble used the newly derived distances of some of the spiral nebulae to show that the observed redshifts were proportional to the galaxies' distances. More distant galaxies have greater redshifts than nearby galaxies. If we interpret this redshift (z) as being due to radial velocity, as in the Doppler effect described in previous lectures, then we conclude that the speed of recession of galaxies increases with distance. The Doppler formula is $z = \Delta\lambda/\lambda_0 = (\lambda - \lambda_0)/\lambda_0 \approx v/c$, or simply $\Delta\lambda/\lambda_0 \approx v/c$, or $z \approx v/c$. At the present time, the more distant galaxies have greater redshifts and, hence, greater speeds of recession than the nearby galaxies. Thus, in general, it appears that more distant galaxies are moving away from us faster than nearby galaxies.

The Hubble diagram, a plot of distance against recession speed, reveals a straight line; thus, speed (v) is proportional to distance (d), and the constant of proportionality is Hubble's constant: $v = H_0 d$, in which H_0 is the current value of Hubble's constant (a constant in space, not time). The value of

Hubble's constant (H) actually changes with time. In the past few years, we have found that $H_0 = 71 \pm 4$ kilometers per second per megaparsec (71 ± 4 km/s/Mpc). Just recently, the measured value of H_0 has been revised to 73 km/s/Mpc, but with a comparable uncertainty of $\pm$ 4 km/s/Mpc. This is not a physical change from 71 to 73 but, rather, a refinement of the determination of the present value of Hubble's constant. However, we will assume for the rest of this course that the Hubble constant is 71 km/s/Mpc, because we have derived various quantities and diagrams based on this value, and in any case, it does not differ statistically from 73 ± 4.

The units of Hubble's constant may initially seem strange: km/s/Mpc. However, the meaning is clear, when one considers **Hubble's law**, $v = H_0 d$. A galaxy's distance in megaparsecs cancels out the 1/Mpc in Hubble's constant. If a galaxy is 10 megaparsecs away, then it is observed to be moving away from us at 710 km/s. A galaxy 20 megaparsecs away is moving away from us at 1420 km/s. According to the equation $v = H_0 d$, we might naively expect that as the distance of a *given* galaxy grows with time, its speed of recession increases. Yet this is not true; the speed doesn't increase with time, because Hubble's constant actually decreases with time in most reasonable universes, and the product $H_0 d$ does not increase. The speed of any given galaxy, at best, remains constant. Or it should slow down with time (in the absence of repulsive effects, such as "antigravity") because galaxies are gravitationally pulling on one another. In any case, Hubble's law alone doesn't imply that the speed of a galaxy should increase or decrease with time. It just says that right now, given a certain Hubble constant, galaxies that are farther away are moving away from us more quickly than galaxies that are nearby.

Historically, there were several competing interpretations for the redshift. Some astronomers wondered if the redshift was gravitational. The problem with that hypothesis is that all parts of a galaxy have essentially the same observed redshift, aside from that caused by spin of the galaxy. If this effect were gravitational, different parts of the galaxy would show different redshifts because of different gravitational field strengths within the galaxy. Moreover, different galaxies in a cluster of galaxies have the same redshift, except for the slightly different radial velocities induced by gravitational interactions among the galaxies. An alternative is that light somehow loses energy on its way toward us, becoming redshifted; this is known as the *tired light hypothesis*. However, the observed brightening and fading of distant supernovae compared with the much quicker time it takes nearby supernovae to brighten and fade is a major strike against this hypothesis. In other words, the observed light curves of distant and nearby supernovae should appear the same if light loses energy as it travels.

If the redshift is caused by expansion of the Universe, however, those distant galaxies and the supernovae within them are moving away from us. From our perspective, their "clocks" should run more slowly than the clocks nearby. In an expanding Universe, time will be dilated for distant supernovae at a significant redshift and not dilated for nearby ones. Because we have observed dilation of the time scale of brightening and fading for the distant supernovae versus the nearby ones, we conclude that the redshift is caused by expansion. Further evidence that the Universe is expanding can be seen in what's called the *surface brightness*—the brightness per unit area—of galaxies as a function of redshift. In an expanding Universe, at a progressively greater redshift, the surface brightness declines quite dramatically, which is what we observe. The surface brightness measurements specifically imply that the correct interpretation of Hubble's law is that space itself expands, rather than galaxies moving away from each other through a preexisting, non-expanding space. This view is also consistent with the underlying general theory of relativity, which is used to theoretically study the expanding universe. Thus, it turns out that the redshifts of galaxies are technically not Doppler shifts, which are produced when objects move through a preexisting space. Nevertheless, the Doppler formula works correctly, at least at low redshifts.

When we say that the Universe is expanding, we mean that the space in between the very distant galaxies is getting larger. We don't mean that the galaxies themselves are getting larger, nor do we mean that planets, planetary systems, stars, star clusters, and clusters of galaxies are expanding. All of these objects are held together by gravitational forces sufficiently strong to overcome the tendency of space to expand. Humans, as well, are not expanding, being held together by **electromagnetic forces**.

The expansion of the Universe is one of the fundamental observations of cosmology. It will play a central role in the remaining lectures of this course. ■

Important Terms

electromagnetic force: One of the four fundamental forces of nature; it holds electrons in atoms.

Hubble's law: The linear relation between the current distance and recession speed of a distant object: $v = H_0 d$. The constant of proportionality, H_0, is called *Hubble's constant*.

Suggested Reading

Ferguson, *Measuring the Universe: Our Historic Quest to Chart the Horizons of Space and Time*.

Ferris, *The Whole Shebang: A State-of-the-Universe(s) Report*.

Hogan, *The Little Book of the Big Bang: A Cosmic Primer*.

Pasachoff and Filippenko, *The Cosmos: Astronomy in the New Millennium*, 3rd ed.

Waller and Hodge, *Galaxies and the Cosmic Frontier*.

Questions to Consider

1. Can you think of possible explanations for the redshifts of galaxies that do not involve the expansion of the Universe? How could you test these hypotheses?
2. What is the distance of a galaxy having a recession velocity of 3000 km/s if the Hubble constant is 71 km/s/Mpc?
3. At what speed is a galaxy 100 million light years away receding from us if Hubble's constant is 71 km/s/Mpc? (Be careful with units!)
4. Do you think there is any fundamental conflict between religion and science (specifically, cosmology)?

Cosmology—The Really Big Picture

Lecture 71—Transcript

We now enter the third major part of this course, cosmology, the study of the Universe as a whole. In this first unit, consisting of eight lectures, I will introduce cosmic expansion as a tool for determining the distances of distant galaxies and allowing us to study the evolution of galaxies, how galaxies change with time. By looking at ones that are very, very far away, we're looking at them as they were in the past. We'll be able to compare them with the galaxies of today and determine how, at least statistically, galaxies evolve. Then later on, I'll move into a study of the structure and evolution of the Universe as a whole, not just of the galaxies within it. Cosmology is one of the grandest subjects, the study of everything, the whole thing, the big shebang. What is the shape of the Universe? What is its size? When was the Universe born, or has it existed forever? What will be its eventual fate, far in the future? I'm sure many of you have heard that the Universe is expanding. Will it expand forever, becoming colder, dimmer, darker, or will it someday re-collapse in on itself, a big crunch to be compared with the Big Bang.

What's going to happen to the Universe? What does it consist of? These are among the deepest and most profound questions that humans have ever asked. They've fascinated humans for, I think, as long as we've had the capacity to have such thoughts. They fascinate us now. People are interested in this stuff. Look at the headlines of major national magazines, *Time* Magazine, "When Did the Universe Begin?" People want to know. This is a fundamental question of our origins. "How the Universe Will End"—that's a bold statement. They don't even have a question mark on this one. "Peering deep into space and time, scientists have just solved the biggest mystery of the cosmos." That was a few years ago when we thought we had the answers. We now are not quite so sure that we know how the Universe will end. Anyway, these are major, major questions that people are interested in.

Another aspect of cosmology is this business of how galaxies form and evolve. Galaxies are the building blocks of the Universe. It behooves us to understand where they came from and where they're going. Look at these grand spiral structures. How did they get there? Were they always this way? Did they just form this way, or did they evolve from something more chaotic-looking? The galaxies, of course, are all over the place. You look everywhere in the sky, and you can see them, with large telescopes. Of course, with your naked eye, you see mostly stars in our Milky Way Galaxy. There are only a few galaxies that you can see outside our Milky Way: the Andromeda Galaxy, the Large and Small Magellanic Clouds, and one additional galaxy, which is sometimes faintly visible.

But if you look with large telescopes, you can see thousands of them. Here's one of my favorite photographs from the Hubble Space Telescope. This is part of the so-called Hubble Ultra-Deep Field. Nearly every splotch of light you see in this image is another galaxy, not a star in our own galaxy. There are a few stars in the Milky Way here. There is one of them, there's another; there's a third one over there. But in fact, I could count them on the fingers of two hands, perhaps. Everything else in this picture is another galaxy. If you count them, there are about 1,000 of them or so. Indeed, astronomers are paid to sit around counting galaxies. It's a really cushy job; it's a lot of fun. But really, we learn about the structure of the Universe, and it's hard work as well. We count the galaxies, and in this picture, there are about 1,000 of them.

This picture is just a tiny fraction of the sky. If you hold a grain of sand at arm's length, just a small grain, and imagine how big it looks, that's the fraction of the sky covered by this photograph. It's just like a grain of sand held at arm's length. Yet this typical fraction of the sky has something like 1,000 galaxies in it. If you extrapolate over the whole sky, you conclude that within the realm of great telescopes like the Hubble Space Telescope, there's something like 50 to 100 billion galaxies, and that's just in the parts of the Universe that we can see. We now have good reasons to think that the Universe extends far, far beyond the parts that we can see. There may even be an infinite number of galaxies. How did they form? How do they evolve with time? These, too, are among the questions of cosmology—the most profound, deepest questions that humans have asked, or at least among the most profound and deep.

These questions represent the epitome of the human intellectual desire to understand our origins. Remember, I said at the beginning of the course that much of astronomy is trying to understand where we came from and where we are going—not in a biological sense, but more in terms of the origin of the elements, of stars, of planets, of the things that led to our existence. Sometimes people think that astronomers, and in particular cosmologists, are somehow trying to eliminate the necessity of God, or somehow determine why He, She, or It created the Universe in the first place. We're not trying to do that. Astronomers and other scientists are not trying to do that. As in the rest of this course, astronomers and physicists are simply trying to discover the fundamental laws of physics and use them to understand how the Universe works. How does it tick?

We are not claiming to determine the purpose of the Universe, whether it had a special creator, why humans exist, or what your moral values should be. These are not the kinds of questions that we're trying to address. Those questions are more in the domain of theology, philosophy, and metaphysics. In general, there is no conflict between science and religion, despite many people thinking that there is somehow. They address different questions, and the rules of the game are different. Religion is based on faith; it's what you believe. Science is based on experimental and observational verification of hypotheses. Scientists are trying to figure out how the Universe works, not why the Universe is here in the first place, or how you should conduct yourself, or whether a special creator created humans in particular, and other animals. That's not what we're trying to understand. Those physicists and other scientists who are religious have seen, in general, no fundamental inconsistency between their religious beliefs and their quest to understand nature, how the Universe works.

Let me end this aside by noting that here, in our study of cosmology, we will see in the end that some of the conclusions, in fact, based on the process of science, by their very nature end up removing themselves from the realm of science because, as far as we can tell, they're not testable, at least not by any techniques that we have so far. Maybe in principle, somehow they're testable, but we haven't figured out any such way to test them. Some of the conclusions, especially near the very end of the course, will have that nature: that they are untestable, yet they were based on the process of science. This can end up blurring the distinction between religion and science. Scientists cannot answer everything. There are many questions we can't answer, and the distinction becomes blurred. I will try to point out the relevant parts of the course, where the conclusions we've come to, by their very nature, at least temporarily, remove themselves from the realm of science. I'll try to be clear when that happens.

Before I move on to cosmology itself, let me point out that there's a lot of confusion among the general public between cosmology, the study of the structure and evolution of the Universe as a whole, and cosmetology, the study of hairdos and facials. They have the same root, cosmos—all that there is, or to make order of. But like astronomy and astrology, which also have the same root and have a common origin, they now have rather distinct meanings. The difference in their spelling is interesting, actually. Cosmetology is cosmology, but with an extra "et". Spell them out; cosmetology is cosmology with an extra "et", like the extra terrestrial. I'm not sure of the significance of that. I think it's just coincidental.

To give you an example of this confusion that I've seen when speaking to people and reading things, here's an ad that a colleague of mine placed in my mailbox some time ago. "Make cosmology your career. Training and supervision in hairstlying, blow-drying, permanent waves, coloring and frosting, scalp treatments, body and skin care, style cuts, [and] basic cuts. For futher information and interviews, call that number." Well, classes started the first Monday in March, but they probably have such classes nearly every quarter or semester at your local university or other institution.

These guys, these ad writers, clearly need a lesson not only on the difference between cosmology and cosmetology, but they need a lesson in spelling and proofreading. In addition to "futher" there, you see "hairstlying" and "colouring" with an "ou." I'll allow the last one because, in fact, it's the British spelling, and my own thesis advisor was British, so I'll allow that. Anyway, if you want to get to the cutting edge of cosmology and make it your career—sorry for the bad pun—as I and many of my colleagues have done, I urge you to take such a course. Next time it's offered, learn all about the Universe and how to figure out how it works on the largest scales.

Joking aside, let's now proceed with some real serious cosmology. The story begins, really, in the 1920s with Vesto Slipher, who noticed that, with few exceptions, the spectra of the spiral nebulae show a redshift. In other words, you see the normal sorts of chemical elements in the spectra that you find in stars in our own galaxy, but their whole pattern is shifted to longer wavelengths or redder colors. There are a few exceptions, like the Andromeda Galaxy is actually blueshifted if you look at its spectrum, and a number of the galaxies in our Local Group are blueshifted. But beyond the Local Group, the spectra are redshifted. If you look at the spectrum, you let the light go through a prism, and you spread it out into its component colors or wavelengths, there are these absorption lines, and there may be emission lines from nebulae that are glowing in the galaxy. If you plot, then, the apparent brightness versus the wavelength, or color, you get this curve, which astronomers and physicists call the spectrum.

At the top here, you see the spectrum of a galaxy, which is not redshifted at all. In other words, all of these absorption lines and emission lines are at the wavelengths at which we observe them in laboratory gases at rest. But the one over here at the bottom is redshifted by a tenth—that is, all of the lines are there in all of the normal patterns, but they've been shifted to redder wavelengths by 10%. A line that normally appears at 6,000 angstroms might be at 6,600 angstroms. Some other galaxy would have the same lines, but again at a different redshift—a different amount to the redder or longer wavelengths of the spectrum, but the patterns are all the same. All the elements are the same; it's just their spectra have been redshifted. That's what Vesto Slipher noticed.

Edwin Hubble took it a step further. In 1929, he used the newly derived distances of some of the spiral nebulae—derived from studies of the Cepheid variable stars and how bright they appear to be and all that. He showed that the observed redshifts of these galaxies are proportional to their distances. The more distant galaxies have greater redshifts of their spectra than the nearby galaxies. All these galaxies, mind you, are so nearby that we are seeing them right now. They might be millions of light years away, so technically we're seeing them as they were millions of years ago, but compared to the 14 billion-year history of the Universe, looking back a few million years is essentially nothing. Let's simply say that the nearby galaxies, we're seeing right now. Later, we'll have to take into account the fact that more distant galaxies we see as they were in the past.

Right now, redshift is proportional to distance. Here's a nearby galaxy; it has a certain redshift. A more distant galaxy, which typically looks fainter and smaller, has a larger redshift. An even more distant set of galaxies, like the Virgo cluster shown here, is even more redshifted. The redshift seems to be proportional to distance. Let me show you some early spectra of galaxies. Here is a galaxy in the Virgo cluster, and here's the spectrum. There's a bunch of comparison lines here due to laboratory gases at rest. This galaxy was thought to be 24 megaparsecs away—that's 24 million parsecs, each parsec being 3.26 light years. The more modern distance is 19 megaparsecs, but this is an old slide. You can see here two absorption lines of ionized calcium, which are called H and K—not to mean that these are hydrogen and potassium, but simply the terminology given for these particular lines of singly ionized calcium. It's weird, I know, but that's the way it goes.

They're shifted a little bit, and this tiny arrow shows the amount of shift redward compared to where those lines appear in a laboratory gas at rest. Here's a galaxy 300 megaparsecs away, and you see the same pair of lines, but they're shifted even farther to the red. Here's another set of galaxies, 780 megaparsecs away, and now the redshift is huge; look at that. The same pair of lines—but way off in the middle of the optical spectrum, not in the ultraviolet or violet part as they were initially. Then finally, here's a galaxy that's over 1,000 megaparsecs away, 1,220 megaparsecs away. These lines, which started out in the violet or near ultraviolet part of the spectrum, have been shifted way off to the bluish-green part of the spectrum.

If you plot the observed brightness of the light versus wavelength, you get the continuum, and you get these absorption lines. Here, I've shown just the hydrogen absorption lines. In a nearby galaxy, the shift is zero, or nearly zero. If the galaxy is more distant, then you see a noticeable redshift, $\Delta\lambda$; $\Delta\lambda$ being the observed wavelength minus the wavelength it would have had if it was produced by a gas at rest in your own laboratory. So, the hydrogen alpha line is shifted, hydrogen beta is shifted, and hydrogen gamma is shifted. What Hubble noticed was that for an even more distant galaxy, this same pattern of lines is shifted even more toward the red part of the spectrum, longer wavelengths, lower frequencies.

Delta lambda becomes bigger, or is seen to be bigger, for the distant galaxies compared with the nearby galaxies. We can define this shift in wavelength, $\Delta\lambda$, divided by the wavelength at which the line would have appeared had it been caused by a laboratory gas at rest. That ratio $\Delta\lambda$ over the rest wavelength, λ_0, we will call z, the redshift. So $z = \Delta\lambda/\lambda_0$; that is, it's the observed wavelength minus the rest or laboratory wavelength, that quantity, divided by the rest wavelength. Suppose the redshift is due to motion along our line of sight. Suppose it's due to a radial velocity. Then we know that the Doppler formula tells us that $\Delta\lambda/\lambda_0$ is V/C. That is, this shift, divided by the rest wavelength, is equal to the speed of the object moving away from you, divided by the speed of light.

Under this interpretation, because redshift increases with distance, you would conclude that speed of recession increases with distance as well for a bunch of different galaxies. At the present time, the more distant galaxies have bigger redshifts, and hence bigger speeds of recession, than the nearby galaxies. Hubble's original plot is shown here, where the speed of recession is along the vertical axis, and the distance in megaparsecs is along the horizontal axis. Interestingly, Hubble had the wrong units on his Y-axis. He gave speeds in terms of kilometers instead of kilometers per second. Nowadays, we would dock students five or 10 points on an exam if they did that, but even the greatest astronomers sometimes make an error, a typo of this sort. Hubble, of course, knew that the units of speed are kilometers per second. You can see here that, with a lot of scatter, in general it does appear to be the case that the more distant galaxies are moving faster than the nearby galaxies. Although there's so much scatter here that it was a little bit of a leap of faith, almost, to conclude that this relation is really real.

Later, Hubble and his assistant, Milton Humason, made a more expanded plot by measuring galaxies that were farther away. Now you can see quite clearly that the more distant galaxies are moving away from us faster than the nearby galaxies. This was the Hubble and Humason plot published in 1931. For comparison, Hubble's original data are in this lower left-hand corner. You can see that now the relationship between distance and velocity is much more convincing. A diagram that a colleague of mine plotted in mid-2003 is given here. It's based on a bunch of distances determined from Type Ia supernovae, in a manner that I'll describe later. It shows this relationship really, really well; that is, the more distant galaxies are moving away faster than the nearby galaxies. The original Hubble plot of 1929 falls in this lower left-hand square down here. We've come a long way in the past many decades. Even more recently in mid-2006, my colleague gave me a revised version of this plot. He was a post-doctoral fellow who worked with me for a while, Saurabh Jha. He now shows this plot with 95 supernovae instead of just 80, and the relationship is even tighter. Now you can see some of the uncertainties plotted in the distances. We don't know the distances of the galaxies perfectly well, but nevertheless, this relationship is very linear and goes very far out, to distances of many hundreds of megaparsecs, billions of light years. If you look at this relationship, speed is proportional to distance, and the constant of proportionality is now known as Hubble's constant, H. Its value right now is given by H_0. The value can actually change with time. H can change with time, so we say that its value now is H_0.

This is the constant of proportionality between the observed speed of recession and the distance. Among the latest values are that the constant of proportionality is something like 71 kilometers per second per megaparsec. I'll define what I mean by that more clearly in a few minutes, but it's 71 kilometers per second per megaparsec. That's what I've assumed for a number of years. Just very recently, the number has been revised to more like 73, perhaps, with a comparable uncertainty, +/- four. I don't want to change all of my graphs and stuff that I've made laboriously with a number equal to 71; and 71 and 73, to within the uncertainties, are about the same anyway. I will assume for the rest of this course that the Hubble constant is 71 kilometers per second per megaparsec. The true value might be closer to 73, but the uncertainty is still four or five.

What do we mean by this? If a galaxy is 10 megaparsecs away, then the Hubble constant, 71 kilometers per second per megaparsec, multiplied by 10 megaparsecs, gives you 710 kilometers per second. That is, a galaxy that's 10 megaparsecs away is observed to be moving away from us at 710 kilometers per second. A galaxy that's 20 megaparsecs away from us is moving away from us with a speed of 20 megaparsecs, multiplied by 71 kilometers per second per megaparsec, or a speed of 1,420 kilometers per second. Usually, you measure the recession speed, and you know the distance somehow, and you determine Hubble's constant. Hubble's constant is this relationship, the quantitative relationship, between the observed speed of recession and the distance of a galaxy right now.

If you look at it, $V = H_0d$, you might think that if Hubble's constant is constant, then as the distance of a given galaxy grows, its speed of recession will increase as well. If $V = H_0d$, and d is increasing, then V should increase as well. Well, it doesn't. Hubble's constant actually changes with time. It actually decreases with time in most reasonable universes. The speed of any given galaxy, at best, remains constant; or you would think maybe it should slow down with time because galaxies are pulling on one another. Hubble's law alone, as this is called, does not itself imply that the speed of a given galaxy should increase, or decrease, or anything. It just says that right now, given a certain Hubble constant, galaxies that are farther away are moving away from us more quickly than galaxies that are nearby.

There are competing interpretations for the redshift. What if it's gravitational? What if light is trying to escape from galaxies and becomes gravitationally redshifted? The problem with that hypothesis is that all parts of a galaxy have the same observed redshift, aside from that caused by the spin of the galaxy. When you take that into account, all parts have the same redshift. If it were gravitational, you wouldn't expect that to be the case because different parts would have different gravitational fields, and there would be a different redshift. Moreover, different galaxies in a cluster have the same redshift, except for the slightly different radial velocities, which Fritz Zwicky measured, and all that. Those are the velocities associated with the gravity of the cluster. If you remove those, all the galaxies in a cluster have the same redshift. If it were gravitational, you wouldn't have expected that.

Another alternative is that light somehow becomes tired on its way toward us, through the vast distances of intergalactic space. Maybe it becomes tired, and loses energy, and becomes redshifted. When you look at the light curves of supernovae—here's a star in a nearby galaxy that brightened and faded—you can plot that observed brightening and fading on a graph, and it shows, as this red line here, the observed brightening and fading of a nearby supernova. If you now observe distant supernovae, you find that they brighten and fade more slowly than the nearby supernovae. What we observe is that distant supernovae brighten and fade more slowly than nearby supernovae. You can measure the distant supernovae with great telescopes like Hubble and some ground-based telescopes. You can accurately measure these light curves, and it is an observed fact that the more distant supernovae take a longer time to brighten and fade.

If light were getting tired along the way, there would not be this observed relationship. That is, the light curve of a supernova should rise and decline just as quickly for a distant supernova as for a nearby one. There should be no difference in the light curves. If, instead, the redshift is caused by expansion of the Universe, those distant galaxies and the supernovae within them are moving away from us. From our perspective, their clocks should run more slowly than the clocks here nearby. A brightening and fading supernova is like a clock; it takes a certain amount of time. In an expanding Universe, that time will be dilated for distant supernovae at a large redshift, and not dilated for the nearby ones. Since we observe dilation of the time scale of brightening and fading for the distant supernovae versus the nearby ones, we conclude that the redshift is caused by expansion, not by tired light. The Universe appears to be expanding.

Finally, you could say, "Well, what if these things really are getting tired, but the supernovae somehow give us the wrong result?" You can look at what's called the surface brightness of galaxies as a function of redshift. The surface brightness is the brightness per unit area. In an expanding Universe, at a progressively greater redshift, the brightness per unit area declines quite dramatically. In the tired light hypothesis, the brightness per unit area declines much less. When you look at galaxies at progressively greater redshifts, and you ask what their surface brightness is, you find that, indeed, the surface brightness appears to be consistent with an expanding Universe, where the redshifts are an indication of recession speed, rather than a Universe where the light is simply getting tired.

Moreover, the quantitative amount of dimming of the surface brightness is consistent not with galaxies zooming like bullets through some pre-existing space, but rather with galaxies that are moving away from us because space itself is expanding. It's not that there's this Universe through which all these pellets, all these galaxies, are zipping. It's that the space in between us and distant galaxies is expanding, and that is what is causing the recession. You can tell that because, again, if you look at the surface brightnesses of galaxies at different redshifts, in an expanding Universe, the surface brightness declines quite a lot with increasing redshift. In a tired-light Universe, it declines very little. If galaxies were simply moving through a pre-existing space, the surface brightness would decline in a manner intermediate between these two extreme examples. What we observe is that the surface brightness declines in a way consistent with expansion of space, not consistent with movement of galaxies through some pre-existing space. So, space itself is expanding.

Finally, you could ask, "Is everything expanding?" No, you aren't expanding, despite what you might think after a big lunch. Maybe you did expand, but that's not the Universe's fault; that's your fault. We're not expanding because we're held together by electromagnetic forces that keep us bound together. Similarly, we are bound to the Earth by gravitational forces, and the Earth itself is bound, and it's not expanding. Individual galaxies are bound by gravity; they're not expanding. Clusters of galaxies are bound together; they're not expanding either. Really, it is the space between very distant clusters of galaxies that is expanding. On smaller scales, clusters, galaxies, stars, planets, people—all those things are held together by forces that overcome the tendency of space itself to expand. So, when we say that the Universe is expanding, what we mean is that the space between extremely distant galaxies, and clusters of galaxies, is what's expanding. On smaller scales, where things are held together by other forces, things aren't expanding. It's this expansion of the Universe that is the fundamental observation of cosmology, and which will play a central role in the remaining lectures of this course.

Expansion of the Universe and the Big Bang

Lecture 72

"Galaxies in our Local Group are gravitationally bound together. … Some might be moving away from us, and some might be moving toward us. … If Edwin Hubble had only looked at the Andromeda Galaxy, he might have concluded that the Universe is collapsing rather than expanding."

In the previous lecture, we learned that the Universe is expanding, one of the most monumental discoveries in astrophysics and an essential theme of cosmology. We continue our discussion of expansion, a tough concept, to help us understand it better. No matter which galaxy we look at, we see that it is moving away from us. At the present time, the nearby galaxies are moving away more slowly than more distant galaxies. There are some exceptions. For the closest galaxies, those in the Local Group, gravity is so strong that it overcomes any tendency for space to expand; instead, the galaxies are gravitationally bound together. Gravity is like a tight spring in this case; it is so tight that it can't be budged by the expansion of space. Some galaxies in the Local Group are moving away from us and others toward us because of gravitationally induced motions. The Andromeda Galaxy, for example, is the nearest big galaxy (about 2.4 million light years away) and a member of the Local Group. It is moving toward the Milky Way Galaxy with a speed of roughly 100 km/s.

Galaxies not strongly bound together but still reasonably close can affect the speed at which each moves. The Virgo cluster, about 60 million light years away and not a member of the Local Group, is receding from us at about 1200 km/s—yet if it weren't for the gravitational effects of the Local Group, the Virgo cluster would recede even faster, by about 150 to 200 km/s. The Virgo cluster is, in a sense, connected to us by a weak spring that is able to stretch by the expansion of space. The *Hubble flow* is the uniform expansion of the Universe. If galaxies were massless and had no reason to attract each other, then all motions other than those caused by universal expansion would be zero. Gravitational attraction produces a *deviation from the uniform expansion of the Universe*, a deviation from the Hubble flow.

These gravitationally induced motions are significant only over relatively short distances, not over distances exceeding a few hundred million light years.

Let's consider uniform expansion of the Universe. Hubble's law is suggestive of an explosion having a well-defined center from which shrapnel is scattered. However, it turns out that the expansion of the Universe is not like a conventional explosion with a unique center. The distribution of galaxies, the number per unit volume as a function of distance from us, implies that there is no unique center to the Universe—at least not in any physically accessible dimension. If our Galaxy were at the center, other galaxies would appear to spread out more the farther away they are from us. Per unit volume, then, there would be fewer galaxies at greater distances, but this is observed not to be the case. A three-dimensional analogy is an expanding loaf of raisin bread. Let the loaf be infinite, or ignore the edges, which are irrelevant to our discussion. Because the yeast is uniformly spread throughout the dough, the dough expands uniformly. The raisins are the galaxies that do not spread themselves. From any one raisin's perspective, the other raisins move away as the bread rises, or expands; that raisin thinks it is at the center of expansion. However, every other raisin also thinks it is at the center of expansion. Thus, there is no unique center. For any given raisin, the more distant raisins recede faster than nearby ones, simply because there is more expanding dough between raisins that were farther apart to begin with. Thus, Hubble's law is satisfied. In the real Universe, Hubble's law is also satisfied: More distant galaxies move away faster than nearby ones. Just as every bit of dough expanded, in the real Universe the more space ("dough") there is between two galaxies to begin with, the more expansion there is and the greater is the apparent recession speed.

Is it possible that the Universe has a unique center that we just can't see? Using an inflated balloon as an analogy of curved space, we have a two-dimensional space; that is, we assume that the laws of physics are constrained to operate only on the surface of the balloon, which curves around itself. In this hypothetical universe, we can move only along the surface of the balloon; we cannot go into the balloon's interior, nor upward out of its surface. This space, though curved, is two-dimensional because we can move forward/backward and left/right and any combination of these two motions, but we cannot move up/down. The center is actually inside the balloon itself, but that dimension is not physically accessible (not being part of the surface), though it is a mathematically real and distinct dimension. The surface curves around this additional mathematical dimension.

To demonstrate the mathematically distinct dimensions, we can look at a one-dimensional circle, defined by the equation $x^2 + y^2 = r^2$, and a two-dimensional sphere, defined by the equation $x^2 + y^2 + z^2 = r^2$, each of which represents a hypothetical universe. *Two* variables, x and y, are needed to define the circle, or a one-dimensional curved path, which means that the path must wrap around a second mathematical dimension. Similarly, *three* variables are needed to represent the surface of a two-dimensional sphere, x, y, and z. The center of the sphere itself is not in the surface, yet clearly, it exists mathematically because there must be three dimensions—x, y, and z—describing this surface. In both the circle and the sphere, the origin—the center—is not part of the circle or sphere. We cannot physically access this point.

"The number density of galaxies, the number of galaxies per unit volume, remains about the same no matter where you look in the Universe."

If we live in a three-dimensional analog of a sphere, where a volume bends around some fourth spatial dimension, the equation for our sphere would be $x^2 + y^2 + z^2 + w^2 = r^2$, w now being this fourth dimension around which we curve. This is called a *hypersphere*, and it's difficult to imagine or describe because we live in a world with three spatial dimensions, not four. To help us visualize *hyperspace* (four-dimensional space as a general class), we can consider the shadow (projection) of a four-dimensional cube, or *hypercube*, onto three-dimensional space. Imagine displacing two cubes along a diagonal and connecting the corners. By analogy, the shadow (projection) of a three-dimensional cube onto two-dimensional space (a sheet of paper, for example) can be drawn by displacing two squares along a diagonal and connecting the corners.

Even if the Universe is infinite (and we don't know whether it actually is), it can still expand; it simply becomes less dense. For example, the counting numbers reach from 1 to infinity. Even if we removed all the odd numbers and just counted with the even numbers, the numbers still keep going infinitely—they are just less dense, in a sense, because the odd numbers have been removed.

Could there be a universe denser than ours? Mathematically, the answer is yes, just as the counting numbers from 1 to infinity can be made denser by adding all the infinite fractions, or *rationals*, that fall in between them. Between 0 and 1, there is an infinite number of fractions, or rationals; this set of rationals, though denser than the counting numbers, can be placed in one-to-one correspondence with the counting numbers. Thus, there are no more rational numbers than there are counting numbers, yet there's clearly a greater density of the fractions. We can demonstrate this mathematical effect by putting all the rational numbers in a table. Though each row and each column has an infinite amount of numbers, all of them can be put in one-to-one correspondence with the counting numbers by tracing along diagonals.

We are convinced that the Universe is expanding. If we extrapolate this observed expansion back in time, we could conclude that the Universe had a definite beginning. If the Universe had a beginning, its matter must have been very dense. For whatever reason, it started expanding from a point called the *singularity*, a point at which everything was possibly infinitely dense (if we temporarily ignore quantum mechanics). We think of the Universe as an expanding gas. When gases expand, they cool, suggesting that the Universe began in a very hot and dense state at time = 0, when the singularity existed. We must determine how far away a distant galaxy is in order to know how far back in time we are looking (our **lookback time**), to understand the origins of the Universe. Recall that the light we see from distant galaxies reveals those galaxies as they were back in time because the light took so long to reach us. We return to Hubble's law to determine the distance of galaxies that are so far away we can't even detect Cepheid variables or any other discrete objects in them. Recall that we can derive the distance of a galaxy by measuring the redshift of its spectrum. If we adopt a certain cosmological model (or history) of the Universe, then for a given redshift, we can calculate the corresponding lookback time given in billions of years. At progressively greater redshifts, the lookback time is a progressively greater number of years, up to about 14 billion years, the age of the Universe.

When we read (in newspapers or magazines) about the distance of a newly discovered galaxy, for example, that quoted "distance" is actually the lookback time—how long it has taken for the light to reach us—not the current distance of the galaxy from Earth. When the galaxy emitted the light we now see, the galaxy was actually closer to Earth than the distance implied by the lookback time; space expanded during the light's journey to Earth. Now, when the galaxy's light finally reaches us, the galaxy is farther away than the distance implied by the lookback time. Thus, the lookback time corresponds to some average, or representative, distance of the galaxy from Earth. Most precisely, it is the amount of time light took to reach us from the galaxy. By observing galaxies at progressively greater distances, we're seeing progressively farther back in time. Thus, we essentially view a movie of the history of the Universe, from which we can learn something about how it and the galaxies within it evolved. ■

Important Term

lookback time: The duration over which light from an object has been traveling to reach us.

Suggested Reading

Dressler, *Voyage to the Great Attractor: Exploring Intergalactic Space.*

Ferris, *The Whole Shebang: A Stat- of-the-Universe(s) Report.*

Hawking, *The Universe in a Nutshell.*

Hogan, *The Little Book of the Big Bang: A Cosmic Primer.*

Kaku, *Hyperspace: A Scientific Odyssey through Parallel Universes, Time Warps, and the 10th Dimension.*

Pasachoff and Filippenko, *The Cosmos: Astronomy in the New Millennium*, 3rd ed.

Smith, *The Expanding Universe: Astronomy's "Great Debate," 1900–1931.*

Questions to Consider

1. Can Hubble's law be used to determine the distance of (a) α Centauri and other stars relatively near the Sun, (b) the center of our Galaxy, (c) the Large Magellanic Cloud, a satellite of our Galaxy, and (d) a cluster of galaxies very far from the Local Group?

2. Why does the recession of galaxies not necessarily imply that the Milky Way Galaxy is at the center of the Universe?

3. Explain how the effective center of expansion can be in an unobservable spatial dimension. Also, what is the Universe expanding into?

4. Can you visualize what our Universe might look like if it's the three-dimensional analogue of the two-dimensional surface of a balloon?

Expansion of the Universe and the Big Bang

Lecture 72—Transcript

In the previous lecture, I introduced the expansion of the Universe—one of the most monumental discoveries of all time, an essential theme of cosmology. Expansion is a tough concept, and to help you understand it better, in this lecture I'll expand on it. No matter which galaxy we look at, we see that it's moving away from us, receding away from us. At the present time, the most nearby galaxies are moving away more slowly than the more-distant galaxies. Here's a relatively nearby galaxy that's moving away with a certain speed. The more-distant galaxies are moving away faster. There are some exceptions. Galaxies in our Local Group are gravitationally bound together, so they're all kind of rumbling around like this. Some might be moving away from us, and some might be moving toward us, but there's no organized recession away from us. For example, the Andromeda Galaxy, the nearest big galaxy to our own, is actually moving toward us at something like 100 kilometers per second.

If Edwin Hubble had only looked at the Andromeda Galaxy, he might have concluded that the Universe is collapsing rather than expanding. You've got to look at other galaxies; Local Group galaxies are bound together. Even galaxies that are pretty far away, but relatively nearby, like the Virgo Cluster shown here, 60 million light years away, are affected by the gravity of the Local Group. So, the Virgo Cluster, though it's moving away from us at something like 1,200 kilometers per second, would have been moving away from us by 150 or 200 kilometers per second more—that is, 1,350 or 1,400 or so kilometers per second—had the Virgo Cluster and the Local Group not been attracting one another. If they were massless particles, massless galaxies, with no reason to attract each other, then they would have been moving away from us more quickly than they're observed to. This is called a deviation from the uniform expansion of the Universe, a deviation from the so-called "Hubble flow." The Hubble flow is the uniform expansion of the Universe. For nearby galaxies, you have to take into account these gravitationally induced deviations.

For the Andromeda, the gravity is so strong that it doesn't even go away at all; it's coming toward us. There are these deviations. For the most nearby galaxies, you can understand that the gravity within, say, the Local Group is so strong that it overcomes any tendency at all for the Universe, for space, to expand. It's like a spring that's so tight that you can't budge it; you can't pull it apart at all. That's the galaxies in the Local Group. They are so bound to each other that we can't pull them apart at all. The Virgo Cluster is, in a sense, connected to us by a spring that's able to be stretched by the expansion of space. That's the kind of analogy you might have. But the spring pulls back a little bit, so it's not being stretched out as quickly as it would have been had the spring not been pulling back in on itself.

Now let's go to this uniform expansion. Let's forget about the deviations. You say, "All right, the nearby galaxies are moving away with a certain speed, and the more distant galaxies are moving away faster." This is kind of like what a bird would see after a few seconds if the bird were to look down at a whole big pile of tennis balls that I had kicked. Suppose I kick a pile of tennis balls. Some of them will have been given a small amount of energy, so they're moving away slowly. Others will have been given a large amount of energy, so they're moving away quickly. After a while, this expanding set of tennis balls look something like this. Here is my foot, where it kicked them, and there are all these balls moving away from me. The more-distant balls are moving faster than the nearby balls because, after all, they were given a bigger kick.

This observed distribution of galaxies and their speeds is suggestive of an explosion with a center—in this case, my foot kicking a bunch of tennis balls or, for example, some fireworks. There's an explosion with a well-defined center, and all the shrapnel go out, for example. Little bits of shrapnel that are given only a small amount of energy move out in a given time by a smaller distance than bits of shrapnel that have a large amount of energy given to them. One characteristic of an explosion with a central point of explosion is that the more distant parts are moving faster than the nearby parts. To have gotten that far away, they had to have been moving faster. The observed recession of galaxies suggests that there was an explosion a long time ago—a Big Bang, let's say—centered on us. Look at the diagram. It looks like it's centered on us.

Well, that's a bit strange, isn't it? I mean why should all the other galaxies be moving away from ours? Are they lactose intolerant? Milky Way Galaxy—okay, sorry, bad pun. Or maybe they just don't like us. Is it something we said, or do we smell? When I give this lecture at my home institution, U.C. Berkeley or Cal, I say are we from Stanford, the big cross-bay rival? A fine institution—just not quite as fine as Cal. Sorry to all of you who might be Stanford grads; anyway, great place. No, we don't think that there's anything like this going on. We don't think that other galaxies don't like us, or whatever. We think that, in fact, there may not be a center to the Universe. Indeed, even this observed recession of galaxies doesn't necessarily imply that there is a center, or that we are at it. Indeed, there's observational evidence that suggests that, to the contrary, there is no center.

If you look at this diagram, and if there were a center, you would notice that near to the center, in a box, say, centered on the center, there are a lot of tennis balls or a lot of galaxies. Farther away, if you draw a box of the same size, there are fewer tennis balls or fewer galaxies. Even farther away, in a box with a given size, there would be even fewer galaxies or tennis balls because a given number of them are spreading out into a progressively bigger volume, and so any given subset of that volume is going to have a smaller and smaller number of galaxies or tennis balls as you go farther out. If our Galaxy were at the center, there would be a clear prediction. Other galaxies spread out more and more as you look farther and farther away; and per unit volume, there should be fewer and fewer galaxies as we look to greater and greater distances.

Is that the case? No. The number density of galaxies, the number of galaxies per unit volume, remains about the same no matter where you look in the Universe, if you spread out over the clusters, and superclusters, and voids, and all that. In other words, take out the known irregularities. Beyond that, then, is there a decrease in the number of galaxies as you get farther away from us? No, they're all about the same. Moreover, we see no other region in the Universe that could call itself the center because we see no other region where the galaxies appear to be concentrated in such a way. In fact, we think that there is no center to the Universe, no unique center, at least not in any of the physically accessible dimensions. I'll have a lot more to say about that later. This is a really packed lecture.

Let's give a couple of demos to maybe clarify what I'm talking about here. Here's a simple one-dimensional universe, where the space is represented by this rubber hose, and the galaxies are the ping pong balls. Space can expand; the galaxies themselves do not because they're gravitationally bound. I expand this space, and from our perspective on this yellow ping-pong ball, all the others are moving away. Indeed, the more-distant ones are moving away faster than the more-nearby ones because there was more rubber between us and the more-distant ones than between us and the nearby ones, and each piece of rubber, each little part of it, expands.

We say that we live in an expanding universe, apparently centered on us. But that conclusion doesn't depend in any way on our choice of which ping-pong ball to view things from. I could choose that one or that one, and from their perspective, all the other ping-pong balls are moving away as well, with a speed that, at any given time, is proportional to their distance. Here's a simple one-dimensional universe that has no unique center, yet each of the galaxies thinks that it is at the unique center. But they're all wrong; there is no unique center. By the way, of course, you could make this into a circular universe like this, to make it finite, if you're bothered by an infinite rubber band, an infinite universe.

Okay, let's now take a three-dimensional example of this phenomenon. If we have a loaf of raisin bread, where the dough is uniformly filled with yeast, and there are raisins that don't expand because they don't have any yeast in them, and you let the thing bake for a while, it'll grow bigger, and all of the raisins will get a greater distance away from all of the other raisins. Here are a bunch of markers. These two are, say, one centimeter apart, and those two are two centimeters and three centimeters. After a while, those same raisins are two, four, and six centimeters apart. From the perspective of any raisin, all the others are moving away from it. It could call up its neighbors and say, "Hey, I'm the center of the Universe." But this one here could say, "Well, no, you're not. You're not the unique center because all the other raisins are moving away from me too." This raisin here would say the same thing.

You might say this raisin is near an edge, and that's weird or something, but let's forget about the edge. The Universe is either infinite, or it wraps around itself in such a way that there is no edge. Here I've drawn an edge, but it's not central to our discussion. In fact, it's confusing. It's hard to draw an infinite universe on a finite sheet of paper. Let's go back to our raisin bread, here. Let's let it bake for an hour. Note the original separation between two raisins, five centimeters and 10 centimeters. We let it bake for an hour, and let's suppose it doubled in size. Now these raisins are separated by 10 centimeters and 20 centimeters.

What do we observe? For the raisin at the left, its final position after one hour is 10 centimeters from us. Its initial position was five centimeters, so it moved 10–five, or by advanced mathematics, five centimeters in one hour—hence an average speed of five centimeters per hour. The one on the right started out at 10 centimeters. After an hour, it's at 20 centimeters. It moved 20 – 10, or 10 centimeters in one hour, for an average speed of 10 centimeters per hour. So, you see, the Hubble law—that is, that the more-distant ones are moving faster than the nearby ones—is satisfied in such a universe. The reason it's satisfied is that every bit of dough, just like every bit of the rubber band, expanded. The more dough there was between two galaxies to begin with, the more expansion there is, and the greater is the apparent recession speed.

You might still ask yourself whether there is a center somewhere that you just can't see. I'd like to illustrate this with another demo, a spherical balloon whose rubber stretches, and on which there are stickers, which represent the galaxies, that don't stretch. Let me expand this balloon, the Universe—not too much because I'll get a little bang. You can see all these stickers moving away from one another. Each sticker sees the others moving away; and indeed, each sticker sees the more-distant ones moving faster than the nearby ones because there's more rubber that's stretching. Each of them thinks that it's at the center. Each of them is wrong. There is no unique center, at least not in the physically accessible dimensions.

What I mean by that is the following. This is a curved universe. It's a two-dimensional space; that is, suppose the laws of physics are constrained to operate only on the surface of the balloon. You can go forwards and backwards, or left and right, or any combination of those two motions, but out of the balloon or into the balloon are forbidden. This is a hypothetical universe. It's two-dimensional—you can only go in these two directions, not up and down—and it curves around itself. The dot creatures could figure that out by, for example, going in one direction and ending up where they began, or drawing large triangles and measuring their geometrical properties and things like that. I'll discuss that more later. But they would figure out that they live in a finite, not an infinite, universe that wraps around itself. They would conclude, after enough thought, that there is a center. But where is it? It's the center of the balloon.

Is that part of this hypothetical universe as I've defined it? No. The center of the balloon is not part of the surface; therefore, it is not part of this universe, or at least of the dimensions that are physically accessible to these dot creatures. It's in a mathematically real and distinct dimension, but that dimension is not physically accessible. That is a really interesting concept. Let's try to take a look at this by drawing some circles. If we look at a circle as a one-dimensional example of the two-dimensional surface I just showed you, a circle is like a one-dimensional universe. I can go one direction or the other direction. But suppose I'm forbidden by the laws of physics from going along the radial direction or the up and down direction. That's like a one-dimensional closed or finite universe.

The equation in Cartesian coordinates for this circle is x^2 (the distance from the origin) + $y^2 = r^2$, the square of the radius. Each of the points along this circle satisfies this equation. That's the equation of the set of points known as the circle. You can also write it in polar coordinates. In polar coordinates, a dot on the circle is some radius, r, from this origin, the center of the circle. The angle from the x-axis is denoted by θ. If you have all points at a given radius, r, and spanning all possible values of θ—0 to 2π, all the way around the circle—again, you will describe the set of points known as the circle. In each case, you can see that two variables were needed to describe this circle, this one-dimensional structure. You needed x and y, or r and θ. Two coordinates were needed to describe a one-dimensional curved path. This means that this one-dimensional path wraps around a second mathematical dimension. In the case of the circle, the origin is not part of the circle. It's not in one of the physically accessible dimensions; rather, it's in the dimension used to mathematically describe that hypothetical universe, the circle.

In the case of the sphere, we can represent the surface of the sphere by an equation, $x^2 + y^2 + z^2 = r^2$. This is just an extension of the Pythagorean theorem, the three dimensions. All dots on this sphere satisfy this equation. The sphere itself is two-dimensional; it's a two-dimensional curved surface. You can only go forward and backward, or left and right. You can't go in and out; remember that. The center of the sphere is not in the surface—yet clearly it exists because there had to be three dimensions (x, y, and z) describing this surface. The origin of coordinates here, $r = 0$, is this point that's not in the sphere. In polar coordinates, you can say that all the points are at radius r, and θ runs the course from 0 to 2π, and φ can go from 0 to π. With this set of points, you describe the full sphere. Once again, one needs three coordinates—r, θ, and φ, or x, y, and z—to describe a fundamentally two-dimensional surface whose center is in a mathematically well-defined, but physically inaccessible, dimension.

If we live in a three-dimensional analog of a sphere, where a volume bends around some fourth spatial dimension, the equation for our sphere would be something like $x^2 + y^2 + z^2 + w^2 = r^2$; w now being this fourth dimension around which we curve. This is called a hypersphere. I can't really imagine what a hypersphere looks like because I live in the normal three dimensions, and I can't put myself into this fourth spatial dimension, from which I could view the hypersphere. It's sort of like the dot creatures. They can't really imagine what their sphere looks like, but they can imagine a circle, which is a one-dimensional counterpart to their two-dimensional surface. In a similar way, the sphere is a two-dimensional counterpart to this three-dimensional volume in which we live, but you'd have to put yourself into the fourth dimension to actually see the hypersphere. I don't know how to imagine it, but I can help you gain some visualization of it by drawing a hypercube in a hypersphere.

Let's first draw a line segment, such as that shown here. If I want to turn that into a square, I just take a copy of it, displace it to the right, let's say, along a dimension perpendicular to the dimension of the line segment, and then join the ends. That gives me a two-dimensional square. If I want to make a cube out of the two-dimensional square, I make a copy of it, displace it in the perpendicular direction, and join up the corners. That gives me a cube. In fact, here's a cube. I just took a square, made a copy, displaced it, joined up the corners, and then sort of filled it in. This we can visualize nicely from our three-dimensional space, but we can also draw its projection onto a sheet of paper. Here's the projection, or shadow, of a three-dimensional cube onto a two-dimensional surface. You take the square and make a copy of it, and instead of displacing it in the perpendicular direction or dimension, which you cannot do in this case because you're confined to the sheet of paper, you displace it along a diagonal and then connect up the corners. That gives you the shadow, or the projection, of a cube onto a two-dimensional sheet of paper.

To make a hypercube, I would need to take a cube, make a copy of it, displace it into this fourth spatial dimension, which is physically inaccessible, and then join up the corners. That would give me a hypercube. But I can't do that because I can't displace it into this other dimension. Where is it? But I can do the next-best thing. I can make a copy of the cube, displace it along a diagonal in this room, and then connect up the corners. That'll give me the shadow, or the projection, of a hypercube onto this three-dimensional room. I hope that helps you to some degree in visualizing hyperspace. I think it helps me a little bit, but I'm not sure.

Let's go back to the expansion of the Universe. I mentioned that the Universe could be finite and curving around itself like a balloon, or a hypersphere or something like that, but it could also be infinite. You might ask, "Does expansion make sense in an infinite space? What is infinity, anyway, and how can we come to some gut feeling for what it means?" It turns out that the Universe can expand even if it's infinite. It just becomes less dense with time. Let me illustrate this with the counting numbers: 1, 2, 3, 4, 5, 6, 7, 8, and so on. They clearly go on for infinity. But now I can strike out all the even numbers, and I have 1, 3, 5, 7, 9, 11, and so on. They clearly go out to infinity as well. There are no fewer of them than there are the counting numbers, but they're, in a sense, less dense. The even numbers are missing.

Now let's strike out all of them except 1, 11, 21, and so on. Those go out to infinity as well. You never stop counting, right? But a bunch of numbers is missing, so in a sense, it's less dense, but they're still infinite. Mathematically, all of these sequences can be put into 1-to-1 correspondence with the counting numbers, and so all of these infinities are mathematically the same size, even though there's a bunch of numbers missing in some of the ones that I wrote down. In a similar way, the Universe can be infinite and "growing"—not becoming a bigger infinity mathematically, but just becoming less dense, like these numerical examples.

Now suppose we thought about universes that might be more dense than ours. Could there be a more-dense universe than ours that's still expanding? Indeed, we can. If we look at the set of fractions, the rational numbers, that is a set of numbers that is more dense than the counting numbers—but, in fact, it's still what we call a "countable infinity." It can be placed into 1-to-1 correspondence with the counting numbers, and so there are no more fractions, rational numbers, than there are counting numbers, yet there's clearly a greater density of the fractions. I mean between 0 and 1, for example, there's 1/2, 1/3, 2/5, 11/13, and so on. Between 1 and 2, there's an infinite number as well. There are all these infinities, but it's still the same size. The rational numbers are more dense than the counting numbers, in this sense.

Let me convince you that I can come up with a scheme to count all the rational numbers and not miss any of them. Let me write them down in a table like this: 1, which is 1/1, 2/1, 3/1, 4/1, and so on—that's just the counting numbers. There's an infinite number of them there. Now let's write 1/2, 2/2—that's the same thing as 1. That's okay; 1 already appeared, but no one said I can't write down the numbers twice or more. Anyway, 3/2, 4/2, 5/2, 6/2, 7/2; there's an infinite number of them there. Now 1/3, 2/3, 3/3, and so on; 1/4, 2/4, ¾, and so on—and so on, and so on, and so forth. Each row has an infinite number of numbers. Each column has an infinite number of numbers. You might think that infinity times infinity gives you a bigger infinity than the counting numbers. It makes sense, right? In fact, it's not a bigger infinity; it's still a countable infinity.

There are a number of schemes you could use to keep track of all these fractions and not miss a single one. Here's one scheme. You start with the number 1, and you go in a loop like this. There's number 2, number 3, number 4, 5, 6, 7, 8, 9, 10, 11, 12—I already counted that one; that's the same thing as 1/2, but that's okay; I can count them more than once—13, 14, 15, 16, 17, 18, 19, 20, 21, and so on along a set of diagonals. Will I miss even a single number? No, I won't. That means that all of these fractions can be put into 1-to-1 correspondence with the set of counting numbers, and that means that the rational numbers are no bigger an infinity than the counting numbers, even though they sure look bigger to me, I must admit.

Given that mathematical infinities are so weird—you have an infinite number of rows and an infinite number of columns, but it's still the same infinity. That whole set is the same infinity as the counting numbers. That's already so weird that, if you can accept that, you surely should be able to accept the concept of a universe that's infinite, but still expanding. That's the way it goes. Infinity is just a hard thing to get a hold of. Later, we'll even see that there are bigger infinities, which will be relevant to some of the things we say about dark energy and the birth of the Universe. But for now, I'm just telling you that even the countable infinities, like the counting numbers and the rational numbers, already are kind of hard to understand.

We're now convinced that the Universe is expanding. If we extrapolate this observed expansion backward in time, we could come to the conclusion that the Universe may have had a beginning, where the matter was very dense, and for whatever reason, started expanding from a point called the singularity—when everything was really dense, possibly even infinitely dense. Although if you include quantum physics, probably it's not infinitely dense; probably it's just really, really dense. The Universe that's expanding is like an expanding gas. When gases expand, they cool. That's what you have with a fire extinguisher. You have all this compressed gas, and it expands and cools. If you compress gas, it heats. This suggests that the Universe began in a very hot and dense state because, as this whole coordinate grid is expanding, as space is expanding, it's also cooling.

If you extrapolate backward in time, you could say that when the Universe was small and very dense, it was also very hot. At time = 0, when this dense, infinitely dense, or very dense, singularity existed, maybe the temperatures, as well, were infinite or nearly infinite—very high, in any case. It was a hot birth to the Universe, a hot Big Bang, where for some reason that we're still trying to understand, the Universe came into existence and has been expanding ever since—growing bigger, less dense, and cooler.

This idea that everything is expanding, and that we can use this expansion, then, to study the history of the Universe, is great. But before we get to that point, I want to now, in the next seven lectures, utilize the expansion of the Universe, Hubble's law, to get the distances of galaxies and to know, then, how far back in time we're looking. The more distant a galaxy is, the longer it has taken for light to reach us—the greater is the so-called "lookback time." Suppose we have some really distant galaxies that are so far away that we can't see the Cepheid variables or anything else in them. Can we derive their distances in some other way? Indeed, we can. We can use the known expansion of the Universe, Hubble's law, with the known Hubble constant—I'll tell you later how we derive it—to get the distance.

If you measure the redshift of a galaxy, z, by looking at its spectrum—that's just $\Delta\lambda/\lambda_0$. You can measure that. You can then compute the recession speed, v, from z = v/c, which at least works at small redshifts. It doesn't work at large redshifts; you have to use a more-correct relativistic formula. But at least at small redshifts, the speed of recession divided by the speed of light is just equal to the redshift. So, now v = zc, but we know from Hubble's law that v = H_0d. Since you've measured the speed of recession, and suppose you know Hubble's constant, you can determine the distance of the galaxy, d. Again, this simple formula works for small redshifts; you need a more complex, relativistic formula for larger redshifts.

Observing galaxies at different redshifts, you can tell their different distances. Since light travels at the speed of light, no faster, we can tell, in fact, how far away those galaxies are, and how long it has taken for light to reach us. Here's a table of redshifts and the so-called lookback time. At a redshift of 0, lookback time is 0; the object is here. At a redshift of 0.1, the lookback time is already 1.3 billion years. At a redshift of 0.2, it's 2.4 billion years. At progressively bigger redshifts, the lookback time is a progressively larger number of years, up to something like 13 billion or 14 billion years, the age of the Universe. In newspaper articles, then, when they tell you the distance of a galaxy that an astronomer has measured, what they're really telling you is the lookback time—how long it has taken for the light to reach us. By looking at progressively larger distances, we're looking progressively farther back in time, and we see a movie of the past history of the Universe, from which we can learn something about how it evolved and how the galaxies within it evolved.

Useful Symbols

In these course notes, the following mathematical symbols are used:

Symbol	**Meaning**
$\sim$	Roughly or around
$\approx$	Approximately equal to (basically synonymous with ~)
$\geq$	Greater than or equal to
$\leq$	Less than or equal to
$\gg$	Much greater than
$\ll$	Much less than
$\propto$	Proportional to

Universe Timeline

(For $t > 10^{14}$ years, assume the Universe expands forever.)

"0" seconds .... The birth of the Universe, perhaps from a quantum fluctuation.

10^{-43} seconds.................................... Space-time foam? Gravity and grand unified force become separate.

10^{-37} seconds? .... Inflation begins.

10^{-35} seconds?.................................. Inflation ends. Strong nuclear and electroweak forces become separate.

10^{-11} seconds..................................... Weak nuclear and electromagnetic forces become separate.

10^{-6} seconds..................................... Matter/antimatter annihilation; slight excess of protons and neutrons.

1 second .... Electrons and positrons annihilate; slight excess of electrons.

10^{2} seconds...................................... Nucleosynthesis of lightest elements from protons and neutrons.

4×10^{5} years (10^{13} seconds) .... Formation of neutral atoms; Universe transparent.

3×10^{6} years (10^{14} seconds) .... First stars begin to form but not in galaxies.

10^{9} years (3×10^{16} seconds)............. Many galaxies form and begin assembling into clusters.

10^{10} years .... Solar System forms (4.6 billion years ago).

1.4×10^{10} years................................ Present age of the Universe.

2×10^{10} years.................................... Sun becomes a red giant and subsequently a white dwarf.

10^{14} years.. Last low-mass stars die.

10^{20} years.. Most stars and planets gravitationally ejected from galaxies.

10^{30} years.. Black holes swallow most of the remaining objects in galaxies.

10^{38} years?.. All objects except black holes disintegrate, due to proton decay.

10^{65} years.. Stellar-mass black holes evaporate due to Hawking process.

10^{100} years.. Largest galaxy-mass black holes evaporate.

10^{110} years?...................................... Positronium atoms (electron-positron pairs) decay, producing photons.

Solar System Timeline

(Given in terms of years ago; 0 = today.)

4.6 billion years............................... Solar System forms.

3.9 billion years............................... Heavy bombardment of Earth by planetesimals subsides.

3.8 billion years............................... Possible formation of primitive life (definitely by 3.5 billion years).

2 billion years.................................. Free oxygen begins to accumulate in atmosphere due to photosynthesis.

600 million years............................. Present atmosphere essentially complete. Multicellular life flourishes.

550 million years............................. *Cambrian explosion*—formation of complex, hard-bodied animals.

240 million years............................. Mesozoic era—earliest dinosaurs appear.

65 million years............................... Extinction of the dinosaurs, along with two-thirds of all living species.

4.5 million years............................... The first *hominids* appear.

160,000 years .... Early *homo sapiens* appear.

3000 years .... Beginning of Iron Age.

250 years .... Industrial Revolution.

100 years .... Radio communication.

Glossary

absolute magnitude: Logarithmic measurement of the luminosity of stars; assumes all the stars to be at the same distance of 10 parsecs from Earth.

absorption line: A wavelength (or small range of wavelengths) at which the brightness of a spectrum is less than it is at neighboring wavelengths.

accelerating universe: The model of the Universe based on recent observations that its expansion is speeding up with time.

accretion: The transfer of matter to the surface of a star or a black hole. When the transferred matter goes into orbit around the object, an *accretion disk* is formed.

active galaxy: A galaxy whose nucleus emits large quantities of electromagnetic radiation that does not appear to be produced by stars. (Radio galaxies are one example of active galaxies.)

adaptive optics: Optical systems providing rapid corrections to counteract atmospheric blurring.

analemma: The apparent figure-8 path made by the Sun in the sky when photographs of the Sun's position taken at a given time of day throughout the year are superimposed on each other.

angstrom (Å): A unit of length commonly used for visible wavelengths of light; 1 Å = 10^{-8} cm.

angular momentum: A measure of the amount of spin of an object; dependent on the object's rotation rate, mass, and mass distribution.

anthropic principle: The idea that given that we exist, the Universe must have certain properties or it would not have evolved so that life formed and humans evolved.

antiparticle: A particle whose charge (if not neutral) and certain other properties are opposite those of a corresponding particle of the same mass. An encounter between a particle and its antiparticle results in mutual annihilation and the production of high-energy photons.

apparent brightness: The amount of energy received from an object per second, per square centimeter of collecting area. It is related to luminosity and distance through the equation $b = L/(4\pi d^2)$, the inverse-square law of light.

archaeoastronomy: The study of the astronomical significance of ancient buildings and other structures.

asterism: A grouping of stars that is not itself a full constellation but is part of a constellation. The Big Dipper is one example.

asteroid: Chunk of rock, smaller than a planet, that generally orbits the Sun between Mars and Jupiter.

astrometry: Measurement of the position and motion of the stars in the plane of the sky.

astronomical unit (AU): The average distance between the Sun and the Earth (1.5×10^8 km).

aurora: The northern or southern lights, caused by energetic particles from the Sun interacting with atoms and molecules in Earth's upper atmosphere, making them glow.

Baily's Beads: During a solar eclipse, the effect of sparkling lights created by sunlight passing through valleys on the Moon's surface.

Big Bang: The birth of the Universe in a very hot, dense state 13.7 billion years ago, followed by the expansion of space.

binary pulsar: A pulsar in a binary system. Often, this term is used for systems in which the pulsar's companion is another neutron star.

binary star: Two stars gravitationally bound to (and orbiting) each other.

bipolar outflow: A phenomenon in which streams of matter are ejected from the poles of a rotating object.

black body: An object that absorbs all radiation that hits it; none is transmitted or reflected. It emits radiation due to thermal (random) motions of its constituent particles, with a spectrum that depends only on the temperature of the object.

black hole: A region of space-time in which the gravitational field is so strong that nothing, not even light, can escape. Predicted by Einstein's general theory of relativity.

brown dwarf: A gravitationally bound object that is insufficiently massive to ever be a main-sequence star but too massive for a planet. Generally, the mass range is taken to be 13–75 Jupiter masses.

charge-coupled device (CCD): A solid-state imaging chip whose properties include high sensitivity, large dynamic range, and linearity.

celestial equator: Projection of Earth's equator onto the celestial sphere.

celestial sphere: The enormous sphere, centered on the Earth, to which the stars appear to be fixed.

centrifugal force: The outward force felt by an object in a rotating frame of reference.

Cepheid variable: A type of pulsating star that varies in brightness with a period of 1 to 100 days.

Cerenkov radiation: Electromagnetic radiation emitted by a charged particle traveling at greater than the speed of light in a transparent medium. The blue light emitted is the electromagnetic equivalent of a sonic boom heard when an aircraft exceeds the speed of sound.

Chandrasekhar limit: The maximum stable mass of a white dwarf or the iron core of a massive star, above which degeneracy pressure is unable to provide sufficient support; about 1.4 solar masses.

chromosphere: Hot, thin layer of gas just below the Sun's corona and above the photosphere.

closed universe: A universe having finite volume.

cold dark matter: Nonluminous matter that moves slowly, such as neutron stars and exotic particles.

collapsar model: Model proposed for some types of gamma-ray bursts, wherein a rotating, massive star collapses and forms two highly focused beams (jets) of particles and light.

comet: An interplanetary chunk of ice and rock, often in a very eccentric (elongated) orbit, that produces a diffuse patch of light in the sky when relatively near the Sun as a result of evaporation of the ice.

constellation: One of 88 regions into which the celestial sphere is divided. The pattern of bright stars within a constellation is often named in honor of a god, person, or animal.

convection: Process by which bubbles of gas or liquid repeatedly heat and expand, rise and give off energy, and fall again; seen in the stars and in Earth's core.

core: In a main-sequence star, roughly the central 10% by mass. In an evolved star, usually refers to the degenerate central region.

corona: The very hot, tenuous, outermost region of the Sun, seen during a total solar eclipse.

cosmic microwave background radiation: Radio electromagnetic radiation that was produced in the hot Big Bang. It now corresponds to $T \approx 3$ K because of the expansion and cooling of the Universe.

cosmic rays: High-energy protons and other charged particles, probably formed by supernovae and other violent processes.

cosmological constant: In Einstein's general theory of relativity, a term (Λ) that produces cosmic repulsion that can counterbalance the attractive force of gravity. Recent evidence suggests that its value is nonzero and somewhat larger than originally postulated by Einstein, causing the observed acceleration of the Universe's expansion.

cosmological principle: The Universe is homogeneous and isotropic (that is, uniform) on the largest scales.

cosmology: The study of the overall structure and evolution of the Universe.

crepuscular rays: Beams of light shining through gaps in clouds, usually best seen near sunset or sunrise.

critical density: The average density of the Universe if it were poised exactly between eternal expansion and ultimate collapse, if the cosmological constant is zero.

dark energy: Energy with negative pressure, causing the expansion of the Universe to accelerate.

dark matter: Invisible matter that dominates the mass of the Universe.

degenerate gas: A peculiar state of matter at high densities in which, according to the laws of quantum physics, the particles move very rapidly in well-defined energy levels and exert tremendous pressure.

deuterium: An isotope of hydrogen that contains one proton and one neutron.

deuteron: A deuterium nucleus.

diffraction: A phenomenon affecting light as it passes any obstacle, spreading it out.

dipole field: The pattern of electric field lines produced by a pair of equal and opposite electric charges or of magnetic field lines surrounding a bar magnet.

Doppler shift: The change in wavelength or frequency produced when a source of waves and the observer move relative to each other. Blueshifts (to shorter wavelengths) and redshifts (to longer wavelengths) are associated with approach and recession, respectively.

$E = mc^2$: Einstein's famous formula for the equivalence of mass and energy.

earthshine: Sunlight illuminating the Moon after having been reflected from the Earth.

eccentric: Deviating from a circle. *Eccentricity* is a measure of this.

eclipse: The passage of one celestial body into the shadow of another or the obscuration of one celestial body by another body passing in front of it.

ecliptic: The path followed by the Sun across the celestial sphere in the course of a year.

electromagnetic force: One of the four fundamental forces of nature; it holds electrons in atoms.

electromagnetic radiation: Self-propagating, oscillating electric and magnetic fields. From shortest to longest wavelengths: gamma rays, X-rays, ultraviolet, optical (visible), infrared, and radio.

electron: Low-mass, negatively charged fundamental particle that normally "orbits" an atomic nucleus.

electroweak force: The unification of the electromagnetic and weak nuclear forces.

ellipse: A set of points (curve) such that the sum of the distances to two given points (foci) is constant.

elliptical galaxy: One of the two major classes of galaxies defined by Edwin Hubble; has a roughly spherical or elliptical distribution of generally older stars, less gas and dust, and less rotation than its spiral counterpart.

emission line: A wavelength (or small range of wavelengths) at which the brightness of a spectrum is more than it is at neighboring wavelengths.

equinox: One of two points of intersection between the ecliptic and the celestial equator, or the time of the year when the Sun is at this position.

escape velocity: The minimum speed an object must have to escape the gravitational pull of another object.

event horizon: The boundary of a black hole from within which nothing can escape.

expansion age: The time estimated for the age of the Universe since the Big Bang, determined by measuring the rate at which galaxies are receding from one another; currently thought to be about 13.7 billion years.

extinction: The obscuration of starlight by interstellar gas and dust.

extrasolar planet: A planet orbiting a star other than the Sun; an *exoplanet*.

eyepiece: A small tube containing a lens (or combination of lenses) at the eye end of a telescope, used to examine the image.

flat (critical) universe: A universe in which the laws of Euclidean geometry hold.

fusion: The formation of heavier nuclei from lighter nuclei.

galactic cannibalism: The swallowing of one galaxy by another.

galaxy: A large (typically 5000 to 200,000 light years in diameter), gravitationally bound system of hundreds of millions (and up to a trillion) stars.

Galilean satellites: The four large moons of Jupiter (Io, Europa, Ganymede, Callisto).

gamma rays: Electromagnetic radiation with wavelengths shorter than about 0.1 Å.

gamma-ray burst (GRB): A brief burst of gamma rays in the sky, now known to generally come from exceedingly powerful, distant objects.

general theory of relativity: Einstein's comprehensive theory of mass (energy), space, and time; it states that mass and energy produce a curvature of space-time that we associate with the force of gravity.

globular cluster: A bound, dense, spherically symmetric collection of stars formed at the same time.

glory: A thin halo of light around the shadow of an object projected on a cloud; caused by the bending of light around and within water droplets.

grand unified theory (GUT): A theory that unifies the strong nuclear ("color") and electroweak forces into a single interaction.

gravitational lens: In the gravitational lens phenomenon, a massive body changes the path of light passing near it so as to make a distorted image of the object.

gravitational redshift: A redshift of light caused by the presence of mass.

gravitational waves: Waves thought to be a consequence of changing distributions of mass.

gravity: The weakest of nature's fundamental forces but the dominant force over large distances because it is cumulative; all matter and energy contribute, regardless of charge.

great circle: The intersection of a sphere with a plane passing through the center of the sphere. The meridian and the celestial equator are both great circles.

green flash: A subtle green glow sometimes visible in very clear skies just as the last part of the Sun is setting (or the first part is rising).

greenhouse effect: The effect by which the atmosphere of a planet heats up above its normal equilibrium temperature because it absorbs infrared radiation from the surface of the planet.

halo (galactic): The region that extends far above and below the plane of the galaxy.

halo (solar or lunar): A circle of light around the Sun or Moon, having a radius of about 22 degrees, formed by light passing through hexagonal ice crystals.

Hawking radiation: According to Stephen Hawking, the thermal radiation emitted by black holes because of quantum effects.

Heisenberg uncertainty principle: One form: In any measurement, the product of the uncertainties in energy and time is greater than or equal to Planck's constant divided by 2π. Another form: In any measurement, the product of the uncertainties in position and momentum is greater than or equal to Planck's constant divided by 2π.

Hertzsprung-Russell (H-R) diagram: A plot of the surface temperature (or color) versus luminosity (power, or absolute brightness) for a group of stars. Also known as a *temperature-luminosity diagram*.

homogeneous: The same (density, temperature, and so on) at all locations.

horizon: The great circle defined by the intersection of the celestial sphere with the plane tangent to the Earth at the observer's location; it is 90 degrees away from the zenith.

hot dark matter: Nonluminous matter with great speeds, such as neutrinos.

Hubble's law: The linear relation between the current distance and recession speed of a distant object: $v = H_0 d$. The constant of proportionality, H_0, is called *Hubble's constant*.

infinity: All numbers. A countable infinity can be put in one-to-one correspondence with the counting numbers, whereas an uncountable infinity cannot.

inflationary universe: A modification of the standard Big Bang theory. Very early in its history (e.g., $t \approx 10^{-37}$ seconds), when the Universe was exceedingly small, it began a period of rapidly accelerating expansion, making its final size truly enormous. Subsequently, the regular Big Bang expansion ensued.

interference: The property of radiation, explainable by the wave theory, in which waves in phase can add (constructive interference) and waves out of phase can subtract (destructive interference).

interferometer: Two or more telescopes used together to produce high-resolution images.

interstellar extinction: *See* **extinction**.

interstellar medium: The space between the stars, filled to some extent with gas and dust.

inverse-square law: Decreasing with the square of increasing distance. For example, the brightness of a star is proportional to the inverse-square of distance, as is the gravitational force between two objects.

ionized: Having lost at least one electron. Atoms become ionized primarily by the absorption of energetic photons and by collisions with other particles.

isotopes: Atomic nuclei having the same number of protons but different numbers of neutrons.

isotropic: The same in all directions (that is, no preferred alignment).

Kelvin: The size of 1 degree on the Kelvin (*absolute*) temperature scale, in which absolute zero is 0 K, water freezes at 273 K, and water boils at 373 K. To convert from the Kelvin scale to the Celsius (centigrade, C) scale, subtract 273 from the Kelvin-scale value. Degrees Fahrenheit (F) = (9/5)C + 32.

Kepler's third law: If one object orbits another, the square of its period of revolution is proportional to the cube of the semimajor axis (half of the long axis) of the elliptical orbit.

Kuiper belt: A reservoir of perhaps millions of Solar-System objects, orbiting the Sun generally outside the orbit of Neptune. Eris and Pluto are the two largest known Kuiper-belt objects, though some astronomers consider them to be planets.

Large Magellanic Cloud: A dwarf companion galaxy of our Milky Way Galaxy, about 170,000 light years away; best seen from Earth's southern hemisphere.

large-scale structure: The network of clusters, voids, and other shapes seen on the largest scales of the Universe.

light curve: A plot of an object's brightness as a function of time.

light year: The distance light travels through a vacuum in 1 year; about 10 trillion kilometers, or about 6 trillion miles.

lighthouse model: The explanation of a pulsar as a spinning neutron star whose beam we see as it comes around and points toward us.

Local Group: The roughly three dozen galaxies, including the Milky Way, that form a small cluster.

lookback time: The duration over which light from an object has been traveling to reach us.

luminosity: Power; the total energy emitted by an object per unit of time; intrinsic brightness.

magnetar: Spinning neutron star with an extraordinarily powerful magnetic field that occasionally releases a burst of gamma rays when the crust of the star undergoes a sudden restructuring (a "star quake").

magnitude: A logarithmic measure of apparent brightness; a difference of 5 magnitudes corresponds to a brightness ratio of 100. Typical very bright stars have mag 1; the faintest naked-eye stars have mag 6.

main sequence: The phase of stellar evolution, lasting about 90% of a star's life, during which the star fuses hydrogen to helium in its core.

massive compact halo objects (MACHOs): Brown dwarfs, white dwarfs, and similar objects that could account for some of the dark matter of the Universe.

merging: The interaction of two galaxies in space, with a single galaxy as a result.

meridian: A great circle passing through the celestial poles and the zenith; the highest point in the sky reached by a star during each day-night cycle.

meteor: The streak of light in the sky produced when an interplanetary rock enters Earth's atmosphere and burns up as a result of friction. If the rock reaches Earth's surface, it is called a *meteorite*.

meteoroid: An interplanetary rock that is not in the asteroid belt.

Milky Way: The band of light across the sky coming from the stars and gas in the plane of the Milky Way Galaxy (our Galaxy).

minor planets: Asteroids. Some astronomers now reserve this term for the largest asteroids and Kuiper belt objects.

mirage: An image of an object, often inverted, formed by light passing through layers of air having different temperatures.

multiverse: The set of parallel universes that may exist, with our observable Universe as only one part.

nebula: A region containing an above-average density of interstellar gas and dust.

nebular hypothesis: Theory of the formation of the Solar System, asserting that spinning clouds of interstellar matter gradually contracted and allowed for the formation of the Sun and the planets.

neutrino: A nearly massless, uncharged fundamental particle that interacts exceedingly weakly with matter. There are three types: electron, muon, and tau neutrinos.

neutron: Massive, uncharged particle that is normally part of an atomic nucleus.

neutron star: The compact endpoint in stellar evolution in which typically 1.4 solar masses of material is compressed into a small (diameter = 20–30 km) sphere supported by neutron degeneracy pressure.

nova: A star that suddenly brightens, then fades back to its original intensity; caused by the accretion of stellar matter from a companion star.

nuclear fusion: Reactions in which low-mass atomic nuclei combine to form a more massive nucleus.

nucleosynthesis: The creation of elements through nuclear reactions, generally nuclear fusion.

Olbers's paradox: The dark night sky; simple arguments suggest that it should be very bright.

open cluster: A loosely bound cluster of stars, usually consisting of young stars that eventually break away from the cluster.

open universe: A universe whose volume is infinite.

parallax: Apparent movement of an object due to a change in the position of the observer. The parallax of a star is defined as the angular distance subtended by 1 AU, the distance between the Earth and the Sun, as seen from the star.

parsec: A unit of distance equal to about 3.26 light years (3.086×10^{13} km).

particle physics: The study of the elementary constituents of nature.

Pauli exclusion principle: Wolfgang Pauli's explanation for the arrangement of electrons in an atom. The quantum mechanical principle states that no two electrons can be in the same "quantum state" (same configuration) in an atom at the same time.

phase transition: The transformation of matter from one phase (e.g., liquid) to another (e.g., solid).

photon: A quantum, or package, of electromagnetic radiation that travels at the speed of light. From highest to lowest energies: gamma rays, X-rays, ultraviolet, optical (visible), infrared, and radio.

photon sphere: A region of space surrounding a black hole at which the curvature of space is so great that it causes light to orbit in circles.

photosphere: The visible surface of the Sun (or another star) from which light escapes into space.

pinhole camera: A hole in an opaque sheet used to project an image of the Sun.

Planck curve: The mathematical formula describing the spectrum of light produced by a perfect thermal emitter.

Planck's constant: The fundamental constant of quantum physics, h; a very small quantity.

planet: A body that primarily orbits a star (so that moons don't count), is large enough to be roughly spherical (typically, larger than about 600 km in diameter), gravitationally dominates its region of space (that is, has largely cleared away other debris), and has never undergone nuclear fusion.

planetary nebula: A shell of gas, expelled by a red-giant star near the end of its life (but before the white-dwarf stage), that glows because it is ionized by ultraviolet radiation from the star's remaining core.

planetary system: A collection of planets and smaller bodies orbiting a star (e.g., our Solar System).

planetary transit: The passage of a planet directly along a star's line of sight, causing a momentary dimming of the star's light; can be used to detect planets in other solar systems.

planetesimals: Small bodies, such as meteoroids and comets, into which the solar nebula condensed and from which the planets subsequently formed.

pole star: A star approximately at a celestial pole (Polaris, in the north).

positron: The antiparticle of an electron.

precession: A conical motion undergone by spinning objects pulled by an external force not directed along the axis. The Earth's precession causes the direction of the north celestial pole to shift gradually with time.

progenitor: In the case of a supernova, the star that will eventually explode.

prograde motion: The apparent motion of the planets when they appear to gradually move from west to east among the stars; *retrograde motion* is the opposite direction.

prominences: Hot plumes of gas streaming from the Sun's photosphere along the lines of the Sun's magnetic fields.

proteins: Molecules consisting of long chains of amino acids.

proton: Massive, positively charged particle that is normally part of an atomic nucleus. The number of protons in the nucleus determines the chemical element.

proton-proton chain: A set of nuclear reactions by which four hydrogen nuclei (protons) combine to form one helium nucleus, with a resulting release of energy.

protoplanetary disks: Also called *proplyds*; concentrations of matter around newly formed or still forming stars out of which planets may form.

protostar: A star still in the process of forming in a cloud of gas and dust, collapsing nearly in free fall.

pulsar: An astronomical object detected through pulses of radiation (usually radio waves) having a short, extremely well-defined period; thought to be a rotating neutron star with a very strong magnetic field.

quantum fluctuations: The spontaneous (but short-lived) quantum creation of particles out of nothing.

quantum mechanics: A 20th-century theory that successfully describes the behavior of matter on very small scales (such as atoms) and radiation.

quark: A fundamental particle with fractional charge; protons and neutrons consist of quarks.

quasar (QSO): A star-like, extremely luminous object, typically billions of light years away. Now thought to be the nucleus of a galaxy with a supermassive black hole that is accreting matter from its vicinity.

quintessence: A new particle or field in physics that can lead to repulsive dark energy.

radial velocity: The speed of an object along the line of sight to the observer.

recombination: Process by which electrons combine with protons and other atomic nuclei to form neutral atoms; believed to have first occurred about 380,000 years after the Big Bang.

red giant: The evolutionary phase following the main sequence of a relatively low-mass star, such as the Sun; the star grows in size and luminosity but has a cooler surface.

redshift: Defined to be $z = (\lambda - \lambda_0)/\lambda_0$, where λ_0 is the rest wavelength of a given spectral line and λ is its (longer) observed wavelength. The wavelength shift may be caused by recession of the source from the observer or by the propagation of light out of a gravitational field.

reflecting telescope: Telescope that uses a mirror instead of a lens to collect light; unlike the refracting telescope, it brings all colors into focus together.

refracting telescope: Telescope that uses a lens to collect light and bring it to a focus.

refraction: The bending of light as it passes from one medium to another having different properties.

relativistic: Having a speed that is such a large fraction of the speed of light that the special theory of relativity must be applied.

resolution: The clarity of detail produced by a given optical system (such as a telescope).

rest mass: The mass of an object that is at rest with respect to the observer. The effective mass increases with speed.

rest wavelength: The wavelength radiation would have if its emitter were not moving with respect to the observer.

retrograde motion: The apparent backward (east-to-west) motion among the stars that planets undergo for a short time each year.

Roche limit: The distance from the center of a planet at which the planet's tidal forces prevent particles from forming a moon through their mutual gravitational attraction.

rotation curve: A graph of the speed of rotation versus distance from the center of a rotating object, such as a galaxy.

Schwarzschild radius: The radius to which a given mass must be compressed to form a nonrotating black hole. Also, the radius of the event horizon of a nonrotating black hole.

second law of thermodynamics: In any closed system, entropy (the amount of disorder) never decreases; it always increases or remains constant.

singularity: A mathematical point of zero volume associated with infinite values for physical parameters, such as density.

solar mass: The mass of the Sun, 1.99×10^{33} grams, about 330,000 times the mass of the Earth.

solstice: The northernmost or southernmost point on the celestial sphere that the Sun reaches, or the time of the year when the Sun reaches this point.

space-time: The four-dimensional fabric of the Universe whose points are events having specific locations in space (three dimensions) and time (one dimension).

special theory of relativity: Einstein's 1905 theory of relative motion, gravity excluded.

spectroscopic binary stars: Binary stars detected by examining the periodically varying Doppler shift in their absorption lines.

spectrum: A plot of the brightness of electromagnetic radiation from an object as a function of wavelength or frequency.

spiral galaxy: One of the two major classes of galaxies defined by Edwin Hubble; made up of a roughly spherical central "bulge" containing older stars, surrounded by a thin disk in which spiral arms are present.

star cluster: A gravitationally bound group of stars that formed from the same nebula.

steady-state theory: A model of the expanding Universe based on the assumption that the properties of the Universe do not change with time. Matter must be continually created to maintain constant density.

Stefan-Boltzmann law: Law stating that, per unit of surface area, an opaque object emits energy at a rate proportional to the fourth power of its surface temperature.

string (superstring) theory: A possible unification of quantum theory and general relativity in which fundamental particles are different vibration modes of tiny, one-dimensional "strings," instead of being localized at single points.

stripped massive stars: Stars that have lost their hydrogen and helium envelopes, either through stellar winds or through transfer of gas to a companion star; thought to be the progenitors for gamma-ray bursts.

strong nuclear force: The strongest force, it binds protons and neutrons together in a nucleus. Actually, it is the residue of the even stronger *color force* that binds quarks together in a proton or neutron.

sundog: A pair of bright spots on the outer edge of the solar halo at roughly the Sun's altitude above the horizon.

sun pillar: A faint pillar of light above the Sun in the sky, best visible after sunset.

sunspots: Cooler regions on the Sun's photosphere that appear as dark blotches.

supercooled: The condition in which a substance is cooled below the point at which it would normally make a phase change.

supergiant: The evolutionary phase following the main sequence of a massive star; the star becomes more luminous and larger. If its size increases by a very large factor, it becomes cool (red).

supernova: The violent explosion of a star at the end of its life. Hydrogen is present or absent in the spectra of Type II or Type I supernovae, respectively.

supernova remnant: The cloud of chemically enriched gases ejected into space by a supernova.

symmetric: Forces that are symmetric act identically. They act differently when the symmetry is broken.

synchronous rotation: The rotation of a body having the same period as its orbit.

terminator: The line between night and day on a moon or planet; the edge of the part that is lighted by the Sun.

terrestrial planets: Rocky, earth-like planets. In our Solar System: Mercury, Venus, Earth, and Mars.

tidal force: The difference between the gravitational force exerted by one body on the near and far sides of another body.

time dilation: According to relativity theory, the slowing of time perceived by an observer watching another object moving rapidly or located in a strong gravitational field.

transverse velocity: The speed of an object across the plane of the sky (perpendicular to the line of sight).

Universe: All that there is within the space and time dimensions accessible to us, as well as regions beyond (but still physically connected to) those that we can see.

variable star: A star whose apparent brightness changes with time.

virtual particle: A particle that flits into existence out of nothing and, shortly thereafter, disappears again.

wavelength: The distance over which a wave goes through a complete oscillation; the distance between two consecutive crests or two consecutive troughs.

weakly interacting massive particles (WIMPs): Theorized to make up the dark matter of the Universe.

weak nuclear force: Governs the decay of a neutron into a proton, electron, and antineutrino.

weight: The force of the gravitational pull on a mass.

white dwarf: The evolutionary endpoint of stars that have initial mass less than about 8 solar masses. All that remains is the degenerate core of He or C–O (in some cases, O–Ne–Mg).

wormhole: A hypothetical connection between two universes or different parts of our Universe. Also: *Einstein-Rosen bridge*.

year: The Earth's orbital period around the Sun.

zenith: The point on the celestial sphere that is directly above the observer.

zodiac: The band of constellations through which the Sun moves during the course of a year.

zodiacal light: A faint glow in the night sky around the ecliptic, stretching up from the horizon shortly after evening twilight and shortly before morning twilight, from sunlight reflected by interplanetary dust.

Biographical Notes

Aristarchus of Samos (roughly 310–230 B.C.). Greek astronomer; measured the Sun-Earth distance relative to the Earth-Moon distance. Realized that the Sun is much larger than the Earth and reasoned that the Sun (rather than the Earth) is at the center of the Universe, predating Copernicus by 1800 years.

Aristotle (384–322 B.C.). The most influential early Greek philosopher. He lectured on a vast range of subjects; however, many or most of his beliefs in physics and astronomy turned out to be wrong. Developed a widely adopted geocentric (Earth-centered) model of the Universe consisting of 55 spheres. Correctly concluded that the Earth is spherical.

Brahe, Tycho (1546–1601). Danish astronomer; measured the positions of planets with unprecedented accuracy, laying the foundations for Kepler's work. Discovered and studied a bright supernova in 1572; thus, the "sphere of fixed stars" is not immutable, in contradiction to Aristotelian and Christian dogma.

Cannon, Annie Jump (1863–1941). American astronomer; classified the photographic spectra of several hundred thousand stars, demonstrating that the spectra depend mostly on the stellar surface temperature. She arranged the spectral types into the sequence OBAFGKM.

Chandrasekhar, Subrahmanyan (1910–1995). Indian-born American astrophysicist. Awarded the Nobel Prize in Physics in 1983 for his work on the physical understanding of stars, especially the upper mass limit of white dwarfs.

Copernicus, Nicolaus (1473–1543). Polish astronomer; proposed the heliocentric (Sun-centered) model of the planetary system. He showed how the retrograde motion of planets could be explained with this hypothesis. His book *De Revolutionibus* was published the year of his death.

Eddington, Sir Arthur (1882–1944). British astrophysicist who studied the physical structure of stars and was an expert on Einstein's general theory of relativity. Through his observations of a total solar eclipse in 1919, he helped to confirm this theory.

Einstein, Albert (1879–1955). German-American physicist, the most important since Newton. Developed the special and general theories of relativity, proposed that light consists of photons, and worked out the theory of Brownian motion (the irregular, zigzag motion of particles suspended in a fluid is due to collisions with molecules). Responsible for $E = mc^2$, the world's most famous equation.

Eratosthenes (276–194 B.C.). Greek geographer who estimated the circumference of the Earth to within 1% accuracy through measurements of the length of a stick's shadow at different locations on Earth.

Galileo Galilei (1564–1542). Italian mathematician, astronomer, and physicist; was the first to systematically study the heavens with a telescope. Discovered the phases of Venus and the four bright moons of Jupiter, providing strong evidence against the geocentric model for the Solar System. After being sentenced by the Inquisition to perpetual house arrest, he published his earlier studies of the motions of falling bodies, laying the experimental groundwork for Newton's laws of motion.

Gamow, George (1904–1968). Russian-American physicist; he suggested that the Universe began in a hot, compressed state and predicted the existence of the cosmic background radiation that was later discovered by Arno Penzias and Robert Wilson. Also devised a theory of radioactive decay.

Guth, Alan (1947–). American physicist; proposed the inflationary theory of the Universe to eliminate some glaring problems with the standard Big Bang model. His perspective was that of an elementary particle physicist, not an astronomer; he was most troubled by the absence of magnetic monopoles.

Hawking, Stephen (1942–). English physicist, best known for his remarkable theoretical work while physically incapacitated by Lou Gehrig's disease (ALS). His prediction that black holes can evaporate through quantum tunneling is an important step in attempts to unify quantum physics and gravity (general relativity). He is Lucasian Professor of Mathematics at Cambridge University, as was Newton.

Hipparchus (c. 160–c. 127 B.C.). Greek astronomer who made the first accurate star catalogue. Refined the methods of Aristarchus of Samos. Determined the length of the year to within six minutes and noticed that the direction of the north celestial pole changes with time.

Hoyle, Fred (1915–2001). English astronomer; proposed the steady-state theory of the Universe, which stimulated much important work in cosmology. Also made fundamental contributions to the understanding of the origin of the chemical elements. Coined the term *Big Bang*.

Hubble, Edwin (1889–1953). American astronomer, after whom the Hubble Space Telescope is named. He proved that "spiral nebulae" are galaxies far outside our own Milky Way and discovered the expansion of the Universe (*Hubble's law*) by recognizing that the redshift of a galaxy is proportional to its distance. He also proposed a widely used morphological classification scheme for galaxies.

Kepler, Johannes (1571–1630). German mathematician and astronomer; was Tycho Brahe's assistant and gained access to Brahe's data after his death. Developed three empirical laws of planetary motion that represent a significant revision of the Copernican model. Studied a very bright supernova in 1604.

Leavitt, Henrietta (1868–1921). American astronomer; demonstrated a relationship between the period and luminosity of Cepheid variable stars. This was done by analysis of Cepheids clustered together and, therefore, at the same distance, so that differences in brightness indicate luminosity differences.

Maxwell, James (1831–1879). Scottish physicist; showed that visible light is only one form of electromagnetic radiation, whose speed can be derived from a set of four equations that describe all of electricity and magnetism. Also investigated heat and the kinetic theory of gases.

Newton, Isaac (1642–1727). English mathematician and physicist; developed three laws of motion and the law of universal gravitation, all published in *The Principia* (1687). Invented the reflecting telescope, determined that white light consists of all colors of the rainbow, and invented calculus. At age 27, became Lucasian Professor of Mathematics at Cambridge University. Became Warden of the Mint in 1696; knighted in 1705.

Ptolemy, Claudius (85–165). Greek astronomer who developed an elaborate model for planetary motions, based on Aristotle's geocentric Universe, that endured for more than 1400 years. Compiled the *Almagest*, a set of 13 volumes that provides most of our knowledge of early Greek astronomy.

Rubin, Vera (1928–). American astronomer; was the first to observationally show that the rotation curves of most spiral galaxies imply the presence of considerable amounts of dark matter. She also obtained early evidence for large-scale *peculiar motions* of galaxies relative to the smooth expansion of the Universe.

Sagan, Carl (1934–1996). American astronomer and the 20th century's most well known popularizer of science, especially astronomy. Among his scientific accomplishments, he demonstrated that Venus suffers from an enormous greenhouse effect. *Cosmos*, his 13-episode public-television astronomy series, has been seen by more than 500 million people. He was an eloquent proponent of unmanned Solar System exploration.

Sandage, Allan (1926–). American astronomer and disciple of Edwin Hubble, he has made fundamental contributions to the determination of globular cluster ages, the distances of galaxies, the Hubble constant, the age of the Universe, and the rate at which the expansion of the Universe is changing.

Shapley, Harlow (1885–1972). American astronomer; correctly deduced that the Sun is not at the center of the Milky Way Galaxy and that the Galaxy is larger than previously believed. Incorrectly concluded that the "spiral nebulae" are within the Milky Way, but most of his reasoning was logically sound.

Zwicky, Fritz (1898–1974). Swiss-American astronomer; proposed that supernovae result from the collapse of the cores of massive stars, producing neutron stars and energetic particles (cosmic rays). Compiled an extensive atlas of galaxy clusters and showed that many such clusters must contain dark matter in order to be gravitationally bound.

Bibliography

Essential Reading:

Adams, Fred, and Greg Laughlin. *The Five Ages of the Universe: Inside the Physics of Eternity*. New York: Free Press, 2006.

Alvarez, Walter. *T-Rex and the Crater of Doom*. London: Vintage, 1998.

Beebe, R. *Jupiter: The Giant Planet*, 2nd ed. Washington, DC: Smithsonian Institution Press, 1996.

Begelman, Mitchell, and Martin Rees. *Gravity's Fatal Attraction: Black Holes in the Universe*. New York: W.H. Freeman, 1998.

Carroll, Bradley W., and Dale A. Ostlie. *An Introduction to Modern Astrophysics*, 2nd ed. New York: Addison Wesley, 2006.

Chaikin, Andrew. *A Man on the Moon*. New York: Time-Life Books, 2001.

Christianson, Gale E. *Isaac Newton and the Scientific Revolution*. New York: Oxford University Press, 1998.

Cristensen, Lars L., Robert A. Fosbury, and M. Kornmesser, *Hubble: 15 Years of Discovery.* Berlin/New York: Springer, 2006.

Davies, Paul. *The Accidental Universe*. Cambridge: Cambridge University Press, 1982.

———. *Superforce: The Search for a Grand Unified Theory of Nature*. New York: Touchstone, 2002.

Dickinson, Terence. *Nightwatch: A Practical Guide to Viewing the Universe*, 4th ed. Buffalo, NY: Firefly Books, 2006.

Dorminey, Bruce. *Distant Wanderers: The Search for Planets beyond the Solar System*. New York: Copernicus Books, 2001.

Drake, Stillman. *Galileo: A Very Short Introduction*. New York: Oxford University Press, 2001.

Dressler, Alan. "The Origin and Evolution of Galaxies," in *The Origin and Evolution of the Universe*. Edited by Ben Zuckerman and Matthew Arnold Malkan. Milwaukee, WI: Gareth Stevens Publishing, 1996.

Ferguson, Kitty. *The Nobleman and His Housedog: Tycho and Kepler: The Unlikely Partnership That Forever Changed Our Understanding of the Heavens*. London: Review, 2002.

———. *Prisons of Light—Black Holes*. Cambridge: Cambridge University Press, 1998.

Ferris, Timothy. *Coming of Age in the Milky Way*. New York: Harper Perennial, 2003.

———. *The Whole Shebang: A State-of-the-Universe(s) Report*. New York: Simon & Schuster, 1998.

Feynman, Richard P., Freeman Dyson (foreword), and Ralph Leighton (ed.). *Classic Feynman: All the Adventures of a Curious Character*. New York: W.W. Norton, 2005.

Filippenko, Alex. "A Supernova with an Identity Crisis." *Sky & Telescope*, Dec. 1993, p. 30.

Gingerich, Owen. *The Eye of Heaven: Ptolemy, Copernicus, Kepler*. Melville, NY: AIP Press, 1993.

———, and James MacLachlan. *Nicolaus Copernicus: Making the Earth a Planet*. New York: Oxford University Press, 2005.

Gleick, James. *Isaac Newton*. London: Vintage, 2004.

Goldsmith, Donald. *Einstein's Greatest Blunder? The Cosmological Constant and Other Fudge Factors in the Physics of the Universe*. Cambridge, MA: Harvard University Press, 1997.

———. *Supernova! The Exploding Star of 1987*. New York: St. Martin's Press, 1989.

———, and Tobias Owen. *The Search for Life in the Universe*, 3rd ed. Sausalito, CA: University Science Books, 2001.

Golub, Leon, and Jay M. Pasachoff. *Nearest Star: The Surprising Science of Our Sun*. Cambridge, MA: Harvard University Press, 2002.

Greene, Brian. *The Elegant Universe: Superstrings, Hidden Dimensions, and the Quest for the Ultimate Theory*. New York: Vintage, 2000.

Gribbin, John, and Martin Rees. *Cosmic Coincidences: Dark Matter, Mankind and Anthropic Cosmology*. New York: Bantam, 1989.

Guth, Alan H. *The Inflationary Universe: The Quest for a New Theory of Cosmic Origins*. Cambridge, MA: Perseus Books Group, 1998.

Harrison, Edward. *Cosmology: The Science of the Universe*. Cambridge: Cambridge University Press, 2000.

———. *Darkness at Night: A Riddle of the Universe*. Cambridge, MA: Harvard University Press, 1989.

Hartmann, William. *The History of Earth: An Illustrated Chronicle of an Evolving Planet*. New York: Workman Publishing, 1991.

———, and Ron Miller. *The Grand Tour: A Traveler's Guide to the Solar System*, 3rd ed. New York: Workman Publishing, 2005.

Hawking, Steven. *A Briefer History of Time*. New York: Bantam, 2005.

Hawley, John, and Katherine Holcomb. *Foundations of Modern Cosmology*. New York: Oxford University Press, 2005.

Hearnshaw, J. B. *The Analysis of Starlight: One Hundred and Fifty Years of Astronomical Spectroscopy.* Cambridge: Cambridge University Press, 1990.

Hirschfeld, Alan W. *Parallax: The Race to Measure the Cosmos*. New York: W.H. Freeman, 2001.

Hogan, Craig J. *The Little Book of the Big Bang: A Cosmic Primer*. Berlin/ New York: Springer, 1999.

Kaku, Michio. *Hyperspace: A Scientific Odyssey through Parallel Universes, Time Warps, and the 10th Dimension.* New York: Oxford University Press, 1995.

Kargel, Jeffrey S. *Mars—A Warmer, Wetter Planet.* Berlin/New York: Springer, 2004.

Katz, Jonathan I. *The Biggest Bangs: The Mystery of Gamma-Ray Bursts, the Most Violent Explosions in the Universe*. New York: Oxford University Press, 2004.

Kaufmann, William J., III. *Black Holes and Warped Spacetime*. New York: Bantam Books, 1981.

Kirkpatrick, Larry D., and Gerald F. Wheeler, *Physics: A World View*, 4th ed. Fort Worth, TX: Harcourt College Publishers, 2001.

Kirshner, Robert P. *The Extravagant Universe: Exploding Stars, Dark Energy, and the Accelerating Cosmos*. Princeton, NJ: Princeton University Press, 2004.

Kitchen, Chris, and Robert W. Forrest. *Seeing Stars: The Night Sky through Small Telescopes.* Berlin/New York: Springer, 1997.

Kowal, Charles T. *Asteroids: Their Nature and Utilization*. New York: John Wiley & Sons, 1996.

Krauss, Lawrence M. *Quintessence: The Mystery of the Missing Mass*. New York: Basic Books, 2000.

Lang, Kenneth R. *The Cambridge Guide to the Solar System.* Cambridge: Cambridge University Press, 2003.

Leslie, John. *Universes*. London: Routledge, 1996.

Littmann, Mark, Ken Willcox, and Fred Espenak. *Totality: Eclipses of the Sun*. New York: Oxford University Press, 1999.

Lynch, David K., and William Livingston. *Color and Light in Nature.* Cambridge: Cambridge University Press, 2001.

Marschall, Laurence. *The Supernova Story*. Princeton, NJ: Princeton University Press, 1994.

McFadden, Lucy-Ann L., Paul Weissman, and Torrance Johnson, eds. *Encyclopedia of the Solar System.* New York: Academic Press, 2006.

Melia, Fulvio. *The Black Hole at the Center of Our Galaxy.* Princeton, NJ: Princeton University Press, 2003.

———. *The Edge of Infinity: Supermassive Black Holes in the Universe.* Cambridge: Cambridge University Press, 2003.

Minnaert, Marcel. *Light and Color in the Outdoors*. Translated by L. Seymour. Berlin/New York: Springer, 1995.

Molnar, Michael R. *The Star of Bethlehem: The Legacy of the Magi*. New Brunswick, NJ: Rutgers University Press, 1999.

Mook, Delo E., and Thomas Vargish. *Inside Relativity*. Princeton, NJ: Princeton University Press, 1991.

The Once and Future Cosmos (special edition of *Scientific American*, 2002).

Pasachoff, Jay M. *Astronomy: From the Earth to the Universe*, 6th ed. Belmont, CA: Thomson Learning, 2002. (This is a longer but somewhat outdated version of *The Cosmos*, listed above.)

———. *A Field Guide to Stars and Planets*, 4th ed. Boston: Houghton Mifflin, 1999.

———. *Peterson First Guide to Astronomy*. Boston: Houghton Mifflin, 1998.

———, and Alex Filippenko. *The Cosmos: Astronomy in the New Millennium*, 3rd ed. Belmont, CA: Thomson Brooks/Cole, 2007.

Pickover, Clifford A. *Black Holes: A Traveler's Guide.* New York: Wiley, 1997.

Randall, Lisa. *Warped Passages: Unraveling the Mysteries of the Universe's Hidden Dimensions*. New York: Harper Perennial, 2006.

Rees, Martin. *Before the Beginning: Our Universe and Others*. Cambridge, MA: Perseus Books Group, 1998.

Rubin, Vera C. *Bright Galaxies, Dark Matters.* Melville, NY: AIP Press, 1996.

Schilling, Govert. *Flash! The Hunt for the Biggest Explosions in the Universe*. Translated by Naomi Greenberg-Slovin. Cambridge: Cambridge University Press, 2002.

Shu, Frank H. *The Physical Universe: An Introduction to Astronomy.* Sausalito, CA: University Science Books, 1982.

Smith, Michael D. *The Origin of Stars.* London: Imperial College Press, 2004.

Smoot, George, and Keay Davidson. *Wrinkles in Time*. New York: Harper Perennial, 1994.

Squyres, Steven. *Roving Mars: Spirit, Opportunity, and the Exploration of the Red Planet.* New York: Hyperion, 2006.

Sutton, Christine. *Spaceship Neutrino*. Cambridge: Cambridge University Press, 1992.

Thorne, Kip S. *Black Holes and Time Warps: Einstein's Outrageous Legacy.* New York: W.W. Norton & Co., 1995.

Waller, William H., and Paul W. Hodge. *Galaxies and the Cosmic Frontier.* Cambridge, MA: Harvard University Press, 2003.

Ward, Peter, and Donald Brownlee. *Rare Earth: Why Complex Life Is Uncommon in the Universe*. Berlin/New York: Springer, 2003.

Weinberg, Steven. *Dreams of a Final Theory: The Scientist's Search for the Ultimate Laws of Nature*. New York: Vintage Books, 1993.

———. *The First Three Minutes: A Modern View of the Origin of the Universe*. New York: Basic Books, 1993.

Will, Clifford M. *Was Einstein Right? Putting General Relativity to the Test.* New York: Vintage, 1994.

Wolf, Fred A. *Taking the Quantum Leap: The New Physics for Nonscientists*. New York: Harper Perennial, 1989.

Wolfson, Richard. *Simply Einstein: Relativity Demystified*. New York: W.W. Norton & Co., 2003.

Zirker, Jack B. *An Acre of Glass: A History and Forecast of the Telescope.* Baltimore, MD: Johns Hopkins University Press, 2005.

Supplementary Reading:

Barrow, John D. *The Origin of the Universe*. New York: Basic Books, 1997.

Beatty, J. Kelly, Carolyn Collins Petersen, and Andrew Chaikin, eds. *The New Solar System*, 4th ed. Cambridge: Cambridge University Press, 1998.

Bell, Jim, and Jacqueline Mitton, eds. *Asteroid Rendezvous: NEAR Shoemaker's Adventures at EROS.* Cambridge: Cambridge University Press, 2002.

Bhatnagar, Arvind, and William C. Livingston. *Fundamentals of Solar Astronomy*. New Jersey: World Scientific Publishing Co., 2005.

Bova, Ben. *The Beauty of Light*. New York: John Wiley & Sons, 1988.

Boyce, Joseph M. *The Smithsonian Book of Mars.* Washington, DC: Smithsonian Books, 2003.

Calle, Carlos. *Superstrings and Other Things: A Guide to Physics*. Washington, DC: Taylor & Francis, 2001.

Chapman, Clark L., and David Morrison. *Cosmic Catastrophes*. New York: Plenum Press, 1989.

Christianson, Gale E. *Edwin Hubble: Mariner of the Nebulae.* Chicago: University of Chicago Press, 1996.

Cohen, I. Bernard. *The Newtonian Revolution*. Cambridge: Cambridge University Press, 1983.

Cohen, Martin. *In Darkness Born: The Story of Star Formation*. Cambridge: Cambridge University Press, 1988.

Cohen, Nathan. *Gravity's Lens: Views of the New Cosmology*. New York: John Wiley & Sons, 1988.

Cooper, W. A., and E. N. Walker, *Getting the Measure of the Stars*. London/New York: Taylor & Francis, 1989.

Croswell, Ken. *The Alchemy of the Heavens: Searching for Meaning in the Milky Way*. Garden City, NY: Anchor, 1996.

———. *Planet Quest: The Epic Discovery of Alien Solar Systems*. New York: Oxford University Press, 1989.

Davies, Paul. *The Last Three Minutes: Conjectures about the Ultimate Fate of the Universe*. New York: Basic Books, 1997.

Drake, Frank, and Dava Sobel. *Is Anyone Out There? The Search for Extraterrestrial Intelligence*. New York: Delta, 1994.

Dressler, Alan. *Voyage to the Great Attractor: Exploring Intergalactic Space*. New York: Knopf, 1994.

Espenak, Fred. *Fifty Year Canon of Solar Eclipses*. Cambridge, MA: Sky Publishing, 1987.

Ferguson, Kitty. *Measuring the Universe: Our Historic Quest to Chart the Horizons of Space and Time*. New York: Walker & Co., 2000.

Filippenko, Alex V. "Stellar Explosions, Neutron Stars, and Black Holes," in *The Origin and Evolution of the Universe*. Edited by B. Zuckerman and M. A. Malkan. Sudbury, MA: Jones & Bartlett Publishing, 1996.

Fischer, Daniel. *Mission Jupiter: The Spectacular Journey of the Galileo Spacecraft*. Berlin/New York: Springer, 2001.

Fraknoi, Andrew, David Morrison, and Sidney C. Wolff. *Voyages through the Universe*, 3rd ed. Belmont, CA: Brooks Cole, 2003.

Goldsmith, Donald. *The Hunt for Life on Mars*. Collingdale, PA: Diane Publishing Co., 1997.

———. *Worlds Unnumbered: The Search for Extrasolar Planets*. Sausalito, CA: University Science Books, 1997.

Greene, Brian. *The Fabric of the Cosmos: Space, Time, and the Texture of Reality*. New York: Vintage, 2005.

Gribbin, John. *In Search of the Big Bang: The Life and Death of the Universe*. New York: Penguin, 1999.

———. *In Search of Schrodinger's Cat: Quantum Physics and Reality.* New York: Bantam, 1984.

Harris, Joel, and Richard Talcott. *Chasing the Shadow: An Observer's Guide to Solar Eclipses*. Waukesha, WI: Kalmbach Publishing Co., 1994.

Hawking, Steven. *The Universe in a Nutshell*. London: Bantam, 2001.

Henbest, Nigel, and Heather Couper. *The Guide to the Galaxy*. Cambridge: Cambridge University Press, 1994.

Hey, Tony, and Patrick Walters. *The Quantum Universe.* Cambridge: Cambridge University Press, 2003.

Hockey, Thomas A. *The Book of the Moon: A Lunar Introduction to Astronomy, Geology, Space Physics, and Space Travel.* Englewood Cliffs, NJ: Prentice Hall, 1986.

Hodge, Paul. *Higher Than Everest: An Adventurer's Guide to the Solar System*. Cambridge: Cambridge University Press, 2001.

Horwitz, Tony. *Blue Latitudes: Boldly Going Where Captain Cook Has Gone Before*. New York: Picador, 2003.

Hoskin, Michael, ed. *The Cambridge Concise History of Astronomy.* Cambridge: Cambridge University Press, 1999.

Hutchison, Robert, and Andrew Graham. *Meteorites*. London: The Natural History Museum, 2001.

Kaku, Michio. *Parallel Worlds: A Journey through Creation, Higher Dimensions, and the Future of the Cosmos*. Garden City, NY: Anchor, 2006.

Kaler, James B. *Stars*. New York: W.H. Freeman, 1998.

Kaufmann, William J., III. *The Cosmic Frontiers of General Relativity.* New York/Boston: Little Brown & Co., 1977.

Kerrod, Robin. *Hubble: The Mirror on the Universe*. Buffalo, NY: Firefly Books, 2003.

Kippenhahn, Rudolf. *100 Billion Suns: The Birth, Life, and Death of Stars*. Translated by Jean Steinberg. Princeton, NJ: Princeton University Press, 1993.

Kuhn, Thomas S. *The Copernican Revolution: Planetary Astronomy in the Development of Western Thought.* Cambridge, MA: Harvard University Press, 1992.

Lederman, Leon, and David N. Schramm. *From Quarks to the Cosmos: Tools of Discovery*. New York: W.H. Freeman, 1995.

Lightman, Alan. *Ancient Light: Our Changing View of the Universe.* Cambridge, MA: Harvard University Press, 1993.

———, and Roberta Brawer. *Origins: The Lives and Worlds of Modern Cosmologists.* Cambridge, MA: Harvard University Press, 1990.

Linde, Andrei. "Future of the Universe," in *The Origin and Evolution of the Universe*. Edited by B. Zuckerman and M. A. Malkan. Sudbury, MA: Jones & Bartlett Publishing, 1996.

Livio, Mario. *The Accelerating Universe: Infinite Expansion, the Cosmological Constant, and the Beauty of the Cosmos*. New York: Wiley, 2000.

Long, Kim. *The Moon Book: Fascinating Facts about the Magnificent, Mysterious Moon*. Boulder, CO: Johnson Books, 1998.

MacLachlan, James. *Galileo Galilei: First Physicist*. New York: Oxford University Press, 1999.

Malin, David. *View of the Universe*. Rohnert Park, CA: Pomegranate, 1997.

Mallove, Eugene F., and Gregory L. Matloff, *The Starflight Handbook: A Pioneer's Guide to Interstellar Travel*. New York: Wiley, 1989.

McKay, Christopher P. "The Origin and Evolution of Life in the Universe," in *The Origin and Evolution of the Universe*. Edited by Ben Zuckerman and Matthew Arnold Malkan. Milwaukee, WI: Gareth Stevens Publishing, 1996.

Osserman, Robert. *Poetry of the Universe*. Garden City, NY: Anchor, 1996.

Overbye, Dennis. *Lonely Hearts of the Cosmos: The Story of the Scientific Quest for the Secret of the Universe*. Boston: Back Bay Books, 1999.

Pagels, Heinz R. *Perfect Symmetry: The Search for the Beginning of Time*. New York: Bantam, 1991.

Pedersen, Olaf. *Early Physics and Astronomy: A Historical Introduction*. Cambridge: Cambridge University Press, 1993.

Petersen, Carolyn Collins, and John C. Brandt. *Hubble Vision: Astronomy with the Hubble Space Telescope*. Cambridge: Cambridge University Press, 1995.

———. *Visions of the Cosmos*. Cambridge: Cambridge University Press, 2003.

Preston, Richard. *First Light: The Search for the Edge of the Universe*. New York: Scribners, 1991.

Rees, Martin. *New Perspectives in Astrophysical Cosmology*. Cambridge: Cambridge University Press, 2002.

———. *Our Cosmic Habitat*. Princeton, NJ: Princeton University Press, 2003.

———, and John Gribbin. *The Stuff of the Universe: Dark Matter, Mankind and the Coincidences of Cosmology*. London: Arrow, 1990.

Rowan-Robinson, Michael. *Ripples in the Cosmos: A View Behind the Scenes of the New Cosmology*. New York: Oxford University Press, 1998.

Sagan, Carl. *Cosmos*. New York: Ballantine Books, 1985.

———. *Pale Blue Dot: A Vision of the Human Future in Space*. New York: Ballantine Books, 1997.

———, and I. S. Shklovskii. *Intelligent Life in the Universe*. Boca Raton, FL: Emerson-Adams Press, 1998.

Silk, Joseph. *A Short History of the Universe*. New York: W.H. Freeman, 1997.

Smith, Robert W. *The Expanding Universe: Astronomy's "Great Debate," 1900–1931*. Cambridge: Cambridge University Press, 1982.

Smolin, Lee. *The Life of the Cosmos*. New York: Oxford University Press USA, 1999.

Sobel, Dava. *Galileo's Daughter: A Historical Memoir of Science, Faith, and Love*. New York: Penguin, 2000.

Sobel, Michael I. *Light*. Chicago: University of Chicago Press, 1989.

Tegmark, Max. "Parallel Universes." *Scientific American*, May 2003, p. 40.

t'Hooft, Gerard. *In Search of the Ultimate Building Blocks*. Cambridge: Cambridge University Press, 1996.

Thoren, Victor E. *The Lord of Uraniborg: A Biography of Tycho Brahe*. Cambridge: Cambridge University Press, 1991.

Tirion, Wil. *The Cambridge Star Atlas*, 3rd ed. Cambridge: Cambridge University Press, 2001.

Trefil, James S. *The Moment of Creation: Big Bang Physics from Before the First Millisecond to the Present Universe*. New York: Dover Publications, 2004.

Velan, A.K. *The Multi-Universe Cosmos*. Berlin/New York: Springer, 1992.

Verschuur, Gerrit L. *Hidden Attraction: The History and Mystery of Magnetism*. New York: Oxford University Press, 1996.

———. *Impact! The Threat of Comets and Asteroids*. New York: Oxford University Press, 1997.

———. *The Invisible Universe Revealed: The Story of Radio Astronomy*. Berlin/New York: Springer Verlag, 1987.

Watson, Fred. *Stargazer: The Life and Times of the Telescope*. Reading, MA: Perseus Books, 2006.

Wheeler, J. Craig. *Cosmic Catastrophes: Supernovae, Gamma-Ray Bursts, and Adventures in Hyperspace*. Cambridge: Cambridge University Press, 2000.

Will, Clifford M. *Was Einstein Right? Putting General Relativity to the Test*. New York: Basic Books, 1993.

Wright, Alan, and Hilary Wright. *At the Edge of the Universe*. New York: Ellis Horwood, 1989.

Zirker, Jack B. *Journey from the Center of the Sun*. Princeton, NJ: Princeton University Press, 2004.

Internet Resources:

Alex Filippenko, astroalex.com.

Astronomical Society of the Pacific (ASP), www.astrosociety.org.

The Astronomy Café, www.astronomycafe.net/.

California and Carnegie Planet Search, www.exoplanets.org (extra-solar planets).

Center for Adaptive Optics, UC Santa Cruz, cfao.ucolick.org/.

Davidson, *Secret Worlds: The Universe Within*, micro.magnet.fsu.edu/primer/java/scienceopticsu/powersof10/.

Eames Office, *Powers of 10*, www.powersof10.com/.

Harvard-Smithsonian Center for Astrophysics, *Supernova*, www-cfa.harvard.edu/oir/Research/supernova/SNlinks.html.

HubbleSite News Center, hubblesite.org/newscenter/.

Institute for Advanced Study, School of Natural Sciences, *John Bahcall*, www.sns.ias.edu/~jnb/ ("popular articles" link).

Light and Optics, ww2010.atmos.uiuc.edu/(Gh)/guides/mtr/opt/home.rxml (rainbows, solar halos, sundogs, and other atmospheric phenomena).

NASA, "Astronomy Picture of the Day," antwrp.gsfc.nasa.gov/apod/.

Nemiroff, *Virtual Trips to Black Holes and Neutron Stars*, antwrp.gsfc.nasa.gov/htmltest/rjn_bht.html.

The Official String Theory Web Site, superstringtheory.com.

Parviainen, P., www.polarimage.fi (solar halos and other atmospheric phenomena).

SETI@Home project, seti.berkeley.edu (download a program that, as a background task, will analyze data from radio telescopes, searching for signs of extraterrestrial life).

Space News, space.com.

Space Telescope Science Institute, hubblesite.org.

Sudbury Neutrino Observatory (SNO), www.sno.phy.queensu.ca/.

Wilkinson Microwave Anisotropy Probe, map.gsfc.nasa.gov.

Young, mintaka.sdsu.edu/GF.

Notes